A249

THE LAW OF
NUCLEAR INSTALLATIONS
AND
RADIOACTIVE SUBSTANCES

AUSTRALIA
LBC Information Services
Sydney

CANADA and USA
Carswell
Toronto ● Ontario

NEW ZEALAND
Brooker's
Auckland

SINGAPORE and MALAYSIA
Thomson Information (S.E. Asia)
Singapore

THE LAW OF
NUCLEAR INSTALLATIONS
AND
RADIOACTIVE SUBSTANCES

STEPHEN TROMANS

AND

JAMES FITZGERALD

FIRST EDITION

LONDON
SWEET & MAXWELL
1997

Published in 1997 by
Sweet & Maxwell of
100 Avenue Road, London NW3 3PF
Typeset by Mendip Communications Ltd, Frome, Somerset.
Printed and bound by Butler and Tanner Ltd, Frome and London.

No natural forests were destroyed to make this product;
only farmed timber was used and replanted

British Library Cataloguing in Publication Data

A CIP catalogue record for this book
is available from the British Library

ISBN 0421-538-805

S. TROMANS & J. FITZGERALD
1997

FOREWORD

The subject of radioactivity has a very mixed reception in the public mind, largely fuelled by the lack of knowledge and the perceived danger of radiation emanating from radioactive sources. This problem is further compounded by the invisible nature of such radiation. This book provides a very good historical survey of radioactivity and its application, via the atom-bomb to the production of energy through nuclear power stations. It also gives a critical survey of the legislation that has been associated with these developments.

One of the main points that arise in any discussion of this subject is related to the high precision with which this class of radiation can be measured. The problem is further compounded by the high values that arise with the use of the *becquerel* as a unit of measurement. Using this unit, relatively small amounts of radiation appear as enormous to the untutored eye and reported statistics of radiation measures may appear to indicate large and apparently dangerous levels.

However, one interesting feature of the ability to measure radiation with such great accuracy is the difference in the problems that arose in the case of contamination of foodstuffs by radiation sources and the current problems that have arisen from BSE in cattle. As a result of fallout from the Chernobyl explosion large areas of sheep grazing land in both England and Wales were contaminated with iodine and caesium. There are still areas of the Lake District where there are restrictions on the sale of sheep due to radioactive contamination. The localisation of the problem arises because of the possibility of detecting the contamination with high precision and the accurate measurement of the radioactive caesium or iodine content of the meat itself. This contrasts with recent problems that have arisen from contamination of meat from cattle which have been affected by BSE. The inability to measure or test for the presence of any contamination of the meat for BSE has led to a total ban on all meat derivatives from cattle in the United Kingdom as this is viewed as providing the only safe and effective means of ensuring no possible danger from consuming meat that may come from contaminated herds.

The advent of the use of nuclear reactors as a potential energy source was initially greeted with considerable enthusiasm by the public and viewed as a "clean" non-polluting source of energy. The occurrence of the Chernobyl incident led to an immediate reassessment of this situation and the unpopularity of nuclear reactors as a solution to the energy problem came to the fore. This has also brought into focus other near accidents such as the incident at Three Mile Island. A difficulty with the nuclear solution to the energy problem was previously noted in the Sixth Report of the Royal Commission on Environmental Pollution which drew attention to the problem of the disposal of nuclear waste at the end of the life-time of the reactor. The waste contains long-lived, highly radioactive materials which are potentially dangerous and the disposal of this waste is still a major problem associated with any development in nuclear reactor programmes. However, in the initial development of the nuclear reactor strategy, this was not fully considered as it was a problem that only arises during decommissioning, a date very much in the future.

In considering the nuclear waste disposal problem it must be recognised that one important difference between nuclear and fossil fuel as an energy source is that in the case of nuclear energy the pollution occurs as a "point source" as opposed to the burning of fossil fuels which provides a "diffuse source" of pollution. However, the magnitude of the problem with the disposal of the nuclear waste is one that cannot be underestimated as far as the potential dangers to human health and environmental damage are concerned. The difficulties in the comparison of the relative dangers to the environment of nuclear or fossil fuel sources of energy arise from the fact that fossil fuel has been used as an energy source from time immemorial and as such the associated dangers have not been fully recognised by the public and have acquired a degree of acceptance within society that is difficult to displace.

It is clear that the development of any energy programme based on nuclear power must be carefully governed by legislation and this has been a point of consideration for many years. The present text gives an excellent overview of the general position of the field in a readable and succinct form. It provides a valuable assessment of a wide range of instruments and legislative procedures that have been introduced and places them in historical context. It also summarises the important areas of liability assessment and insurance risk. These are two areas that are of prime consideration for the development of any form of legislation but are also fraught with difficulties.

There is little doubt that the debate on the future use of nuclear and radioactive substances will continue into the distant future. The short-term pollution problems of fossil fuel and the introduction of the additional hazards associated with carbon dioxide production and global warming have to be set against the longer-term problems of nuclear waste disposal from reactor sites. Recent discussion on the disposal of low and intermediate-level radioactive waste by NIREX illustrates the general concern of the public. A clear understanding of the relative problems and the associated legislation is essential for a balanced solution to this problem. This book goes a long way towards providing this background information.

Lord Lewis

April 10, 1997

PREFACE

This book's genesis lies in practical work undertaken by the authors for H.M. Government relating to nuclear licensing and liabilities and radioactive waste management, specifically in the context of the Atomic Weapons Establishment and Devonport Royal Dockyard. In providing that advice, it soon became apparent that there was no up to date legal text or commentary on the key pieces of legislation, the Nuclear Installations Act 1965 and the Radioactive Substances Act 1993. We therefore rashly set out to write one and, some eighteen months later, this book is the end result. The earliest writing was done in snatched moments on a visit by the first-named author to the USA in Autumn 1995, and perhaps few law textbooks can claim, as this one can, to have been commenced sitting on a bench overlooking the camel enclosure at Salt Lake City Zoo.

The period during which the book was written has been an active one in relation to many aspects of its subject matter, and events after the book went to press indicate that there is little danger of nuclear issues lessening in terms of pubic controversy. Such developments include major civil unrest in Germany in protest at transport of radioactive waste to its storage site at Gorleben, Lower Saxony (March 1997). Nearer to home, in October 1996 the Irish Supreme Court ruled that a group of Dundalk residents were entitled to sue British Nuclear Fuels plc in the Irish High Court in relation to alleged endangerment and consequent mental distress caused by the THORP thermal oxide reprocessing plant at Sellafield. Concern was also expressed in November 1996 over levels of the isotope technetium-99 in seafood close to the site, possibly linked to the operation since 1994 of the new enhanced actinide removal plant (EARP). In December 1996 it was reported that the Environment Agency was to initiate proceedings over the dumping of radioactive waste (low-grade uranium U-238) in a field near Rushden, Northamptonshire in 1994. Revelations of accidents involving aircraft loaded with nuclear weapons at US and RAF military bases in the 1950s have fuelled further public concern of possible radioactive contamination (August 1996).

Amidst such anxieties, the job of the regulators – the Nuclear Installations Inspectorate and the Environment Agency – is not likely to get any easier. The NII is having to scrutinise proposals within the privatised nuclear industry for significant jobs cuts, with their attendant possible implications for safety at nuclear sites: this involves the implementation of appropriate "management for change" arrangements (HSE, *Nuclear Safety Newsletter*, February 1997). The Environment Agency in January 1997 announced for consultation a proposed agreement with the nuclear waste company Nirex as to the provision of information on investigations which might lead to an application to develop a deep waste repository for intermediate and low-level waste.

Among these developments, three are particularly noteworthy and require special mention here, production constraints have precluded their inclusion in the text. These are the revised Basic Safety Standards Directive, the High Court decision in *Blue Circle Industries plc v. The Ministry of Defence*, and the decision of the Secretary of State in relation to the proposed rock laboratory of Nirex.

The revised Basic Safety Standards Directive – Council Directive 96/29/EURATOM (O.J. L159) lays down basic safety standards for the protection of the health of workers and the general public against the dangers arising from ionising radiation. With effect from May 13, 2000, the date on which it is required to be implemented by Member States, it will replace the current Basic Safety Standards Directive 80/836/EURATOM, which is referred to in the text and described particularly at paragraph 1–47. Since the new Directive will not take effect for three years, it was felt that it might be unduly confusing to analyse its provisions in detail in the text. As the Preamble to the new Directive makes clear, it reflects the development of scientific knowledge concerning radiation protection, as expressed in particular in Recommendation No. 60 of the International Commission on Radiological Protection. The structure of the Directive remains basically similar, though the drafting is improved and clarified in a number of respects, for example in relation to the general principles such as justification (Art. 6). The most notable numerical change is the reduction of the effective dose limit for members of the public from 5 to 1 mSv a year (Art. 13).

Blue Circle Industries plc v. The Ministry of Defence (*The Times*, December 11, 1996) – This decision of Carnwath J. concerned a claim under the Nuclear Installations Act 1965 against the Ministry of Defence following the escape of radioactive material in liquid form from the Atomic Weapons Establishment, Aldermaston. The resulting contamination of the plaintiff's commercial estate led to the breakdown of negotiations for sale of the estate to a prospective purchaser for £10 million. It was held, in contrast to the case of *Merlin v. British Nuclear Fuels plc* (see para. 3–26), that the contamination involved a physical change to the property requiring major engineering remedial works, and so constituted "damage to property" within section 7(1) of the 1965 Act. The assessment of damages for loss of the sale was subject to the ordinary rules governing remoteness (see para. 3–27). On such principles, the plaintiff was awarded damages in excess of £5 million. The decision is fully analysed in the Sweet & Maxwell *Environmental Law Bulletin* Number 31, January 1997, p. 40.

The Nirex decision – The decision of the Secretary of State on the planning appeal by Nirex against refusal of permission by Cumbria County Council for its proposed rock characterisation facility at Gosforth, Cumbria, was announced on March 18, 1997. As explained in Chapter 5, the facility was intended as an underground laboratory to establish the suitability of the site for the storage or disposal of intermediate level waste. Early reports of the decision indicate that the Secretary of State was critical of "poor design, layout and arrangements for access" to the site, close to the Lake District National Park. More fundamentally, he also appears to have criticised the process of site selection, the adequacy of the applicant's environmental statement, and "scientific uncertainties and technical difficulties" in the proposal. The decision was immediately hailed as an "historic victory" by environmentalists.

In writing this book, we have found it necessary to understand and explain, however briefly, some of the history and science behind the development of nuclear energy. Much more expert analysis, for those interested, can be found in a number of the papers contained in *Radiation Protection Dosimetry. Becquerel's Legacy: A Century of Radioactivity* (Vol. 68, Nos. 1/2 1996). Our focus has been on the law relating to nuclear installations and radioactive waste in the U.K., and whilst we have dealt with the Conventions underlying that issue, we have not sought to address the wider issues of international law relating to matters such as nuclear weapons testing. Such issues have been much in the news in relation to the French underground nuclear tests at Mururoa and Fangataufa: for a useful collection of essays, see *Reciel,* Vol. 5 Issue 3, September 1996.

Finally, we must acknowledge the kind contribution of a number of people within the nuclear industry or involved with the management of radioactive substances and waste, who have taken the time to read and comment on various parts of the book in draft. The Environmental Law Department at Simmons & Simmons has been an encouraging and conducive place within which to write, and we are grateful to various colleagues who have assisted in numerous ways, both directly and indirectly, to lighten our load as authors.

Stephen Tromans and James FitzGerald
Simmons & Simmons,
London, EC2M 2TX

March 20, 1997

CONTENTS

CHAPTER ONE

NOVEL TRANSFORMATIONS-NUCLEAR INSTALLATIONS AND RADIOACTIVE SUBSTANCES IN CONTEXT

CHAPTER TWO

THE LICENSING OF NUCLEAR INSTALLATIONS

CHAPTER THREE

LIABILITY AND INSURANCE

CHAPTER FOUR

RADIOACTIVE SUBSTANCES: CONTROLS OVER THEIR USE AND TRANSPORT

USE OF RADIOACTIVE SUBSTANCES

TRANSPORT

CHAPTER FIVE

RADIOACTIVE WASTE

EVOLUTION OF POLICY ON RADIOACTIVE WASTE

INTERNATIONAL ASPECTS

CONTROL UNDER THE RADIOACTIVE SUBSTANCES ACT 1993

TABLE OF CASES

(All references are to paragraph numbers)
(All references with prefix A, B or C are to Appendices)

UNITED STATES

EUROPEAN COMMUNITY

TABLE OF STATUTES

(All references are to paragraph numbers)
(All references with prefix A, B or C are to Appendices)

UNITED STATES

TABLE OF STATUTORY INSTRUMENTS

(All references are to paragraph numbers)
(All references with prefix A, B or C are to Appendices)

TABLE OF EUROPEAN AND INTERNATIONAL LEGISLATION

(All references are to paragraph numbers)
(All references with prefix A, B or C are to Appendices)

EUROPEAN AND INTERNATIONAL TREATIES AND CONVENTIONS

REGULATIONS

DIRECTIVES

COUNCIL DECISIONS, RESOLUTIONS AND RECOMMENDATIONS

LIST OF ACRONYMS

ACSNI . . . . . Advisory Committee on the Safety of Nuclear Installations
AEC . . . . . . Atomic Energy Commission
AGR . . . . . . Advanced Gas-cooled Reactor
ALARA . . . . . As Low As Reasonably Achievable
ALARP . . . . . As Low As Reasonably Practicable
BNDC . . . . . British Nuclear Design and Construction
BNFL . . . . . British Nuclear Fuels Limited
BPEO . . . . . Best Practicable Environmental Option
BPM . . . . . . Best Practicable Means
BSL . . . . . . Basic Safety Limit
BSO . . . . . . Basic Safety Objective
CCGT . . . . . Combined Cycle Gas Turbine
CEGB . . . . . Central Electricity Generating Board
COMARE . . . Committee on Medical Aspects of Radiation in the Environment
DRAWMOPS . . Decommissioning and Radioactive Waste Management Operations
EBRD . . . . . European Bank for Reconstruction and Development
EURATOM . . European Atomic Energy Community
HSC . . . . . . Health and Safety Commission
HSE . . . . . . Health and Safety Executive
IAEA . . . . . . International Atomic Energy Agency
ICRP . . . . . . International Commission on Radiological Protection
ILW . . . . . . Intermediate-Level Waste
IMF . . . . . . International Monetary Fund
IRAC . . . . . . Ionising Radiations Advisory Committee
ISO . . . . . . International Standards Organisation
ITP . . . . . . Integrated Toxic Potential
LAWDC . . . . Local Authority Waste Disposal Company
LLW . . . . . . Low-Level Waste
LTSR . . . . . . Long Term Safety Review
MoD . . . . . . Ministry of Defence
MOX . . . . . . Mixed Oxide Fuel
MPC . . . . . . Maximum Permissible Concentration
NEA . . . . . . Nuclear Energy Agency
NFFO . . . . . Non-Fossil Fuel Obligation
NHL . . . . . . Non-Hodgkins Lymphoma
NII . . . . . . Nuclear Installations Inspectorate

NIREX . . . . Nuclear Industry Radioactive Waste Executive
NNC . . . . . National Nuclear Company
NPC . . . . . Nuclear Power Company
NRPB . . . . National Radiological Protection Board
NSD . . . . . Nuclear Safety Division
NUSSAG . . . Nuclear Safety Standards Advisory Group
OECD . . . . Organisation for Economic Co-operation and Development
OEEC . . . . Organisation for European Economic Co-operation
PPI . . . . . Paternal Pre-conception Irradiation
PSA . . . . . Probabilistic Safety Analysis
PSR . . . . . Periodic Safety Review
PWR . . . . . Pressurised Water Reactor
RADREM . . . Radioactivity Research and Environmental Monitoring Committee
RADWASS . . . Radioactive Waste Safety Standards
REC . . . . . Regional Electricity Company
RPV . . . . . Reactor Pressure Vessel
RWMAC . . . Radioactive Waste Management Advisory Committee
SAP . . . . . Safety Assessment Principles
SDR . . . . . Special Drawing Rights
SEPA . . . . Scottish Environment Protection Agency
SGHWR . . . Steam Generating Heavy Water Reactor
THORP . . . . Thermal Oxide Reprocessing Plant
TNPG . . . . The Nuclear Power Group
TRCL . . . . The Radiochemical Centre Limited
UAEC . . . . European Atomic Energy Committee
UKAEA/AEA . . United Kingdom Atomic Energy Authority
UNSCEAR . . . United Nations Scientific Committee on the Effects of Atomic Radiation

CHAPTER ONE

NOVEL TRANSFORMATIONS—NUCLEAR INSTALLATIONS AND RADIOACTIVE SUBSTANCES IN CONTEXT

Attitudes to radioactivity   1–01

"There are few subjects in the field of environmental pollution to which people react so emotionally as they do to radioactivity".

Royal Commission on Environmental Pollution, Sixth Report, *Nuclear Power and the Environment*, Cmnd. 6618 (1976).

Twenty years have passed since these words were written, yet over those two decades nothing has diminished the deep-seated fears of ordinary people as to the potential harm they perceive may be caused to them, their children, and the environment, by ionising radiation. On the contrary, those fears have, if anything, been increased by two very serious nuclear accidents, at Three Mile Island, Pennsylvania in March 1974, and Chernobyl near Kiev in April 1986, together with the tendency of the media to focus on, and play up, nuclear risks.

It was clear to the Royal Commission that one reason for the emotional reaction to radioactivity was its association with the purely destructive use of nuclear energy in the bombs which destroyed and maimed so many at Hiroshima and Nagasaki; as explained below, the development of nuclear reactors to generate energy from nuclear power was inextricably linked with, and stemmed from, the drive to create an atomic bomb during the Second World War. Apart from the horrific effect on those exposed to the immediate release of energy from a nuclear bomb, the effects of radioactive emissions from military and peaceful uses of nuclear energy are potentially the same. It is not surprising, therefore, that the public at large make the link between the consequences of the military use of a nuclear weapon and those from sources of nuclear/radioactive material utilised for peaceful purposes. The nature of these consequences add a further layer of unease to the public perception; the horrors of unseen forces that can deleteriously affect cells, promote birth defects and cause cancers (perhaps many years later) promote understandable fears in people's minds. Then there is the environmental dimension; the creation of waste products of varying degrees of radioactivity, some of which will still present hazards hundreds if not thousands of years on.

Few subjects are as capable of generating lurid headlines as radioactivity. Consider the following selection from recent months: "Nuclear Alert over Forgotten Dump" (*The Independent*, December 27, 1995); "Food Fears as N-Waste Survives Sea Burial" (*Sunday Telegraph*, December 31, 1995); "MoD Facing Humiliation over Nuclear Accident Lies" (*The

Guardian, August 12, 1996); and "Costain's Newbury Bypass in New 'Radioactive' Row (*Reuters News Service*, August 14, 1996).

Yet radioactivity has enormously useful—indeed now essential—applications. As well as the generation of electricity from nuclear reactors, the medical and industrial uses of radioactive substances are of great benefit to society. It is therefore vital to have in place legal controls in which the public can have full confidence to safeguard them against the risks which they fear, and to avoid serious, long-term or irreversible environmental consequences. In the United Kingdom, control over the keeping and use of radioactive substances and the accumulation, discharge or disposal of radioactive waste is provided by the Radioactive Substances Act 1993. The installation and operation of nuclear reactors and certain other types of nuclear apparatus are subject to licensing under the Nuclear Installations Act 1965. The same Act provides a system of compensation for injury and damage caused by nuclear occurrences at licensed sites.

Both these major Acts are the subject of detailed commentary in this book, and are reproduced in their entirety as Annexes. However, it may be helpful, in order to put that legislation into context, to consider in this first Chapter, the historical circumstances of the development of nuclear power, the relevant policy and institutional structures within the United Kingdom, and some of the relevant major international developments.

It is salutary to compare the brief span of time during which mankind has begun to understand the energetic possibilities of the atom, with the enormous timescales over which radioactive waste products will present hazards, as discussed by the Royal Commission in its 1976 Report. The half-life of radioisotopes such as plutonium, americium, and neptunium, means that they could remain hazardous for "tens or hundreds of thousands of years" (para. 384). By contrast, it is just over 100 years since the German physicist Wilhelm Röntgen discovered a new and strange radiation (X-rays) emitted from a cathode-ray tube. Forty years later in 1935, Lord Ernest Rutherford, the pre-eminent Cambridge atomic physicist, was reported in *The Times* under the heading "The Neutron. Novel Transformations" as having concluded that the "transformation of atoms ... was a very poor and inefficient way of producing energy, and anyone who looked for a source of power in the transformation of the atoms was talking moonshine."[1] Within a decade this statement had been proved wrong, such was the pace of atomic technology.

1–02 Exploring the atom

As mentioned above, in November 1895 Röntgen discovered X-rays. In 1897, the director of the Cavendish Laboratory in Cambridge, J. J. Thomson, used the cathode-ray tube to demonstrate the existence of the

[1] Quoted in: Richard Rhodes, *The Making of the Atomic Bomb* (Penguin, 1986), p. 27.

electron, and verify the particulate theory of matter. In the meantime, the French physicist Henri Becquerel had, by experiments with uranium salt and photographic paper, observed the phenomenon of naturally occurring radioactivity. Marie and Pierre Curie, working in their draughty attic laboratory, then isolated the radioactive elements, polonium and radium, from the uranium ore, pitchblende. It was Rutherford who identified two different types of radiation, alpha and beta particles (the first being readily absorbed, the latter of a more penetrative character). A third type, high energy gamma radiation, was later discovered by the French scientist, P. V. Villard.

Rutherford, working with Frederick Soddy, a young chemist, also observed the spontaneous transmutation of the radioactive element, thorium, into the non-radioactive elemental gas, argon. They thereby concluded that each different radioactive element had a characteristic half-life which governed its rate of decay or transformation into another element, or variant of the same element. Soddy coined the term "isotope" to describe such variants. It was also discovered that such "radioactive change" set free energy on a grand scale, although, at this time, the potential destructive power of such energy was the subject only of speculation by prescient writers such as H. G. Wells. More immediately, attention focused on the medical applications of X-rays and radium; medicines containing radium were prescribed and "radium spas" became popular in some parts of Europe.

Research was continued by a number of exceptionally gifted young physicists during the 1920s—German universities, such as Göttingen, were particular centres of excellence. Many key scientists such as Max Born, Leo Szilard and Enrico Fermi moved to London or the United States with the onset of anti-semitism in Germany and Italy, and in due course were to make an enormous contribution to the development of the atomic bomb for the Allies.

The first nuclear reactor   1–03

Fermi in particular was a vital figure in nuclear research. It was he who was responsible, in November 1942, for construction of the world's first nuclear reactor, assembled in a doubles squash court at Stagg Field, a sports-stand in Chicago University. The "pile", as it was known, was constructed by teams of men working in shifts over 17 days, stacking layers of graphite and uranium on a wooden frame, with control rods consisting of cadmium sheet nailed to 13-foot wooden strips which had to be inserted and removed by hand. On December 2, 1942 the last control rod was painstakingly removed by six-inch increments, allowing a self-sustaining nuclear reaction to take place for 4.5 minutes. One description of this gigantic step in the development of nuclear power highlights the relationship between the new technology and legal responsibilities in an interesting way:

"For once Compton [Arthur Compton, the Nobel prize-winning experimental physicist who was a major co-ordinator of the atomic bomb project] made a quick decision; with control seemingly assured, he allowed Fermi to build CP-1 [Chicago Pile Number One] in the west stands. He chose not to inform the President of the University of Chicago, Robert Maynard Hutchins, reasoning that he should not ask a lawyer to judge a matter of nuclear physics. 'The only answer he could have given would have been—no. And this answer would have been wrong. So I assumed the responsibility myself.' The word *meltdown* had not yet entered the reactor engineer's vocabulary—Fermi was only then inventing that specialty—but that is what Compton was risking, a small Chernobyl in the midst of a crowded city, except that Fermi was, as he knew, a formidably competent engineer."[2]

1–04 The bomb

Fermi's experiment was a vital stage in the Allied project to create the atomic bomb, and it was later in December 1942 that President F. D. Roosevelt authorised the "Manhatten Project" which involved full-scale plants to produce enriched plutonium (near Hanford, Washington) and enriched uranium (Oak Ridge, Tennessee), as well as the huge research laboratories at Los Alamos, New Mexico. As is well known the project reached its culmination in the dropping of the uranium bomb "Little Boy" on Hiroshima on August 6, 1945, which resulted in 200,000 deaths by 1950, either immediately, or from radiation-induced illness. "Fat Man", a plutonium bomb, was dropped on Nagasaki on August 9, 1945, causing an ultimate total of 140,000 deaths.

1–05 Nuclear power in the United States

With the end of the war, the peaceful applications of nuclear power began to be developed. In the United States, the Atomic Energy Act of 1946 transferred authority over atomic energy from military to civilian control, creating the five-member Atomic Energy Commission (AEC) with a monopoly over all aspects of civil and military research into atomic energy. However, the efforts of the AEC were still concentrated on weapons production. It was not until the Atomic Energy Act of 1954 allowed private companies to enter the field under licence from the AEC, that real progress in the field of civil nuclear energy began to be made, thereby fulfilling the commitment made in 1953 by President Eisenhower to the General Assembly of the UN that the United States would "... devote its entire heart and mind to find the way by which the miraculous

[2] Richard Rhodes, *The Making of the Atomic Bomb* (Penguin, 1986), pp. 432–433.

inventiveness of man should not be dedicated to his death, but consecrated to his life."[3] In order to encourage rapid technological development within a non-existent industry, a "virtually unique" regulatory scheme gave the AEC broad and non-prescriptive responsibility (*Carstens v. NRC*,[4] quoting *Siegel v. AEC*[5]). The first reactor in the United States to be connected to the commercial power grid was built on the Ohio River at Shippingport, a joint venture by the AEC, the Duquesne Electric Power Company and the Westinghouse Corporation. The reactor design, incidentally, was based upon the pressurised water reactor technology used in U.S.S. *Nautilus*, the first nuclear-powered submarine. The second U.S. commercial reactor, at Detroit, led to a Supreme Court decision which confirmed the validity of the staged process of permitting used by the AEC, allowing it "to defer a definitive safety ruling until operation is actually licensed" (*Power Reactor Dev. Co. v. International Union*).[6]

The Price-Anderson Act of 1957 removed a further obstacle to civil nuclear power by providing indemnities to reactor licensees and AEC contractors in the event of an accident. However, the perceived lack of return on nuclear reactors as investments constituted a further brake on new projects until the creation of pooling arrangements among electricity generators in the mid-1960s allowed surplus power to be sold, and thus changed the economics. The market for new reactors then took off with a vengeance in 1966, so much so that by the end of the 1960s the position of nuclear power had been strengthened enormously. However, this position was reversed during the 1970s with renewed uncertainties over the economics of reactor projects, and growing public concern as to safety, particularly after the accident at Three Mile Island in 1979 (see para. 1–33 below).

The post-war activities of the U.S. nuclear industry, particularly the AEC and Department of Energy, have left an appalling legacy of contamination of soil and groundwater, for example at the Hanford Site in eastern Washington State and at Rocky Flats in Colorado. Cost estimates of clean-up efforts vary between the conservative sum of $200 billion, to as high as half a trillion dollars; in any event, the cost will be at least as high as that of producing the U.S. nuclear weapons arsenal in the first place. Blame for this situation has been attributed to the system of self-regulation, by which the AEC and, after it, the Department of Energy "regulated" themselves with little or no external accountability.[7]

---

[3] Cited in: Campbell-Mohn, Breen and Futrell (eds.), *Sustainable Environmental Law* (St. Paul, Minn., 1993), p. 889.
[4] 742 F.2d 1546, 1551 (D.C. Cir, 1984), cert. denied 471 U.S. 1136, 105 S.Ct. 2675.
[5] 400 F.2d 778, 783 (D.C. Cir, 1968).
[6] 367 U.S. 396, 407; 81 S.Ct. 1529, 1534.
[7] see: A. Caputo, "A Failed Experiment", *Environmental Forum* (1996), Vol. 13, No. 1, p. 17.

1–06 Post-war development in the United Kingdom

The position in the United Kingdom may be compared with that in the United States. There are many parallels although, fortunately, these do not include in the United Kingdom environmental damage on the massive scale referred to above.

From 1946 to 1954, the responsibility for nuclear matters lay with the Ministry of Supply, an administrative relic of the Second World War. The Atomic Energy Act 1946 placed a general duty on the Minister of Supply "to promote and control the development of atomic energy" (s.1) and he was given wide powers to do so, including compulsory acquisition. By section 10, the Minister of Supply was enabled to make orders prohibiting the use of uranium, thorium, plutonium, neptunium and any other prescribed substances, for the production or use of atomic energy, or for research into such matters. No orders were in fact ever made under the section.

During the period until 1954, the Ministry developed nuclear energy for military purposes. It oversaw the establishment of the Atomic Energy Research Establishment at Harwell, and production facilities for plutonium and uranium at Windscale and Springfields. It also commenced the Calder Hall project, an installation which produced weapons-grade plutonium, with electricity as a by-product.

1–07 Creation of the United Kingdom Atomic Energy Authority

Following representations to Government that a new entity was needed to take responsibility for nuclear energy, a committee of inquiry was set up, the outcome being the creation of the United Kingdom Atomic Energy Authority by the Atomic Energy Authority Act 1954. This Act constituted the Authority which, like its U.S. counterpart, had very broad powers and was singularly unconstrained by Parliamentary or other controls, save in respect of weapons production or the searching for, or working of, minerals in the United Kingdom. The powers of the Authority under subsection 2(2) were as follows:

(a) to produce, use and dispose of atomic energy and carry out research into any matters connected therewith;

(b) to manufacture or otherwise produce, buy or otherwise acquire, store and transport any articles which in the opinion of the Authority are, or are likely to be, required for, or in connection with, the production or use of atomic energy or such research as aforesaid, and to dispose of any articles manufactured, produced, bought or acquired by them;

(c) to manufacture or otherwise produce, buy or otherwise acquire, treat, store, transport and dispose of any radioactive substances;

(d) to do all such things (including the erection of buildings, and the execution of works and the searching for and working of minerals) as appear to the Authority necessary or expedient for the exercise of the foregoing powers;

(e) to make arrangements with universities and other institutions or persons for the conduct of research into matters concerned with atomic energy or radioactive substances and, with the approval of the Lord President of the Council and the Treasury, to make grants or loans to universities and other institutions or persons engaged in the production or use of atomic energy or radioactive substances or in research into matters connected with atomic energy or radioactive substances; and

(f) to distribute information relating to, and educate and train persons in matters connected with, atomic energy or radioactive substances.

The unique position of the Agency has been described in the following terms:

"The UKAEA was unlike any other agency in Britain. Its financial and administrative powers were substantial; its control by Parliament was limited and tenuous. It was financed by a direct "vote" of public funds, under conditions that offered little opportunity for MPs to find out what would be done with the money voted, either before or after it was spent. The very first estimate of the annual budget of the UKAEA put the sum likely to be required from Parliament at £53 million, at 1954 prices—a staggering sum to be found within an economy still trying to right itself after a devastating war."[8]

Civil nuclear power   1–08

The original focus of the Agency was intended to be weapons, but it has been described as: "Over the years ... the Adam from whose ribs a number of different organisations have been created."[9] A number of these organisations related to the civil applications of nuclear energy.

One of the Agency's first acts in this respect was to assist in the formulation of a White Paper published in 1955, *A Programme of Nuclear Power*.[10] The White Paper, with the benefit of hindsight, appears wonderfully optimistic about the future of nuclear energy, describing it as "the energy of the future" (para. 1) and suggesting that a large nuclear power station might take as much as "five or more years to complete, including finding the site, designing the station and building it" (para. 3). The programme proposed by the White Paper was for a number of

[8] Walter C. Patterson, *Going Critical* (Paladin, 1985), pp. 4–5.
[9] Royal Commission on Environmental Pollution, Sixth Report, *Nuclear Power and the Environment*, Cmnd. 6618 (1976).
[10] Cmd. 9389 (1955).

nuclear power stations, generating 1,500–2,000 megawatts, to be ordered by 1965. These power stations were to be of the "Magnox" design—so called because of the magnesium alloy used to clad the uranium fuel-rods. Five consortia were set up, each led by a major manufacturer of heavy electrical plant, to construct the power stations. Contracts for the first Magnox power stations went to various of these consortia: Berkeley, Gloucestershire and Bradwell, Essex to be constructed for the Central Electricity Authority (later the Central Electricity Generating Board); and Hunterston, Scotland for the South of Scotland Electricity Board.

However, as in the United States, the economics of nuclear power remained dubious when compared with increasingly efficient fossil-fuel power stations. Lack of orders led to a gradual consolidation among the five consortia, ultimately into two: The Nuclear Power Group (TNPG), and British Nuclear Design and Construction (BNDC). Nine Magnox stations were built in all—the last at Wylfa in Anglesey—but the limitations of the design in terms of efficiency and the cost of materials meant that attention turned to new designs: the Advanced Gas-Cooled Reactor (AGR), and later, the U.S. water-cooled designs. The AEA was throughout this period actively involved in providing design information and training to the various consortia. Orders were placed for a number of AGR stations during the late 1960s—Dungeness B, Hinkley Point B, Hunterston B and Hartlepool—but this was a troubled period of delayed projects, serious technical problems, and financial difficulties for the participants of some consortia.

1–09 Division of Atomic Energy Authority

In 1971, the AEA was split into three parts by the Atomic Energy Authority Act 1971. Part of the Authority's undertaking was transferred to British Nuclear Fuels Limited (BNFL) (s.1) which took over all fuel-service activities, namely uranium processing at Springfields, fuel enrichment at Capenhurst, and fuel reprocessing and plutonium manufacture at Windscale. The activities included, as had those of the Authority, the provision of weapons-grade plutonium for defence purposes. Section 2 of the 1971 Act transferred another part of the Authority's undertaking to The Radiochemical Centre Limited (TRCL), which manufactured and sold radioisotopes for use in industry and medicine and which was subsequently privatised as Amersham International plc. The shares of both BNFL and TRCL were held by the Authority. The third aspect of the Authority's work, the manufacture of nuclear weapons at Aldermaston, Berkshire, passed to the Ministry of Defence (see para. 1–30 below).

The two remaining commercial consortia for nuclear power station construction, TNPG and BNDC, were amalgamated in 1973, somewhat against their will. The National Nuclear Company (NNC) was created, with a minority Government shareholding; NNC in turn held all the shares in the Nuclear Power Company (NPC), which shared out construction and design activities. Fifteen per cent of shares in NNC were held by the Authority, 35 per cent by British Nuclear Associates (a group of seven companies which had been involved in the old consortia) and 50 per cent by GEC. These shareholdings were subsequently adjusted, GEC reducing its stake.

Nuclear power in difficulties   1–10

The 1970s were a difficult time generally for nuclear power in the United Kingdom. There was the hard-hitting Royal Commission Report (the "Flowers Report") in 1976 (discussed in various contexts below), then problems with the construction of AGRs, and subsequently the abandonment in 1978, after much effort and expenditure, of the proposed steam generating heavy water reactor (SGHWR). An indication of the lack of activity is provided by the fact that Heysham B station, an AGR, when the contract between the CEGB and NNC was signed in 1981, represented the first contract for a new station that NNC had received since it had been set up in 1974, and the first order for a nuclear power station by the CEGB since 1970.

The PWR stations   1–11

A new chapter for nuclear power in Britain began in 1983 with the proposal to construct a pressurised water reactor (PWR) at Sizewell B in Suffolk. Consent was given following a mammoth public inquiry of extremely wide-ranging scope, a decision which probably prevented the break up of the United Kingdom nuclear plant design industry.[11]

The PWR design used for Sizewell B followed that adopted in the United States and in France, and involved collaboration between the CEGB and the U.S. Corporation, Westinghouse. The proposal was approved[12] and the station is now operating. Further proposals followed, and consent was granted for a further nuclear power station at Hinkley Point C.[13] Subsequently, Nuclear Electric applied for a site licence for Sizewell C, a twin-reactor near-replica version of Sizewell B. However, in December 1995, it was announced that Nuclear Electric would not be proceeding with either Sizewell C or Hinkley Point C; the reason advanced was that the future of United Kingdom energy prices was not sufficiently certain for the investment to take place in the short term, and that the priority should be to concentrate on privatisation of the industry in 1996. The Sizewell C project would have cost about £3 billion, and Hinkley Point C about £1.9 billion; some commentators suggested that

[11] O'Riordan, Kemp and Purdue, *Sizewell B—An Anatomy of the Inquiry* (MacMillan Press, 1988), p. 19.
[12] see: *The Sizewell B Public Inquiries—A Report by Sir Frank Layfield Q.C.* (HMSO, 1987).
[13] see: *The Hinkley Point Public Inquiries—A Report by Michael Barnes Q.C.* (HMSO, 1990).

the real concern was the possible reaction of the City to such investment, coupled with overcapacity in the electricity market. Some analysts expect British Energy (the holding company of Nuclear Electric and Scottish Nuclear) to diversify into cheaper gas-fired generation plant after privatisation (see para. 1–24 below).

1–12 Nuclear reprocessing and fuel services

During the operation of a reactor the fissile content of the nuclear fuel providing the power changes. Such fuel is said to be irradiated and at some stage this fuel must be replaced so that a chain reaction (through which controlled releases of energy are made available for the generation of electricity, for example) can be maintained. The effective life of nuclear fuel in a reactor depends upon the reactor type, the nature of the fuel and the cladding material. Irradiated fuel, removed from a nuclear reactor, is intensely radioactive and capable of generating high temperatures. All irradiated fuel is cooled prior to reprocessing, some of it in water-filled ponds. Following either treatment the irradiated fuel is in a state in which it may be reprocessed. It consists of a mixture of unused uranium, plutonium and other elements formed by transmutation and fission products. The original object of the first reactors at Windscale was to produce plutonium for nuclear weapons, and methods were developed for it to be extracted chemically from irradiated fuel. The extraction of plutonium is still an important reason for reprocessing nuclear fuel; however, there are other reasons, such as the recovery of uranium for recycling, and to facilitate management of the waste by reducing its volume.

The reprocessing itself involves relatively simple chemical processes, but these are complicated by the radioactive nature of the materials, and by the precautions necessary to avoid a state of criticality or uncontrolled reaction. The reprocessing facilities at Windscale were expanded significantly during the 1960s to accommodate the needs of fuel reprocessing for the Magnox reactors. Additionally, the Atomic Energy Authority as operator was beginning to provide a variety of fuel services, *i.e.* fuel manufacture, uranium enrichment and reprocessing, to a number of foreign customers. In 1965, the fuel service activities of the AEA moved onto a commercial basis, with separate accounts and a "trading fund". Together with the AEA's small reprocessing plant at Dounreay in Northern Scotland, Windscale took fuel for reprocessing from such countries as Japan, Italy, Canada, Denmark and Germany.

On the break-up of the AEA in 1971 (see para. 1–09 above), BNFL took over the fuel service operations of the Authority at Windscale, together with the uranium and fuel manufacturing plant at Springfields, Lancashire, the enrichment plant at Capenhurst, Cheshire and the Calder Hall and Chapelcross reactors which served the dual purpose of producing weapons materials and generating power. During the first half of the

1970s, potentially serious difficulties caused the long-term shutdown of some of the reprocessing facilities, leading to a worrying accumulation of spent fuel awaiting reprocessing.[14]

At the same time, attention was turning to the reprocessing needs of fuel from the newer AGR stations; this fuel was made of ceramic uranium dioxide which is more durable than uranium metal. Additionally, AGR fuel could be irradiated for a longer time and so could become more radioactive than an equivalent mass of Magnox fuel. BNFL began to formulate plans for an oxide reprocessing plant to service those United Kingdom power stations using oxide fuel, and also Japanese nuclear power stations. This led to the coining of the term "radioactive dustbin" by Friends of the Earth, and later in the same year the lurid article and headline in the *Daily Mirror* of October 21, 1975: "Plan to Make Britain World's Nuclear Dustbin". At the opposite extreme of the spectrum of academic reputability, the 1976 Sixth Report of the Royal Commission on Environmental Pollution,[15] voiced serious doubts as to the dangers and civil liberties implications of the creation and transport of plutonium.

The Windscale Inquiry   1–13

BNFL's application for the construction of new processing facilities at Sellafield was considered by the local planning authority, Cumbria County Council, in 1976; by November, the planning committee of the Council announced that they were minded to approve the application subject to suitable planning conditions. It appeared at first that the Secretary of State for the Environment, Peter Shore, would not call in the application for his own determination; however, he announced his intention to do so on December 22, 1976. The inquiry into the thermal oxide reprocessing plant (THORP) was held in 1977, chaired by the High Court judge, Sir Roger Parker, and is generally known as "The Windscale Inquiry". The Inquiry was given wide terms of reference and examined policy matters of safety, national interest, and nuclear proliferation.

The Report of the Inquiry was published in 1978.[16] The Report supported the construction of THORP and dismissed the various objections with little discussion. A novel procedure was developed to deal with the Report's findings. The Secretary of State announced that he would formally reject the application, though he found the Report completely persuasive; there would then be a Parliamentary debate on the Report, following which he would lay before Parliament a special development order authorising construction, which would in turn be debated. Ultimately, the matter was voted on in division, and in both votes

[14] Walter C. Patterson, *Going Critical* (Paladin, 1985), pp. 102–103.
[15] Royal Commission on Environmental Pollution, Sixth Report, *Nuclear Power and the Environment*, Cmnd. 6618 (1976).
[16] *The Windscale Inquiry—A Report by Sir Roger Parker* (HMSO, 1978).

the proposal was approved, though with significant numbers of votes against, from across the political spectrum.

1–14 THORP

The THORP project, approved in 1978, was intended to be in operation by the late 1980s, an assumption which proved to be optimistic. Two particular matters gave rise to concern about its ultimate viability—one was the possibility of long term dry storage of oxide fuel as an alternative management strategy, the other was an increase in the supply of uranium from new mines, leading to a drop in price.

Detailed planning permission was eventually applied for and obtained by BNFL in 1983, and construction work began. The project was subject to further legal controversy at the stage of authorisation for the discharge of radioactive wastes as liquid and gaseous effluents from the reprocessing plant, with challenges to the relevant decisions brought by Greenpeace in judicial review actions. These proceedings, and the legal issues raised, are discussed in detail at paras. 5–30 to 5–32 below.

THORP began operation in March 1994. BNFL's published accounts for the year to March 31, 1995 suggest that revenue from the plant reached £484 million during the first year of operation, of which £287 million came from overseas.

1–15 The future of BNFL

The Government's 1995 Nuclear Review (see para. 1–23 below) confirms that BNFL will continue to offer the full range of nuclear fuel-cycle services "so long as the market continues to demand them" (para. 8.31), and recognises that the challenge for BNFL is to develop its business in overseas markets where growth is strongest. In order to preserve its position, BNFL is constructing a Mixed Oxide Fuel (MOX) plant at Sellafield. This is to accommodate the desire of many customers to have the plutonium separated from the spent fuel during reprocessing returned as MOX fuel, a form of plutonium and uranium. The plant is due for completion this year (1997).

BNFL also operates a nuclear fuel transport service through its subsidiary company Pacific Nuclear Transport Limited, and its associate company Nuclear Transport Limited. During 1993/94 they carried some 400 tonnes of spent nuclear fuel from Japan to Europe.

Although the Government's Nuclear Review focused principally upon privatisation of the nuclear generators, the Government has not ruled out privatisation of BNFL as a longer term aim, and is satisfied that its proposed restructuring of the industry will not close off that option (para. 8.35). Whether the prospects of future privatisation will be affected depends to a large degree on the extent to which BNFL is required to shoulder the costs relating to the older Magnox stations, which were excluded from the initial privatisation process (see para. 1–24 below).

Radioactive waste management   1–16

A key legacy of the use of nuclear or radioactive material is the generation of radioactive waste. The identity of the elements in this waste will determine the half-life of the various components and, because some radionuclides have a very long half-life, provision for isolation of some radionuclides for unimaginable lengths of time needs to be incorporated in any effective waste management programme. At present, the process by which isolation can be achieved most effectively has yet to be successfully demonstrated but the industry believes that options exist and research on those options continues.

The issue of radioactive waste management is, of course, a highly emotive one. The Royal Commission on Environmental Pollution, in its Sixth Report,[17] was firmly of the view that there should be no commitment to a large programme of nuclear fission power until it had been demonstrated beyond reasonable doubt that a method existed to ensure the safe containment of long-lived, highly radioactive waste for the indefinite future (para. 33). The legal and policy issues relating to radioactive waste disposal are dealt with in Chapter 5, but it may be helpful to outline below the main types of waste involved in practice.

In July 1995, the Government published its *Final Conclusions on its Review of Radioactive Waste Management Policy*.[18] These conclusions are considered in detail in Chapter 5.

Gaseous waste   1–17

The principal radioactive discharges to the atmosphere from nuclear reactors are of inert gases and iodine. Iodine is a highly volatile product of fission with a short half-life and can present dangers if ingested (for example through milk) or inhaled. Iodine-131 would be one of the main sources of danger to the public in the event of a large accidental discharge of radioactivity. There are other radioactive isotopes which may be of concern at different types of site, *e.g.* tritium, carbon 14 and krypton 85.

---

[17] Royal Commission on Environmental Pollution, Sixth Report, *Nuclear Power and the Environment*, Cmnd. 6618 (1976).
[18] Cm. 2919 (1995).

1-18 Liquid waste

The output of aqueous radioactive effluent from nuclear power stations is normally quite small, for example plant washings and laundry effluents. However, more serious problems can arise if cooling pond water becomes contaminated by reason of the corrosion of fuel element cladding.

Apart from relatively short-lived liquid wastes arising from medical uses of radioactive materials, the main source of liquid effluent in the United Kingdom is the reprocessing carried out at BNFL's Sellafield Works, which is discharged by pipeline to the Irish Sea and is subject to complex limits. In 1976, the Royal Commission found there was insufficient evidence to require the discharges of plutonium as a component of this effluent to be reduced, but expressed some concern as to the potential consequences if new exposure pathways emerged in years to come for the plutonium accumulated in sediment of the Ravenglass estuary.[19]

1-19 Low-level solid waste

Large amounts of solid waste with low levels of radioactivity (approximately 10,000 cubic metres) are disposed of in the United Kingdom each year. This low-level waste consists largely of paper towels, discarded laboratory clothing, lightly contaminated plant items and equipment arising from the operation of nuclear facilities. Very low-level radioactive waste may be disposed of at landfill sites or by incineration depending upon the nature and quantity of the waste while BNFL operates a disposal facility at Drigg which receives most of the low-level waste arising in the United Kingdom; small volumes are also disposed of at Dounreay.

1-20 Intermediate-level solid waste

Such waste cannot be accepted for land burial, but on the other hand does not require the special cooling which high-level waste does in order to remove fission decay heat. Intermediate-level waste arises mainly from nuclear power stations. It comprises irradiated metal components and instruments removed from the reactor core, and miscellaneous incombustible materials such as filters, valves, ash from incineration of waste, and dusts. Such waste is accumulated at the power stations, and produced under strict conditions. When the Royal Commission reported in 1976, they found a "lack of clarity" about where responsibility rested for determining the best strategy for dealing with such waste.[20] Intermediate-

---

[19] Royal Commission on Environmental Pollution, Sixth Report, *Nuclear Power and the Environment*, Cmnd. 6618 (1976), paras. 353–354.
[20] Royal Commission on Environmental Pollution, Sixth Report, *Nuclear Power and the Environment*, Cmnd. 6618 (1976), para. 364.

level waste in different forms arises from reprocessing, particularly the fuel element cladding which is removed from the irradiated fuel.

High-level liquid wastes   1–21

The high-level wastes contain fission products left after uranium and plutonium have been extracted from irradiated fuel. They must be isolated from the environment for extremely long periods, and will initially generate so much heat as to require active steps to remove the heat and thus ensure the integrity of containment. The waste is in the form of an acid solution, containing numerous fission products, including the elements known as actinides (*e.g.* thorium, uranium, neptunium, plutonium, americium and curium) many of which are extremely radio-toxic. The composition of the waste will depend on the type of fuel reprocessed.

The acid is generally allowed to boil for a period under its own radioactive heat to reduce its volume. It is kept in elaborate stainless steel storage tanks with special cooling. Removal of decay heat over periods of a few hundred years is essential to avoid the solution boiling dry and releasing volatile materials into the atmosphere. Whilst such storage is acceptable as an interim measure, a longer term solution is obviously required; this could involve the vitrification of the waste by heating together with glass-making materials. Development of the process began in the 1950s, but was seriously delayed, which the Royal Commission found strange, given the importance of the technology for safe waste management.[21]

The question of ultimate disposal raises very difficult issues as to whether irretrievable disposal is to be preferred to indefinite storage. In 1976, the main options which appeared to the Royal Commission to exist were disposal to geological formation on land, disposals to the deep oceans, or disposal in the deep ocean bed. Three other possibilities of dramatic, if not science-fictional aspect were suggested but rejected as fundamentally unsound: removal by rocket to space, subduction into the earth's mantle between the edges of tectonic plates in ocean trenches, and deep burial in Antarctic or Greenland ice-sheets.

Privatisation of the electricity supply industry   1–22

The Government's intention to privatise the electricity supply industry in the United Kingdom was announced in the 1988 White Paper, *Privatising Electricity*.[22] The original intention was for nuclear power stations to be

---

[21] Royal Commission on Environmental Pollution, Sixth Report, *Nuclear Power and the Environment*, Cmnd. 6618 (1976), para. 381.
[22] Cm. 322 (1988).

privatised along with the fossil fuel stations. This decision was taken against the background of the Government's view that the existing nuclear stations should continue to operate in the interests of diversity of supply and reduction of polluting emissions. Furthermore, the cost of continued generation was low compared with the high "back end" costs, including the cost of spent fuel reprocessing, radioactive waste management and disposal, and decommissioning.

It soon became apparent that potential investors in a privatised nuclear industry were seriously concerned on various counts, including the "back end" costs referred to above, the relatively poor performance of some AGR stations, and the costs of financing the number of new PWR power stations which at that time were proposed. A report of the National Audit Office in June 1993 drew attention to the high costs of decommissioning old nuclear power stations and other facilities (estimated at some £18 billion), and cast doubt on the adequacy of financial provision being made. Advice to the Government was that potential investors would be unwilling to take on the risks associated with those issues in the absence of unprecedented guarantees and underwriting by the Government.

The Government's judgment was that the provision of such guarantees would not be in the interests of taxpayers, and in November 1989 it was announced that the Government would retain all U.K. nuclear power stations within the public sector, in a decision described at the time by some commentators as a "humiliating withdrawal".[23] The nuclear power stations operated by the CEGB were accordingly vested in a new company, Nuclear Electric, and the SSEB stations were vested in Scottish Nuclear.

Arrangements were put in place to secure the short-term position of existing nuclear power stations. In England and Wales, this took the form of the non-fossil fuel obligation (NFFO), orders made under the Electricity Act 1989 by the Secretary of State requiring public electricity suppliers to contract for specified amounts of generating capacity from non-fossil fuel sources, including nuclear power. Regional electricity companies (RECs) fulfilled this obligation by contracting with Nuclear Electric for all its capacity and output to 1998. The additional costs incurred by the RECs in meeting this obligation (which was later applied not only to nuclear power but also to various renewable sources) was reimbursed through the fossil fuel levy, charged as a fixed percentage of the value of electricity produced and sold in England and Wales.

In relation to Scotland, these arrangements took the form of the Nuclear Energy Agreement (NEA) between Scottish Nuclear and the two Scottish public electricity supply companies; by this agreement, Scottish Nuclear was required to sell all its output to Scottish Power and Hydro-Electric, who were required to take that output. The NEA covers the period up to the year 2005.

Both Nuclear Electric and Scottish Nuclear are underwritten by the Government, by way of arrangements in respect of funds required to meet

---

[23] *The Independent,* November 10, 1989.

their nuclear liabilities as they fall due, and which these companies would not otherwise be able to provide.

The future of nuclear power
1–23

At the time of electricity privatisation, the long-term future of nuclear power in the United Kingdom was uncertain. The Government decided, therefore, to impose a moratorium on further new nuclear construction in the public sector at least until 1994, pending review of the prospects for nuclear power. Existing nuclear power stations were to continue to operate, as described above, and the construction of existing projects was to be completed.

The Government published the conclusions of its nuclear review in a White Paper in May 1995 entitled *The Prospects of Nuclear Power in the U.K.*[24] This involved consideration, with the help of a number of professional advisers, of the feasibility of privatisation of the nuclear generating companies, the prospects for introducing private sector finance for the nuclear industry, the most appropriate way of managing nuclear liabilities, and the strength of the case for new nuclear construction.

The White Paper referred to a number of "significant developments" within the nuclear industry since electricity privatisation, including an apparent improvement in operating performance of the AGR stations, and an improved commercial environment. The latter consisted of new long-term contractual arrangements between Nuclear Electric, Scottish Nuclear, and BNFL, relating to fuel fabrication and processing, which embodied significant cost reductions and enhanced the competitive position for nuclear power.

The White Paper also suggested that the actual costs of decommissioning some of the early Magnox stations had confirmed (at least for the early stages) that cost assumptions made in 1989 were pessimistic. The improved productivity of nuclear plant (by as much as 40 per cent over projections for the year 2000) has been ascribed to improved performance of the AGRs and to tighter application of commercial disciplines.[25]

The Government's policy towards nuclear power, as expressed in the 1995 White Paper, is essentially non-interventionist. The Government's aim is to secure diverse and sustainable supplies of energy at competitive prices; an aim best achieved, in the Government's view, through the operation of open, competitive markets. The Government therefore proposes to leave the market to take decisions about the relative merits of new electricity generating projects; the Government deems that it would not be appropriate to apply any less stringent test to proposals for nuclear power stations built with public money than the market would apply to private projects. The Government's position, expressed at para. 3.43 of

---

[24] Cm. 2860 (1995).
[25] see: DTI, *Energy Paper 65* (March 1995); and *Climate Change: The U.K. Programme—Progress Report on Carbon Dioxide* (HMSO, 1995).

the White Paper, is that nuclear power plays a key role in meeting the U.K.'s energy needs and should continue to do so, "provided it maintains its current high standards of safety and is competitive".

On the issue of competitiveness, Chapter 4 of the White Paper considers the commercial case for new nuclear power stations. The test adopted by the Government is that a new nuclear power station may only be considered genuinely commercial if it is capable of attracting investment on the same terms as comparable projects in the private sector (the current comparable projects being gas-fired combined cycle gas turbine (CCGT) power stations). On the basis of information provided by the nuclear industry, the Government concluded that the best economic case for a nuclear power station in the short term would be a twin PWR at Sizewell C, and that in current market conditions and at current prices for gas, it was unlikely that such a project would provide a rate of return competitive with a CCGT station.

The Government emphasised in the White Paper that this was simply "a snapshot of the position at this particular point in time" and that various factors could change that balance. The Government was also advised that private finance for a new nuclear station was unlikely to be forthcoming under current conditions, in the absence of a transfer of nuclear risks away from private investors to some other party in the project (*i.e.* Nuclear Electric or the Government). Nonetheless, the Government wishes to encourage private sector operators to investigate the construction of nuclear power stations on a fully commercial basis, whilst emphasising that public sector support for new nuclear stations could only be justified on the basis of compelling strategic needs. Whether such needs exist is considered in Chapters 5 and 6 of the White Paper.

On strategic environmental issues, the Government's conclusion is that new nuclear stations are not required in the near future on emissions abatement grounds. Coming to this conclusion, the Government takes due account of existing nuclear power stations which play a significant role in helping to meet the United Kingdom's current commitments regarding the limitation of carbon dioxide emissions. It also recognises that there could be problems in meeting such targets if existing nuclear power stations are not replaced by other generating facilities which do not emit carbon dioxide when they reach the end of their working lives (some time after 2010).

In the longer term, a case for new nuclear stations could be strengthened if the Government was obliged to achieve further substantial reductions in carbon dioxide emissions; to some extent, therefore, the future of the nuclear industry depends upon international and domestic policies relating to climate change. The White Paper also concludes that the rigorous regulatory regime within which nuclear power operates, supported by the safety record of United Kingdom nuclear power stations, indicates that there are no disproportionate environmental disbenefits from nuclear power. It should be emphasised, however, that operating safety is only part of the issue, and that the longer-term issues are those relating to decommissioning and waste disposal.

Another strategic issue is that of diversity of fuel supply; here the Government's view is that there is no reason to suppose that the market will not of its own accord provide an appropriate level of diversity, and that there is currently no case for public financial support for new nuclear power stations on diversity grounds.

A further strategic issue relates to the maintenance of a viable nuclear industry, to keep the nuclear option open, and to provide benefits in terms of export potential and technology spin-offs. The Government concludes that it may reasonably assume that existing technology will not be lost, even in the absence of new nuclear stations being constructed, and that the option to build nuclear power stations will be available for "some time to come".

Export opportunities are seen as focusing predominantly in the areas of nuclear fuel cycle services, decommissioning, and consultancy for refurbishment of nuclear plant. The Government sees the international market for construction of new nuclear power plant as being fragmented and problematic in any event, but suggests that the United Kingdom nuclear industry will remain well placed to win orders for plant and equipment for overseas nuclear power stations as a sub-contractor.

Proposals for privatisation of the nuclear industry   1–24

The 1995 White Paper also considered the case for privatisation of the public nuclear generating companies. The Government accepted the view of the industry that it would not be practicable to privatise the nine old Magnox stations, essentially because they would not generate enough cash over their remaining lifetime to meet the accrued liabilities for spent fuel reprocessing, nuclear waste disposal, and decommissioning.

The Government's intention had been to transfer full responsibility for the Magnox stations and their associated liabilities to BNFL, in order to benefit for vertical integration within BNFL's existing Magnox fuel cycle activities. However, BNFL made it clear to the Government that it would not take the Magnox stations without proper financial provision for decommissioning and other liabilities; the Government therefore re-organised the Magnox stations into a stand-alone company, Magnox Electric, to remain state-owned at least for the time being. The liabilities of Magnox Electric have been estimated at some £8 billion. The intention is that after privatisation of the nuclear industry the Magnox stations will be transferred to BNFL, subject to agreement on financial arrangements. At the time when BNFL gave evidence to the House of Commons Trade and Industry Committee (November 1995), talks on the detail of those terms had not yet commenced.

By contrast, the Government believed that privatisation of the AGR and PWR stations was viable, and would bring significant benefits to the nuclear industry, taxpayers and electricity consumers. The intended way forward was to create and then privatise a single holding company with Nuclear Electric and Scottish Nuclear as wholly-owned subsidiaries. It was subsequently announced that the holding company's name would be British Energy Plc, with headquarters in Edinburgh. It was hoped that the retention of the actual operation of power stations within the existing companies holding site licences would avoid the need for a protracted wholesale re-licensing exercise. However, the Nuclear Installations Inspectorate in its evidence to the House of Commons Trade and Industry Select Committee on Nuclear Privatisation (Memorandum, para. 15), indicated that as a consequence of the proposed restructuring, the Inspectorate had determined that the changes were sufficient to require re-licensing of all 16 Nuclear Electric and Scottish Nuclear sites (see also para. 1–28).

Doubt as to whether the sale would in fact take the form of a flotation emerged in February 1996, when it was revealed that the Government had been in discussion with the U.S. nuclear operator, Duke Power, as to a possible trade-sale of the assets.

1–25 Provision for long-term liabilities on privatisation

Privatisation of nuclear power generation, of course still leaves the issue of long-term nuclear liabilities in relation to the AGR and PWR power stations, currently estimated at about £6 billion. The Government's expressed aim is to ensure that the privatised companies will meet these, and all other obligations, in full. So long as nuclear generation remains in public ownership the creation of a segregated fund to meet nuclear liabilities provides no additional assurance. However, the licensing of private sector nuclear operators would raise starkly the question of assured availability of funds to meet long-term liabilities. The Government's view was that adequate assurance in this respect can be achieved by means of creating segregated funds, and that the detailed implications of such an approach should be considered as privatisation progressed.

The Government indicated in evidence to the Commons Select Committee on Trade and Industry in November 1995, that a funding company would be created, owned by an independent trust, operated by actuaries, and working under contract with the privatised nuclear companies. Money provided by the nuclear companies was to be invested by the trust company, subject to actuarial advice. The fund will not, however, as presently envisaged, cover the cost of reprocessing accumulated spent fuel, or long-term management of nuclear waste. The Government subsequently announced that the fund would be set up as from March 31, 1996. In March 1996, it was announced that British Energy would contribute a lump sum of £225 million into the fund, topped up by payments of £15 million a year for the next 30 years. Subsequently, in responding to the Commons Select Committee Report, in May 1996, these figures were revised to an initial contribution of £228 million (£157 million by Nuclear Electric and £71 million by Scottish Nuclear) and an annual contribution of £16 million over 40 years. Investment income from such a fund will not be trading income capable of being set-off against trading losses under the Income and Corporation Taxes Act 1988 (see *Nuclear Electric plc v. Bradley (Inspector of Taxes))*.[26]

Privatisation and U.K. Nirex   1–26

Another aspect of long-term liabilities relates to the activities of United Kingdom Nirex Limited (the company jointly owned by Nuclear Electric, Scottish Nuclear, BNFL and the United Kingdom Atomic Energy Authority), which is responsible for constructing a deep underground repository for the disposal of intermediate level radioactive waste. The issue of radioactive waste disposal is discussed in more detail in Chapter 5, but the Government's review considers whether any change is warranted in the ownership of Nirex, particularly since nuclear privatisation would result in the creation of new entities producing and holding radioactive waste requiring disposal.

The Government's view is that the existing broad structure of Nirex is capable of incorporating such new entities by the integration of any new privatised generating company into its existing structure, with appropriate share holdings. The various shareholders in Nirex are currently bound by a shareholders' agreement which establishes the basic operating regime for Nirex and sets out the various duties of the shareholders; these include provision of the necessary funding, based upon the volumes of waste produced by each shareholder.

It is intended that the President of the Board of Trade will continue to hold a special share in Nirex on behalf of the Government and, as such, his consent would be required to any changes to the shareholders' agreement. The current shareholders have agreed in principle to include within the shareholders' agreement specific undertakings to follow Government policy on radioactive waste management and disposal, as set out from time to time.

Privatisation and financial support structures   1–27

In relation to the existing financial support structures, the Government's Review concluded that there are no convincing arguments in favour of extending the nuclear non-fossil fuel obligation, nor the nuclear levy, beyond 1998. It was also stated that there would be a corresponding withdrawal of support by way of the premium price payable under the Nuclear Energy Agreement in Scotland. However, in

---

[26] [1995] STC 285; [1995] STC 1125; *The Times*, March 29, 1996.

December 1995, it was announced by the Office of Electricity Regulation that the levy would not be abolished altogether on privatisation (which had been considered as a possibility), but would remain at 10 per cent for 1996, and would then be reduced on privatisation.

1–28 Privatisation: the outcome

The offer prospectus for British Energy was published on June 26, 1996, indicating an unprecedented wide range of values for the company, from £1.26 to £1.96 billion—reflecting the pricing difficulties involved and contrasting with the £2.6 billion which the Government had in 1995 anticipated that the privatisation would raise. Both the main and pathfinder prospectuses referred to the possibility of economic performance being affected by tighter regulation. The general perception of investment advisers was that the company would be a good investment in the short-term, but with somewhat uncertain long-term prospects. The long-term uncertainty related to the absence of any plans for expansion of the nuclear power industry and the forthcoming costs of decommissioning. In the medium-term the outlook for electricity prices, and their implications for profitability, were also regarded as important.

There appeared to be greater interest amongst private investors in the offer than within the financial institutions. Closing of the public offer on July 10, 1996 coincided with the shutdown by British Energy, for safety reasons, of Hinkley Point B and Hunterston B reactors to check for weld cracks, an irony not lost on the press, and one prompting a Stock Exchange investigation as to whether any announcement was necessary. A few days later it was suggested that difficulties had also been encountered with leaks in fuel pins at the Sizewell B reactor. By July 15, shares were priced at 203p, valuing the company at £1.4 billion, less than half the cost of building Sizewell B. The value of shares fell sharply in trading on their first day, described by the *Financial Times* (July 16, 1996) as "... the worst debut by a newly privatised company in nearly a decade." British Energy subsequently confirmed the safety of the reactors at Hinkley Point B and Hunterston B, following which the share price improved somewhat. The unfortunate episode however served to illustrate the volatility of share prices in the industry to output and safety issues.

1–29 The future of the Atomic Energy Authority

Whilst the AEA was, as explained above, initially involved in designing and operating nuclear facilities, it became clear during the 1960s that the research undertaken for nuclear purposes could have potentially lucrative non-nuclear applications. From 1965, the AEA was authorised to embark upon non-nuclear research and development, on a commercial basis. With the decline in orders for new nuclear power stations which has occurred since the 1960s, greater emphasis has had to be placed on developing the AEA's non-nuclear business. In May 1992, the AEA was subject to a report by the Monopolies and Mergers Commission,[27] which considered in detail the structure, organisation, and activities of the Authority. By 1995, the AEA employed about 7,000 employees based at six main sites in the United Kingdom—the largest at Harwell in Oxfordshire—and, after reorganisation in 1993, was structured into three operating divisions as follows:

1. COMMERCIAL DIVISION — (known as AEA Technology) which comprises all the AEA's commercial activities.

2. GOVERNMENT DIVISION — which is responsible for decommissioning the AEA's former sites and for managing AEA's liabilities. The Government Division holds the nuclear site licences relating to AEA nuclear facilities. The relevant sites are mainly the experimental nuclear reactors at Harwell, Dounreay, Winfrith, and Windscale.

3. SERVICES DIVISION — dealing with facilities management services. These have been disposed of to the private sector, or to the other two divisions, or closed.

The Atomic Energy Act 1986 put the AEA onto a trading-fund basis, so that it was no longer vote-funded, but was expected to raise all its income from commercial trading. In 1992, the review of the AEA by the Monopolies and Mergers Commission concluded that the Authority's commercial business activities rested uneasily in the private sector and should be removed from it so far as was practicable.

It was on this basis that the Government introduced in 1995 the Atomic Energy Authority Bill, to enable the privatisation of AEA Technology. This part of AEA was described by the Minister for Energy and Industry (Mr Tim Eggar) in moving the Second Reading of the Bill as:

> "... an international science and engineering business ... It operates in four closely related areas: plant and process performance optimisation, product improvement, safety and risk management and environmental and waste management."[28]

The Atomic Energy Authority Act 1995 thus contains provision for the AEA, if so directed by the Secretary of State, to make a scheme or schemes providing for the transfer of property, rights and liabilities to any person (subs.1(1)). Such a transfer scheme is the precursor to privatisation. By subsection 1(3), no transfer scheme may provide for the transfer of a nuclear site licence, or of the freehold of land comprised in a site in

[27] *U.K. Atomic Energy Authority: A Report on the Service Provided by the Authority,* Cm. 1947 (1992).
[28] *Hansard,* H.C. Vol. 256, col. 701.

respect of which a site licence held by the AEA is in force; this reflects the fact that licences will continue to be held by the Government Division. The transfer scheme may apportion property, and create mutual rights between the AEA and transferee (Sched. 1, paras. 2–5). This is important, since the AEA Government Division and the new companies will, in many cases, be operating at the same sites. A transfer scheme was made by the AEA on March 7, 1996, transferring certain assets to a publicly-owned company, AEA Technology plc, pursuant to a direction by the Secretary of State.[29]

The Government has been at great pains to emphasise that these new arrangements do not involve any transfer of nuclear liabilities from the public sector. However, the Government and AEA are committed to increasing competition for decommissioning and radioactive waste management operations work (rejoicing in the acronym, DRAWMOPS). This involves putting work out to tender to AEA Technical Division among other contractors. The intention is not only to achieve value for money, but also to transfer some risk to the private sector.[30]

This strategy appears to be having some effect in reducing costs, when coupled with the separation of functions within AEA. The Government Division of AEA reported in August 1995 that its estimated total projected costs for decommissioning the facilities for which it is responsible had been cut from £8.2 billion to £7.5 billion. The task of the Government Division in cutting costs has been assisted by the wish of a number of companies to get into the decommissioning business at an early stage, resulting in highly competitive pricing.

1–30 Weapons production

When the United Kingdom Atomic Energy Authority was reorganised (see para. 1–09 above), its Weapons Group was transferred to the Secretary of State for Defence by the Atomic Energy Authority (Weapons Group) Act 1973. Subsection 1(1) of that Act provided that the Weapons Group would cease to form part of the Authority and that, from the appointed day, it should be for the Secretary of State (and not the Authority, except under contract to the Secretary of State, or by his direction, or with his approval) to carry on any activities which before that day were activities of the Group, and involved working on explosive nuclear devices. The appointed day for this purpose was April 1, 1973 (S.I. 1973 No. 463).[31]

These activities were carried out at establishments at Aldermaston and Burghfield in Berkshire, Foulness in Essex, and Cardiff. As activities

---

[29] see also: AEA Technology plc (Capital Allowances) Order 1996 (S.I. 1996 No. 2101), determining certain writing-down allowances and capital allowances available to the company as successor.

[30] see: H.C. Standing Committee D, First Sitting, col. 26 (March 23, 1995).

[31] Atomic Energy Authority (Weapons Group) Act 1973 (Appointed Day) Order 1973 (S.I. 1973 No. 463).

carried out by the Group, they were not required to be licensed under section 1 of the Nuclear Installations Act 1965. Following the Government's decision to consider contracting out the management of these establishments, the Atomic Weapons Establishment Act 1991 made provision for designated activities to be carried out under contract.

Section 1 of the 1991 Act allows the Secretary of State to designate activities and premises for the purposes of the Act. Any activities may be designated if they are connected with the development, production, or maintenance of nuclear devices, or with research into such devices and their effect. The Act then applies if the Secretary of State makes arrangements: (a) for a company formed under the Companies Act 1985 to carry on designated activities at the premises under contracts with him, and (b) for that or another company to become the employer of employees from the undertaking, and to acquire rights in respect of the premises. Designation of the four establishments was effected by the Atomic Weapons Establishment (Designation and Appointed Day) Order 1992 (S.I. 1992 No. 2743), giving March 31, 1993 as the appointed day for the purpose of determining the relevant employees. The Act provides for the jurisdiction of the Ministry of Defence Police in respect of the relevant premises (s.4), and makes provision as to the application of the enactments referred to in the Schedule (s.3). Under this Schedule, the position so far as the Nuclear Installations Act 1965 is concerned, is that the use of a site in designated premises by the contractor is treated as use by a Government department, and accordingly does not require a nuclear site licence (Sched., para. 6(1)). However, the contractor is subject to the provisions of the 1965 Act on liability and insurance as if he was a licensee (para. 6(2)). For the purposes of section 13 of the Radioactive Substances Act 1993 (dealing with authorisations for the disposal of radioactive waste), a nuclear site within designated premises is treated as if a nuclear site licence were in force.[32] It should be noted that these provisions may be amended by ministerial order under subsection 3(2) so as to remove or curtail, but not extend, any privilege or immunity for the time being provided by the Schedule. It is anticipated that the immunity from nuclear site licensing will be removed in this way.

The new arrangements took effect at the establishments on April 1, 1993. A new company, AWE PLC, is the employer of the relevant employees, and a contract for management of the establishments was awarded to a consortium, Hunting-BRAE.[33] Shortly thereafter, the Health and Safety Executive announced its intention to carry out a comprehensive review of health and safety management systems and standards of risk control at all four establishments.[34] This review resulted in a partially critical report published in October 1994.[35] The review was described by

---

[32] Para. 10A, inserted by the Radioactive Substances Act 1993, s.49(1) and Sched. 4, para. 10.
[33] *Hansard*, H.C. Vol. 183, col. 183.
[34] *HSE News Bulletin—E75:93* (May 1993).
[35] *The Management of Health and Safety at Atomic Weapons Establishment Premises—Review by the HSE* (HSE Books, 1994).

the Director General of the HSE as "one of the most comprehensive exercises the HSE had ever conducted". Whilst not criticising the new management, the HSE found that they had taken over long-standing problems in terms of inadequate safety systems, records, equipment, and emergency procedures. Some 65 deficiencies were identified, requiring rectification within a year; 19 required immediate action.

The HSE issued a prohibition notice under its general powers in the Health and Safety at Work, etc. Act 1974, thereby preventing operations in the part of the plant where enriched uranium was machined; conditions there, the HSE found, could present the danger of a runaway nuclear chain reaction. The HSE also strongly recommended that immunity from nuclear site licensing requirements should be removed so as to allow for effective control.

1–31 Reactor safety

The issue of reactor safety and siting was addressed by the Royal Commission in Chapter 6 of its Sixth Report.[36] It noted that an accident leading to an uncontrolled release of radioactive material could have highly unpredictable consequences depending on a number of factors including the substances released, and the wind direction at the time. Most concern, the Commission noted (para. 262), centred on two issues: whether in the event of a potentially dangerous situation developing, the reactor could be shut down quickly; and whether the residual decay from the fission products could then be removed:

> "If the safeguards provided to cover these contingencies were to fail, the temperature might rise to values at which the fuel would melt and interactions involving the molten fuel within the core could develop pressures that would be sufficient to rupture the containment, so allowing a release of the gaseous and more volatile fission products."

To date worldwide, there have been three serious reactor accidents, although only one involved the worst scenario outlined by the Royal Commission in the extract above. These were the Windscale fire in the United Kingdom, Three Mile Island in the United States, and Chernobyl in the former Soviet Union. Each is briefly described below.

1–32 Windscale

The two reactors (or "piles") at the site named Windscale on the Cumbrian Coast were constructed after the Second World War in a disused ordnance factory and were intended, like the corresponding

---

[36] Royal Commission on Environmental Pollution, Sixth Report, *Nuclear Power and the Environment*, Cmnd. 6618 (1976).

installations in the United States, France and the Soviet Union, for the production of weapons grade plutonium. They were fuelled by natural uranium, clad in aluminium and positioned in horizontal channels in a core of graphite. They were cooled by air, blown through cooling channels by powerful fans, and discharged into the air through a 126 metre stack.

A serious (potentially catastrophic) fire occurred in Number 1 Pile in October 1957. The fire resulted from the decision of the engineer to boost power during the course of a routine operation known as a "Wigner release". During this operation the uranium thermocouples recording heat were not at the hottest part of the core, which was hotter than the engineer realised. The boost of energy led to the ignition of the graphite within the reactor. By the time the fire was detected, two days later, the fuel rods were red-hot and could not be removed. There was a real risk that the heat could have affected the integrity of the concrete shielding. An unsuccessful attempt was made to extinguish the fire with carbon dioxide and eventually, in the early hours of October 11, a decision was made to call in the emergency services. Fire hoses were coupled to fuel channels and by this means the fire was brought under control, and finally extinguished. This, in itself, was an extremely risky course of action, given the possibility of an oxidising and explosive reaction between the water and the molten metal.

It became clear that the fire had released a cloud of radioisotopes, some of which had been discharged from the stack. Studies at the time (referred to in Cmnd. 302, *Accident at Windscale No. 1 Pile*), identified iodine 131 among these; this was hazardous because of its high activity and its potential involvement in the functioning of the thyroid gland. Milk produced over a 500 square kilometre area (about 2 million litres) was poured away; farmers were paid about £50,000 in compensation.

The No. 1 Pile was written off, and No. 2 Pile was closed while enquiries were carried out. It became clear that it would be prohibitively expensive to modify No. 2 Pile to reduce the risk of a recurrence of the fire; both reactors were therefore sealed with concrete.

Three Mile Island   1–33

The accident of March 1979, at the Three Mile Island (TMI) power station at Harrisburg, Pennsylvania, was both serious and costly. It attracted particular attention in the United Kingdom in the context of the public inquiry as to the construction of a PWR at Sizewell in Suffolk. The episode provides an all too graphic demonstration of the uncertainty and panic which may attend a serious nuclear incident. The initial problem was a failure of pumps providing feed-water to the steam generators of the reactor. This led to the turbine being automatically shut down, which in turn resulted in an increase in both temperature and pressure within the primary cooling water of the reactor. A relief valve was opened automatically to reduce the pressure, but remained open undetected when it should have closed, thereby causing coolant to be lost. Within two minutes, the reactor pressure fell so as to trigger emergency high pressure pumps providing cooling water. However, the operators, in the belief that the water level in the reactor was rising too fast, switched off this flow. The coolant began to boil and the level fell below the top of the reactor core; the temperature of the exposed fuel rods soared and began to melt, resulting in a chemical reaction which produced a bubble of the potentially explosive gas, hydrogen.

Meanwhile, the water which was escaping through the open relief valve overflowed the tank into which it was meant to drain. Sealing discs were ruptured, and 600,000 gallons of coolant water, contaminated by radioactivity from the damaged fuel, flooded the reactor sump and the basement of the building. Radioactive fission products entered the ventilation system and were released as gas into the atmosphere.

The relief valve was eventually identified as open and was shut immediately. Concern then focused on whether the hydrogen formed inside the reactor vessel might explode, breaking the containment building. In fact, a smaller explosion had already occurred within the reactor, a fact that was only discovered later. While experts from the Nuclear Regulatory Commission agonised over this risk, the Governor of Pennsylvania took the decision to recommend the evacuation of children and pregnant women within a five mile radius of TMI. In fact, a far greater number, approximately 11,000 people, left their homes.

Whilst the feared explosion never happened, the costs of the accident at TMI were enormous. Operations to assess the damage, clean up the huge quantities of contaminated water, purging radioactive gas, and defuelling the reactor, took many years. It was not possible to view the reactor core, even through a remote control camera, for three years, when the entire core was found to be damaged, even reduced to rubble in some parts. Tests confirmed that temperatures had exceeded 2,800°C during the incident, and that fuel had in fact melted. The recovery operation was estimated to cost $1,000 million; material damage to the reactor was estimated at $200 million. Two U.S. nuclear pools provided cover, subject to a limit of $300 million, which was fully expended. The pools also provided third-party insurance; claims under a class action for loss of earnings and business interruption were settled in 1981 for approximately $20 million, and a fund of $5 million was set aside to provide long-term medical surveillance for local inhabitants and their children. Further claims were made for psychological harm and for expenses incurred by municipalities. The total claims to be met from liability insurance have been estimated to be around $50 million.[37] By 1996, about $56 million had been expended in settlements and legal expenses. Some 2,100 claims were pending in 1996 for punitive damages,

---

[37] Figures taken from: James C. Dow, *Nuclear Energy and Insurance* (1st ed., 1989), p. 411.

following a ruling of the U.S. Third Circuit Court of Appeals that Federal nuclear legislation does not prohibit such claims.[38]

The accident also had wider repercussions, representing a substantial setback for the nuclear industry in the United States and in Sweden, and encouraging the concerns of the powerful anti-nuclear lobby in Germany.

Chernobyl   1–34

The accident at Unit 4 of the Chernobyl power station near Kiev in the Ukraine, in April 1986, eclipsed both Windscale and TMI in terms of severity. The nuclear plant was massively damaged, deaths resulted, and there was serious environmental damage caused by fallout, affecting not only the local population, but also having consequences in other countries.

The accident resulted from an experiment undertaken to test for arrangements on external power failure; this involved reducing reactor power to about 25 per cent of normal. Contrary to operating rules, this was maintained for 24 hours, leading to dangerous instability in the reactor because of its inherent design. When the power increased rapidly as a result, the engineers did not (as they should have) shut the reactor down, but rather attempted to bring the power down rapidly. This caused a phenomenon called "xenon poisoning" to the reactor system, which made control difficult. Coolant flow to the reactor became insufficient, resulting in the reactor becoming "super-critical". The dramatic result was a 100-fold increase over normal operating power, which occurred within 4 seconds. The phenomenal energy released ruptured the fuel elements, which led to an interaction of molten fuel and coolant, causing massive steam explosions, blowing the lid off the reactor vessel and destroying the cooling circuits. A further explosion projected burning molten nuclear fuel to a great height. Secondary fires then caused radionuclides to be released for up to the next 10 days.[39]

A major and heroic operation succeeded ultimately in bringing the fires under control, and stopping the nuclear reaction within the red-hot core. This resulted in 31 immediate deaths from radiation exposure, and 203 casualties. The work involved helicopters dropping 800 tons of dolomite, 40 tons of boron carbide, 1,800 tons of clay, and 2,400 tons of lead into the core, and local miners digging beneath the reactor to strengthen foundations and prevent collapse.[40]

The population of the nearest town, Pripyat, and a number of other villages were evacuated, a total of 135,000 people. It is too soon to say what

[38] see: DYP, *Liability Risk and Insurance—No. 64* (1995).
[39] see further: *The Chernobyl Accident and its Consequences—U.K. Atomic Energy Authority* (2nd ed., HMSO, 1988).
[40] Peter Gould, *Fire in the Rain: The Democratic Consequences of Chernobyl* (Polity Press, Oxford, 1990), pp. 17–18.

the health consequences of exposure will be, but fatalities from all types of cancer due to the incident have been predicted at 1,000. A massive relief and decontamination operation followed, together with the construction of new settlements to house the evacuees, the removal of contaminated soil and vegetation, and the creation of a 30 km controlled zone around the reactor.

Whilst Unit 4 was obviously a complete write-off, the other three units have continued to operate. A first priority was to provide a decontaminated strip of road through the surrounding area to allow access by operating staff. Unit 4 was sealed with clay, sand, boron, lead, and dolomite, with concrete reinforcement, and is now effectively entombed in a "sarcophagus". The initial estimated cost to the Soviet economy was £2,000 million; this is now estimated at nearer £8,000 million.[41] The power station was state-owned and uninsured, and since the Soviet Union was not a party to the Vienna Convention on Liability for Nuclear Damage, there was no clear mechanism for third-party claims.

Many other countries were affected by the radioactive cloud resulting from the incident. These included Poland, Germany, Italy, France, and Greece. The United Kingdom was less seriously affected, but sales of lamb and other produce from affected areas have been subject to restriction for some years by orders under section 1 of the Food and Environment Protection Act 1985. These restrictions have been gradually removed on the basis of annual monitoring results. There remains a divergence of view regarding the continuing effects of the Chernobyl accident on farms in the United Kingdom. Following the derestriction order in January 1996, according to official sources some 13 agricultural holdings, with about 21,000 sheep post-lambing, remained subject to restriction.[42] However, a newspaper article (*The Independent*, April 22, 1996) under the headline "Nuclear Cloud Hangs Over The Hills" claimed that 390 farms, with 220,000 sheep, remained "under the Chernobyl cosh".

Foreign claimants, not compensated by their own Governments, would face the difficult prospect of claims against the Soviet government under private international law. Claims have been made against the European Commission and Council by KYDEP, a Greek agricultural co-operative, under Article 215 of the E.C. Treaty in respect of the Community recommendations and rules as to maximum tolerances of radioactivity in products following the incident. The claim was subject to an unfavourable opinion from Advocate General Van Gerven,[43] on the basis that there was no provision of Community law obliging the Commission to adopt measures providing for compensation for the affected agricultural sector. Also, in the opinion of the Advocate General, there had been no infringement of the principle of proportionality, in that the Commission

---

[41] James C. Dow, *Nuclear Energy and Insurance* (1st ed., 1989), pp. 414–415.

[42] *Ministry of Agriculture, Fisheries and Food* (MAFF) *News Release 15/96* (January 1996).

[43] Case C–146/91, *KYDEP v. Council and E.C. Commission:* [1994] E.C.R. I–4199; [1995] 2 C.M.L.R. 540.

in establishing uniform maximum tolerances had to take account of divergent approaches within different Member States; factual and scientific information was sparse at the time, and there was no evidence that the figures adopted were unreasonable, or were stricter than necessary.

Since the accident, western Governments, including the United Kingdom, have funded safety upgrading measures at various former Soviet, and East European reactors. A fund, initiated by the G7 Group of countries, is managed by the European Bank for Reconstruction and Development (EBRD). The fund is intended to finance urgent safety improvements, concentrating on the older RBMK-Chernobyl type, and VVER Model 230 PWR reactors, which are perceived as presenting the greatest risks.[44] In December 1995, a Memorandum of Understanding was signed between the G7 countries and the Ukraine on a comprehensive programme of measures designed to close the remaining parts of the Chernobyl plant by the year 2000, with some $500 million in grant assistance, and $1.8 billion in projected investments by the international financial institutions.[45]

Hazards of radioactivity   1–35

Those who discovered radioactivity were not long in realising that it had potentially harmful properties. First of all, radiation can burn. Becquerel suffered such a burn through carrying a vial of radium in his pocket. In 1903, Ernest Rutherford visited Marie and Pierre Curie in Paris; in their garden they demonstrated the brilliant luminosity created by radium, causing zinc sulphide to fluoresce. The resulting light was bright enough to show Rutherford Pierre Curie's hands, "in a very inflamed and painful state due to exposure to radium rays".[46] Radium burns were also an unfortunate by-product of the early medical treatments with the substance.

Next to emerge were the delayed adverse effects of exposure to radiation. One early example was the case of women employed in watch factories, making watches with dials made luminous by zinc oxide and radium. Through repeated licking of the brushes to achieve a fine point, a large proportion of the women suffered bleeding gums and anaemia, and later bone cancer. Another example was provided by the miners of uranium in mines such as the former silver mine at Joachimsthal, who were prone to *Bergkrankheit*, a form of lung cancer identified in the 1930s as being caused by inadequate ventilation in the mines. This awareness of the dangers seemed to cause no great public concern at the time—as one commentator has written:

---

[44] see: *DTI Press Notices P/93/56* and *P/93/332*.
[45] see: *DTI Press Notice P/95/903*.
[46] Richard Rhodes, *The Making of the Atomic Bomb* (Penguin, 1986), p. 45.

"It was as if the controversy over radiation had an even longer latency period than the disease."[47]

1-36 Hazards of ionisation

Radioactive substances may cause harm to living organisms if they are ingested or inhaled, or if they enter the body through a break in the skin. However, certain types of radiation—gamma radiation for example—can easily pass through matter, living or otherwise. Neutron radiation can penetrate matter to a considerable depth before being absorbed. Radiation may have the property of being ionising, a process whereby electrons are dislodged from atoms, creating positively charged atoms or molecules (ions) in the matter through which it passes. When ionising radiation passes through material it causes atomic and molecular structural changes releasing energy; the effect will depend upon how much energy is released into a given amount of material. Where the material in question comprises living cells, the disruptive effect may have serious consequences. These may occur to somatic cells, affecting the living organism itself, or to reproductive cells, potentially affecting offspring or descendants.

Somatic effects may include damage to the digestive or central nervous systems, damage to skin, cataracts on the eyes, cancers such as leukaemia, or loss of fertility. Small exposures of reproductive cells to radiation may give rise to congenital problems, including birth defects or delayed cancers in the offspring.

1-37 Measurement

The hazards of radiation may be measured in various ways according to the purpose of measurement and the requirements of various branches of science—resulting in a confusing variety of units which has been described as "*une belle salade radioactive*".

One method of measurement is simply the level of emitted radioactivity. The standard unit of measurement is the *becquerel* (Bq), defined as one emission of a particle (one transformation) per second. Another unit (now obsolete) is the *curie*, which is 37,000 million (37 billion) transformations per second, the radioactivity of one gram of radium. Such measures do not indicate the impact of the radioactivity, simply the amount.

Other units may be used to describe the effects of radioactivity on living cells. One measure is the amount of radiation absorbed by the cells, *i.e.* the amount of energy deposited in relation to the amount of matter. This unit is the *gray* (Gy) which corresponds to the amount of radiation which

---

[47] Walter C. Patterson, *Nuclear Power* (2nd ed., Penguin, 1983), p. 105.

will cause 1 kg of matter to absorb one joule of energy. The gray superseded as an international unit the *rad* (radiation absorbed dose), which equates to one-hundredth of a gray.

This measure does not take account of the differing responses of cells and tissues to different forms of radiation. Alpha particles and neutrons will have a more disruptive effect on cells for the same amount of energy discharged; they are more "biologically effective". The current unit for this purpose is the *sievert* (Sv), which replaced the *rem* in 1977. Doses for occupational purposes will often be expressed in *millisieverts* (mSv), *i.e.* one-thousandth of a sievert. The *rem* (Roentgen equivalent man) was introduced in 1954 by the International Commission on Radiological Protection (ICRP) to allow for the differing responses to exposure. To give some indication of the nature of the unit, a dose of 4 Sv over the whole body is likely to kill 50 per cent of adult recipients; the background, naturally-occurring, radiation to which people are typically exposed is 2 mSv per year. A dose of between 10–50 Sv will cause failure of the gastro-intestinal tract, extensive internal bleeding, and death within a few days.

The adverse effects of radiation will also depend upon which part of the body is exposed. Very high doses of radiation can kill cells, but lower doses can prevent a cell dividing, or cause it to divide abnormally. The tissues that are most susceptible to such damage are those containing a high proportion of dividing cells; examples are bone marrow, reproductive and foetal tissues. Another risk is the accumulation of radioactivity in bone structures; all inhabitants of the planet have small amounts of radioactive material in their skeletons as a result of fall out from past atmospheric atomic tests.

The International Commission on Radiological Protection in its Recommendations of 1990, lays down what the Commission regards as acceptable dose limits of radiation for those occupationally exposed, and dose limits for members of the public. The issue of radiological protection is considered in more detail below (see para. 1–40 below). The higher figure for workers is attributable to the fact that they are voluntarily exposed, and will be strictly supervised. Dose limits can be, and are, set for various different parts of the body; these may be allocated different weighting factors.

Toxicity   1–38

Some radioactive substances are also extremely toxic, irrespective of their other hazardous properties. In its 1976 Report on *Nuclear Power and the Environment*,[48] the Royal Commission on Environmental Pollution noted that an individual dose of "only a few milligrams" of plutonium would be sufficient, if inhaled, to cause massive fibrosis of the lungs, and death

---

[48] Royal Commission on Environmental Pollution, Sixth Report, *Nuclear Power and the Environment*, Cmnd. 6618 (1976).

within a few years (para. 322); it was the risk of this appallingly dangerous material falling into terrorist hands which led the Commission to question the security measures necessary in a "plutonium economy", and their potential repercussions for civil liberties.

1–39 Environmental levels

The hazards of radioactive substances mean that they need to be monitored carefully in the environment, or when released to the environment. The ICRP's maximum dose levels referred to earlier are supplemented by maximum permissible concentrations (MPCs) of individual isotopes in air and water. These can be used to check the environment of radiation workers.

Where radioactive wastes are released into the environment, a different approach is needed, since various substances may follow a particular pathway, leading to specific groups of people being exposed to undue risk. The approach is to identify a group which is likely to have the greatest exposure, and then to restrict the discharge so as to keep the dose to that group at an acceptable level. This "critical group" may change over time; for example, in relation to aqueous discharges from Windscale into the Irish Sea, the Royal Commission instanced the change from an original critical group of persons eating laver bread made from contaminated seaweed, through to salmon fishermen in the Ravenglass estuary, then to consumers of fish and shellfish caught in the Irish Sea. This change of emphasis was due to changes in dietary habits, and changes in the composition of the actual discharge over time.

1–40 Principles of radiological protection

The process of assuring protection for the public from radiation involves three stages:

    (1) deciding on what should be the maximum allowable dose of radiation, whether to the whole body or some particular part, or organ;

    (2) discovering the connection between discharge of a particular radioisotope at a certain rate and the radiation dose which will result for the most exposed member of the public, or a typical member of a "critical group" (see para. 1–39 above); and

    (3) deciding what discharge will be permitted, taking account of all other discharges and what monitoring and reporting system will be required to ensure that the permitted level is not exceeded.[49]

The setting of radiological standards is not purely a domestic matter, but

---

[49] Royal Commission on Environmental Pollution, Sixth Report, *Nuclear Power and the Environment*, Cmnd. 6618 (1976), para. 220.

involves a number of international agencies, which are examined in the following paragraphs.

The International Commission on Radiological Protection (ICRP)   1–41

The ICRP comprises a group of eminent scientists chosen by the International Congress of Radiology. It was originally established in 1928 as the International X-ray and Radium Protection Committee, and was restructured and renamed in 1950. It is not a part of any inter-governmental organisation, and is accountable to the world's professional bio-radiologists, meeting in congress. Its publications and recommendations are used as an authoritative basis for protection standards, though their status is really that of opinion of eminent experts, based on conclusions from available scientific fact. Its recommendations only become effective through adoption as national law, which may first involve their adoption by one or more of the international agencies referred to below.

Since 1958, the ICRP has promulgated defined doses to members of the public which should not be exceeded, founded on the principle that any dose, however small, is regarded as harmful to some degree; all doses should, therefore, be kept as low as is reasonably achievable.

The 1990 Recommendations of the ICRP identifies three types of exposure: exposure at work, medical exposure as part of diagnosis or treatment, and public exposure (all other forms). The system recommended by the ICRP is based on the following general principles:

(1) no practice involving exposures to radiation should be adopted unless it produces sufficient benefit to the exposed individuals, or to society to offset the radiation detriment it causes (the justification of a practice);

(2) in relation to any particular source of exposure, the magnitude of individual doses, the number of persons exposed and the likelihood of exposure should all be kept as low as reasonably achievable, while taking account of economic and social factors; to offset the unfairness which would inevitably result from taking decisions on economic and social factors on a case-by-case basis, the approach should be constrained by uniform restrictions on doses to individuals and/or risk to individuals (known as the optimisation of protection);

(3) exposure of individuals resulting from a combination of all relevant practices (*i.e.* human activities increasing exposure) should be subject to dose limits, or to some control of risk in the case of potential exposures (individual dose and risk limits). Such limits are aimed at ensuring no individual is exposed to radiation risks that are judged unacceptable in any normal circumstances.

The Recommendations then refer to the three possible types of exposure. For occupational exposure, the Commission recommends a limit on effective dose of 20 mSv per year, averaged over 5 years (100 mSv in total), with the further provision that in no single year should the dose exceed 50 mSv. No specific limits are recommended for medical exposure; the issue there being the compatibility of the dose with the direct medical benefit to the patient.

In relation to public exposure, the Commission recommends control at source (rather than in the environment) by procedures of optimisation and the use of prescriptive limits. The recommendation is that the limit for public exposure should be expressed as an effective dose of 1 mSv in a year, though in special circumstances a higher value of effective dose could be allowed in a single year, provided the average over five years does not exceed 1 mSv per year. The Commission points out that higher doses may be received from natural sources, *e.g.* ambient radon gas, but that the Commission's recommendation is concerned only with exposure as a result of deliberate practices which are a matter of human choice.

1–42 The UN Scientific Committee

The UN Scientific Committee on the Effects of Atomic Radiation (UNSCEAR) was set up by the General Assembly in 1955, and is the prime scientific review body on a global basis. It meets annually, and reports on the effects of radiation from different sources and on the scientific evidence of their effects.

1–43 The International Atomic Energy Agency (IAEA)

The IAEA grew out of President Eisenhower's famous speech of 1953 on "Atoms for Peace", and was founded in 1957 as a UN Agency to promote the peaceful uses of atomic energy; it remains an autonomous organisation. The founding statute of the IAEA sets out various functions, including assisting research, encouraging exchange of expertise in the peaceful uses of nuclear energy, and establishing and administering safeguards to avoid the diversion of nuclear matter to non-peaceful ends. Its sixth authorised function is to establish and adopt standards of safety for protection of health, and minimisation of danger to life and property. It is based in Vienna and has a Board of Governors and permanent staff. The IAEA is the Agency responsible for the Vienna Convention on Civil Liability for Nuclear Damage of 1963 (see para. 3–11 below).

The IAEA publishes numerous materials on all aspects of safety and radiological protection, generally based upon ICRP recommendations (see para. 1–41 above), which include the IAEA's *Basic Standards for Radiation Protection (Safety Series No. 9)*. In its 1982 revision of this document, the IAEA laid stress on the requirement to keep doses as low as reasonably achievable, over and above the basic dose limits. These standards, which are written in the form of regulations, are intended to be used by the competent authorities of Party States, and are thus something more than simply general guidance. The Standards distinguish between foreseeable exposures which can be controlled at source, and abnormal exposure conditions, where exposure can only be limited (if at all) by remedial action. The approach of the ICRP (see para. 1–41 above) is followed in that doses, including exposure to ionising radiations or to radioactive substances, are to be restricted by a system of dose limitation which shall include:

(a) justification of the practice;
(b) optimisation of radiation protection; and
(c) annual dose equivalent limits.

One of the requirements of the Standards is that there should be a system of regulatory control by way of notification, registration and licensing, making it possible for appropriate protection requirements to be imposed. IAEA Safety Series No. 89 lays down principles as to when radiation sources and practices may be exempted from regulatory control, *e.g.* low levels of activity, and naturally-occurring sources. The Standards are approved not only by the IAEA, but were approved or adopted also by the Nuclear Energy Agency of the OECD (see para. 1–44 below), the International Labour Organisation, and the World Health Organisation.

Another important IAEA document is *Safety Series No. 77* on *Protection of the Public and the Environment*, which provides *Principles for Limiting Releases of Radioactive Effluents* during the normal controlled operations of nuclear installations. These follow the basic tenets referred to earlier, *i.e.* practices must be justified by their benefits, and control of releases must be optimised by keeping resulting doses as low as reasonably achievable, while taking account of economic and social factors.

Assessment of doses under these principles involves various stages: identification of the nature of the source and the amount, composition and timing of release; using mathematical models to analyse the possible environmental pathways to man and using models to estimate doses with assumptions as to age, sex and living habits of potential recipients. This process involves the concept of the "critical group" (see para. 1–39 above).

Individual doses can arise from various sources, and the IAEA *Principles* refer to "upper bounds" of individual dose; these are the fractions of the dose limit for members of the public allocated to the various sources and practices which may gave rise to exposure. This provides a formalised way of ensuring that contributions to individual dose from all sources, present and future, are taken into account when setting limits for releases from a particular source. IAEA *Safety Series No. 92* takes this issue further by reference to the sources which may give rise to global doses, relying on the information collected and reviewed by UNSCEAR (see para. 1–42 above).

The IAEA is also working to prepare a number of publications under its Radioactive Waste Safety Standards (RADWASS) programme. These include a safety fundamentals document, *The Principles of Radioactive Waste Management,* and a Safety Standard on *Establishing a National System for Radioactive Waste Management.* Further details on the approach of the IAEA to the issue of radioactive waste management are given at para. 5–13 below.

As well as its various advisory documents, the IAEA operates through Conventions; the Vienna Convention on liability has already been referred to above, while the IAEA's Convention on Nuclear Safety was opened for signature in September 1994, and negotiations are in hand for a Convention on the Safety of Radioactive Waste Management.

1–44 The Nuclear Energy Agency (NEA)

The NEA, created in 1956, is a regional grouping, broadly comprising the Western European States, together with the USA, Canada, Japan, Australia, and New Zealand. It was established within the framework of the OECD, and, as with the IAEA, has the general purpose of furthering the development of peaceful uses of nuclear energy; its objects include contribution by the responsible national authorities to promotion of the protection of workers and the public, and of the preservation of the environment, as well as the promotion of the relative safety of nuclear installations. Another object is contribution to promoting a system of third-party liability and insurance with respect to nuclear damage; this contribution takes the form of the Paris and Brussels Supplementary Conventions (see paras. 3–04 to 3–10 below). The NEA, like the OECD, bases its standards and principles on ICRP recommendations.

The Radioactive Waste Management Committee of the NEA publishes collective opinions on various issues; for example, its 1991 *Opinion on Current Scientific Methods for Conducting Safety Assessments of Radioactive Waste Disposal Systems.*

The NEA has its own stable of publications on nuclear safety, radiological protection, and radioactive waste management, as well as general nuclear science. These include a useful comparative study of legislation on compensation for nuclear damage.[50]

1–45 EURATOM: The Treaty

The European Atomic Energy Community (EURATOM), unlike the IAEA and NEA, has power to set and enforce binding radiation protection standards. The background to the creation of EURATOM has

---

[50] NEA, *Liability and Compensation for Nuclear Damage—An International Overview* (1994).

been usefully outlined in an article by Dr Linda Spedding.[51] The article points out that EURATOM was created by treaty to assist the development of a civil nuclear industry in Europe, and to "ensure for Europe a place in the energy revolution". In 1957, when EURATOM was created, atomic power was considered the means by which national differences in economic growth could be minimised, and the gap between domestic energy supplies and increasing demand for energy within Member States could be reduced.

The creation of the Community was also intended to promote development of the peaceful uses of atomic energy through co-operation, given that the resources of individual Member States might be insufficient in this respect.

EURATOM must also be seen within the context of the development of the other European Communities; its creation followed the conclusion of the European Coal and Steel Community Treaty in 1951, and coincided with the creation of the European Economic Community by the Treaty of Rome in March 1957. The three Treaties were supplemented in 1965 by a Merger Treaty which brought together the Councils of the three Communities. Whilst each Treaty therefore contains different objectives, the three Treaties should be considered as a whole.

The Treaty establishing the European Atomic Energy Community was signed in Rome on March 25, 1957 (the same date as the Treaty establishing the European Economic Community). Article 1 states that it is the task of the Community to contribute to the raising of the standard of living in Member States, and to the development of relations with other countries, by creating the conditions necessary for the speedy establishment and growth of nuclear industries. In order to perform this task Article 2 requires the Community to:

(a) promote research and ensure the dissemination of technical information;

(b) establish uniform safety standards to protect the health of workers and the general public and ensure they are applied;

(c) facilitate investment and ensure, particularly by encouraging ventures on the part of undertakings, the establishment of the basic installations necessary for the development of nuclear energy in the Community;

(d) ensure that all users in the Community receive a regular and equitable supply of ores and nuclear fuels;

(e) make certain, by appropriate supervision, that nuclear materials are not diverted to purposes other than those for which they are intended;

(f) exercise the right of ownership conferred in respect of special fissile materials;

---

[51] *Law Society Gazette—No. 29* (1989), p. 23.

(g) ensure wide commercial outlets and access to the best technical facilities by the creation of a common market in specialised materials and equipment, by the free movement of capital investment in the field of nuclear energy, and by freedom of employment for specialists within the Community; and

(h) establish with other countries and international organisations such relations as will foster progress in the peaceful uses of nuclear energy.

Under Article 3, the tasks entrusted to the Community are to be carried out by the European Parliament (assisted by the Economic and Social Committee acting in an advisory capacity), the Council, the Commission, and the Court of Justice. By Article 7, it is the responsibility of the Commission to promote and facilitate nuclear research in the Member States, and to complement it by carrying out a Community research and training programme. Article 8 required the Commission to establish a Joint Nuclear Research Centre, in order to ensure that the research programmes and other tasks assigned to it by the Commission were carried out. Article 9 empowered the Commission, after obtaining the opinion of the Economic and Social Committee, and within the framework of the Joint Nuclear Research Centre, to set up "schools for the training of specialists, particularly in the field of prospecting for minerals, the production of high-purity nuclear materials, the processing of irradiated fuels, nuclear engineering, health and safety and the production and use of radioisotopes."

This emphasis on research and the development of a healthy nuclear industry also appears in Chapter II of the Treaty. This deals with the dissemination of information and the exchange of research results, allowing Member States and undertakings to obtain non-exclusive licences in relation to patent rights owned by the Community. This Chapter of the Treaty also requires Member States to ask applicants for domestic patents, relating to specifically nuclear subjects, to agree that the contents of the application be communicated to the Commission. Other economically-related provisions of the Treaty include Chapter IV which deal with investment, and Chapter V, which makes provision for joint undertakings.

Other aspects of fundamental importance to the development of the nuclear industry are Chapter VI, relating to the supply of ores, fissile materials and source materials, on the principle of equal access to sources of supply; Chapter VIII, dealing with property ownership of fissile materials; Chapter IX, with the nuclear common market; and Chapter X, dealing with external relations.

EURATOM: Treaty requirements on health and safety   1–46

By contrast with the economic provisions mentioned above, the provisions of the Treaty dealing with "Health and Safety" (Chapter III) are relatively brief. The main articles in this respect are as follows:

ARTICLE 30 — this Article requires basic standards to be laid down within the Community for the protection of the health of workers and the general public against the dangers arising from ionising radiations. In this context, the expression "basic standards" is defined to mean the maximum permissible doses compatible with adequate safety, maximum permissible levels of exposure and contamination, and the fundamental principles governing the health surveillance of workers. By Article 31, these basic standards are to be worked out by the Commission, having obtained the opinion of a group of experts appointed by the Scientific and Technical Committee, and having obtained the opinion of the Economic and Social Committee. Following a proposal from the Commission, the Council, after consulting with the European Parliament, establishes the basic standards acting by a qualified majority.

ARTICLE 33 — this Article requires each Member State to lay down the appropriate provisions, by legislation, regulation, or administrative action, to ensure compliance with the basic standards established under Article 31.

ARTICLE 34 — each Member State is required to consult the Commission about obligatory additional safety measures when it intends to perform a particularly dangerous experiment in its territory. The assent of the Commission is required where the experiment is liable to affect the territories of other Member States. It has been argued that this requirement could have applied to the controversial French nuclear weapons tests at Mururoa, Polynesia, in 1995.[52] However, in Case T–219/95R, the European Court of First Instance denied interim relief to three Tahitean residents who were seeking to challenge the Commission's decision not to intervene in respect of the testing: the plaintiffs were held not to have standing to bring such a challenge as they were not "individually concerned".

ARTICLE 35 — each Member State is required to establish the facilities necessary to carry out continuous monitoring of the level of radioactivity in the air, water, and soil, and to ensure compliance with the basic standards. The Commission is to have the right of access to such facilities and may verify their operation and efficiency.

ARTICLE 37 — this important article requires each Member State to

---

[52] S. Diemann and G. Betlem, "Nuclear Testing and Europe" (1995) New L.J. 1236. Cm. 2485 (1995).

provide the Commission with such general data relating to any plan for the disposal of radioactive waste as will make it possible to determine whether the implementation of the plan is liable to result in the radioactive contamination of the water, soil, or air space of another Member State. The Commission must deliver its opinion within six months, having consulted the group of experts appointed under article 31. This Article is more extensively discussed below in the context of radioactive waste disposal (see para. 5–15).

ARTICLE 38 — the Commission is required to make recommendations to Member States with regard to the level of radioactivity in the air, water and soil; in cases of urgency the Commission must issue a direction requiring the Member State concerned to take all necessary measures within a period laid down by the Commission to prevent infringement of the basic standards, and to ensure compliance with the regulations. In default of compliance with such a direction, the Commission or the Member State concerned may bring the matter before the Court of Justice.

The Council has adopted numerous directives under the Treaty to provide basic standards. The key regulations and directives include measures relating to waste movement, notification of accidents, emergency procedures, and maximum permitted levels of radioactive contamination in foodstuffs following nuclear accidents or radiological emergencies.[53] However, the most important measure relates to basic standards generally, and is described at para. 1–47 below.

1–47 EURATOM: Basic Safety Standards Directive

Basic standards for the protection of workers and the public were laid down initially by directives in 1959; the most recent major amendment of these directives was by Directive 80/836/EURATOM.[54] This Directive applies to the production, processing, handling, use, holding, storage, transport, and disposal of natural and artificial radioactive substances, and to any other activity which involves a hazard arising from ionising radiations.

Subject to specified exceptions in Article 4, these activities must, by Article 3, be subject to prior authorisation in cases decided by each

---

[53] see, for example: Regulation 3954/87/EURATOM: [1987] O.J. L371/11 laying down maximum permitted levels of radioactive contamination of foodstuffs and of feedingstuffs following a nuclear accident or any other case of radiological emergency, as amended; Directive 92/3/EURATOM: [1992] O.J. L635/24 on the supervision and control of shipments of radioactive waste between Member States and into and out of the Community; Directive 89/618/EURATOM: [1989] O.J. L457/31 on informing the general public about health protection measures to be applied and steps to be taken in the event of a radiological emergency.
[54] [1980] O.J. L246.

Member State. By Article 5, irrespective of the degree of danger involved, prior authorisation must be applied in respect of the administration of radioactive substances for medical or research purposes, the use of radioactive substances in toys, and the addition of radioactive substances to foodstuffs, medicinal products, cosmetics, and products for household use.

Article 6 requires the limitation of individual and collective doses resulting from controllable exposures to be based on three general principles, which in turn follow the approach of the ICRP:

(a) every activity resulting in exposure shall be justified by the advantages which it produces;

(b) all exposures shall be kept as low as reasonably achievable; and

(c) the sum of doses received and committed doses (*i.e.* the dose over a period of 50 years) shall not exceed the dose limits specified for exposed workers, apprentices, students and members of the public. This final principle does not apply to exposure of individuals as a result of medical examination and treatment.

Dose limits are given for whole and for partial body exposure. The figure for whole body exposure for workers is 50 mSv per year (13 mSv a quarter for women of reproductive capacity, and 10 mSv during any period of declared pregnancy). For members of the public the figure is 5 mSv per year. These figures accord with the IAEA recommendations, but are higher than the most recent recommendations of the ICRP (see para. 1–41 above).

The Directive has been revised to take into account the changes in radiological protection criteria recommended by the ICRP.[54a]

U.K. legislation on radiological protection   1–48

The main piece of U.K. primary legislation dealing with radiological protection is the Radiological Protection Act 1970. This statute created a public authority, the National Radiological Protection Board (NRPB) with the following functions:

(a) by means of research and otherwise, to advance the acquisition of knowledge about the protection of mankind from radiological hazards; and

(b) to provide information and advice to persons, including government departments, with responsibilities in the United Kingdom in relation to protection from radiation hazards.

The NRPB has power to provide technical services in this respect, and to

[54a] see further the note in the Preface to this text.

make charges for such services and for providing information and advice (subs.1(2)). The NRPB, in this respect, assumed certain responsibilities previously undertaken by the Atomic Energy Authority (AEA), and by the Medical Research Council (subs.1(3)), as well as the functions of the Advisory Committee established under section 6 of the Radioactive Substances Act 1948 (subs.1(5)).

The Board consists of a chairman and between 9 and 12 other members, who are appointed by the Health Minister after consultation with the AEA and the Medical Research Council (subs.2(1)). The Board is a body corporate, and may appoint a secretary and other officers and employees (Sched. 1, paras. 1 and 13).

The NRPB is a substantial organisation operating from Chilton, Oxfordshire, and from regional centres in Leeds and Glasgow, with an income for the year ended March 31, 1995, of over £6 million from government grants, and £7.5 million from charges for its activities. Its work is multi-faceted, including contribution to the work of the international agencies referred to above, as well as to the work of the United Kingdom's Radioactive Waste Management Advisory Committee, and the Advisory Committee on Safety of Nuclear Installations. The NRPB also publishes its own series of documents providing information and advice. It acts as radiation protection advisor to many users of radiation, including industry, hospitals and clinics, research institutions, local authorities, and airports, and similar facilities using radioactive means of luggage and cargo inspection. It maintains an extensive research programme on environmental radioactivity, including radon.

The NRPB produces various publications, including the six-volume *Documents of the NRPB*, dealing with issues such as radon, electromagnetic fields, dose limits, and restrictions on food and water following radiological incidents. It also produces reports in its *Guidance on Standards* series, *Guidance Notes* on radiological protection, numerous Scientific and Technical Reports, and various publications for the general public in its *At a Glance* series.

1–49 The Ionising Radiations Regulations 1985

The Ionising Radiations Regulations 1985 (S.I. 1985 No. 1333) are made under the Health and Safety at Work, etc, Act 1974. They constitute the main Regulations dealing with radiological protection and implementing in part, as respects Great Britain, the EURATOM basic safety standards directive (80/836/EURATOM). It has been held judicially that these Regulations do not constitute the entirety of United Kingdom law implementing the requirement under that directive for the justification of radiation exposure.[55]

The 1985 Regulations supersede a number of previous regulations,

---

[55] see: *R. v. Secretary of State for the Environment, ex p. Greenpeace* [1994] 4 All E.R. 352.

namely the Ionising Radiations (Unsealed Radioactive Substances) Regulations 1968 (S.I. 1968 No. 780), the Ionising Radiations (Sealed Sources) Regulations 1969 (S.I. 1969 No. 808), the Radioactive Substances (Road Transport Workers) (Great Britain) Regulations 1970 (S.I. 1970 No. 1827), and the Employment Medical Advisory Service (Factories Act Order, etc. Amendment) Order 1973 (S.I. 1973 No. 36).

It is not proposed to analyse the 1985 Regulations in detail, but the following main points may be noted:

1. REGULATION 4 requires co-operation between employers where work with ionising radiations undertaken by one employer is likely to give rise to exposure to ionising radiations to the employee of another employer; clearly, this may be relevant in cases where work is carried out by contractors.

2. REGULATION 5 requires the notification of certain types of work involving ionising radiation to the Health and Safety Executive, who must be furnished with the particulars specified in Schedule 4 to the Regulations.

3. PART II of the Regulations deals with dose limitation. Regulation 6(1) requires every employer, in relation to work with ionising radiations, to take all necessary steps to restrict, so far as reasonably practicable, the extent to which his employees and other persons are exposed to those ionising radiations. The words "reasonably practicable" here may be compared with the formulation "as low as reasonably achievable" frequently used in the context of radioactive waste disposal practices (see para. 5–54 below). This obligation is further amplified by reference to the provision of engineering controls, design features, and warning devices, as well as suitable personal protective equipment.

4. REGULATION 7 requires every employer to ensure that his employees and other persons are not exposed to ionising radiations to such an extent that any dose or limit specified in Schedule 1 for each such employee or person is exceeded. Schedule 1 contains dose limits for the whole body and dose limits for various individual body parts. There is a greater restriction on dose limits applied to the abdomen of women of reproductive capacity, and pregnant women at work. Current dose limits for the whole body are 50 mSv in any calendar year for employees aged 18 and over, 15 mSv for trainees aged under 18 years, and 5 mSv for any other person. These figures relate to the effective dose equivalent from external radiation, and the committed effective dose equivalent from that year's intake of radionuclides.

5. PART III of the Regulations deals with the regulation of work with

ionising radiations, including the requirement for designation of controlled and supervised areas for such work, the designation of classified persons (employees likely to receive a dose of ionising radiations exceeding three tenths of any relevant dose limit), the appointment of radiation protection advisers and supervisors and qualified persons, the making of written local rules of work and general information, instruction and training.

6. PART IV of the Regulations deals with dosimetry and medical surveillance, requiring assessments to be made of all significant doses of ionising radiations received by employees, and arrangements with approved dosimetry services for the making of systematic measurements and assessments of such doses. Regulation 16 applies to classified employees (see 5. above), other employees who have received over-exposure, and employees who for health reasons should only be engaged in ionising radiations work under certain conditions. This regulation requires the provision of medical surveillance and the keeping of health records for at least 50 years from the date of the last entry made in them. By Regulation 17, the Health and Safety Executive may serve notice requiring an employer to make approved arrangements for the protection of certain employees.

7. PART V of the Regulations deals with arrangements for the control of radioactive substances. There is a general requirement under Regulation 18 that where a radioactive substance is used as a source of ionising radiations, the employer should ensure that, whenever reasonably practicable, the substance is in the form of a sealed source (defined as a radioactive substance bonded wholly within a solid inactive material, or encapsulated within an inactive receptacle of, in either case, sufficient strength to prevent any dispersion of the substance under reasonably foreseeable conditions of use). Regulation 19 requires employers to take such steps as are appropriate to account for, and keep records of, the quantity and location of, radioactive substances. Other regulations under this Part deal with the keeping of radioactive substances which are not in use, the transport and movement of radioactive substances, the provision of adequate washing and changing facilities, and the provision and maintenance of approved personal protective equipment.

8. PART VI of the Regulations deals with the monitoring of ionising radiations and requires every employer undertaking work with such radiations to take such steps as are requisite (otherwise than by the use of assessed doses of individuals) to ensure that levels of ionising radiations are adequately monitored for each controlled area or supervised area designated, in order to ascertain the efficacy of the methods used in those areas for restricting exposure of persons to

ionising radiations. The relevant examinations and tests of the monitoring equipment must be carried out by qualified persons.

9.   PART VII of the Regulations deals with assessments and notifications. An employer must not carry on working with ionising radiations unless he has made an assessment adequate to identify the nature and magnitude of the radiation hazard to employees and other persons likely to arise from the work, in the event of any reasonably foreseeable accident, occurrence or incident. Where the assessment reveals radiation hazards, the employer must take all reasonably practicable steps to prevent any such accident, occurrence or incident, to limit its consequences, and to provide employees with the information, instruction, training, and equipment necessary to restrict their exposure. Regulation 26 deals with special hazard assessment reports required in the case of work with ionising radiations involving more than certain quantities of radioactive substances. Other provisions of this Part deal with the making of contingency plans to respond to accidents, occurrences or incidents (including consultation with public and emergency services), the investigation of cases where employees are exposed to ionising radiations to an extent that three-fifths of the annual dose limit is exceeded for the first time in any calendar year, the investigation and notification of over-exposure, and additional restrictions in relation to employees who have suffered over-exposure.

10.   PART VIII of the Regulations deals with the safety of articles and equipment. In the case of work equipment, obligations are placed on manufacturers to ensure that design and construction are such as to restrict exposure so far as is reasonably practicable. In relation to equipment used for medical exposure, the employer must ensure that the equipment is designed, installed and maintained so as to be capable of restricting, so far as is reasonably practicable, the exposure of the patient to an extent that is compatible with the potential clinical benefits. Regulation 34 contains a general prohibition on intentionally or recklessly misusing, or without reasonable excuse interfering with, any radioactive substance or radiation generator.

HSC policy statement on radiation protection   1–50

Since the Ionising Radiations Regulations 1985 are made under the Health and Safety at Work, etc. Act 1974, responsibility for their implementation falls to the Health and Safety Commission and Health and Safety Executive. The HSC issued a *Policy Statement on Radiation Protection* in August 1993, based upon a review carried out by the Working Group on Ionising Radiations (the predecessor of IRAC; see para. 1–51 below).

49

The HSC restates its belief that radiation protection in the United Kingdom is best served by a policy based on the recommendations of the ICRP, as incorporated in directives made under the EURATOM Treaty; the HSC saw no need to amend the 1985 Regulations until such time as the renegotiation of the EURATOM Basic Safety Standards Directive was complete. The significance of this decision was that in Case C–376/90, *Commission of the European Communities v. Kingdom of Belgium,*[56] the European Court of Justice had held that there could be situations that justified a Member State setting stricter levels of protection than those contained in the Directive; it would, therefore, have been open to the United Kingdom to revise the 1985 Regulations in advance of completion of the review of the Directive.

1–51 The Ionising Radiations Advisory Committee (IRAC)

IRAC was appointed by the Health and Safety Commission in 1995 to advise on ionising radiations issues. The terms of reference of IRAC, which supersedes the HSC's Working Group on Ionising Radiations, are:

> "To consider all matters concerning protection against exposure to ionising radiations that are relevant to the work of the Health and Safety Commission and are referred to the Committee by the Commission or by the Health and Safety Executive, and to advise the Commission and Executive."

IRAC consists of a chairman nominated by the HSE, and 18 other members nominated by various public and professional bodies.

1–52 Radioactive Waste Management Advisory Committee (RWMAC)

RWMAC was originally set up in response to a recommendation of the Sixth Report of the Royal Commission on Environmental Pollution regarding *Nuclear Power and the Environment.*[57] Its terms of reference were revised in 1991 and are:

> "To advise the Secretaries of State for the Environment, Scotland and Wales on the technical and environmental implications of major issues concerning the development and implementation of an overall policy for all aspects of the management of civil radioactive waste, including research and development; and on any such matters referred to it by the Secretaries of State."

---

[56] [1992] O.J. C333/17; [1993] 2 C.M.L.R. 513.
[57] Royal Commission on Environmental Pollution, Sixth Report, *Nuclear Power and the Environment*, Cmnd. 6618 (1976).

Radioactive waste management is addressed in Chapter 5, but RWMAC is mentioned here as its activities are clearly relevant to the issue of radiological protection. The Committee publishes annual reports, as well as special reports on relevant issues.

Committee on Medical Aspects of Radiation in the Environment (COMARE)   1–53

This Committee is an independent expert advisory committee offering medical and scientific advice to the Government on the health effects of both ionising and non-ionising radiations, both natural and man-made. The Secretariat of the Committee is provided jointly by the Department of Health, and the National Radiological Protection Board. The terms of reference of COMARE are:

> "To assess and advise Government on the health effects of natural and man-made radiation in the environment and to assess the adequacy of the available data and the need for further research."

The Committee has published a number of reports on epidemiological issues, including the incidences of cancer in West Cumbria (1986 and 1996), Dounreay (1988), and Aldermaston (1989). COMARE collaborates with the RWMAC on certain reports, such as that into the potential health effects and possible sources of radioactive particles found in the vicinity of Dounreay.

The development of legislation on radioactive substances and nuclear installations   1–54

The foregoing material has hopefully provided a context in which to consider and understand the U.K. legislation dealing with radioactive substances and nuclear installations, which is described in detail in the ensuing chapters. It may be helpful, however, to close this chapter with a brief chronology of how the primary legislation has developed.

1946   Atomic Energy Act 1946 placed general duties on the Minister of Supply to promote and control the development of atomic energy.

1948   Radioactive Substances Act 1948 introduced controls over import, export, sale and supply of radioactive substances and gave power to make safety regulations with regard to use of radiation substances and disposal of radioactive waste.

1954   Atomic Energy Authority Act 1954 constituted the United Kingdom Atomic Energy Authority (UKAEA) and conferred on it wide powers.

1959　Nuclear Installations (Licensing and Insurance) Act 1959 introduced controls over nuclear installations and a general framework for liability and insurance.

1960　Radioactive Substances Act 1960 instituted a two-fold system of registration for use of radioactive materials and authorisation for accumulation and disposal of radioactive waste.

1965　Nuclear Installations (Amendment) Act 1965 amended the 1959 Act so as to implement Britain's international obligations under the Paris and Brussels Supplementary Conventions. The 1959 and 1965 Acts were consolidated by the Nuclear Installations Act 1965.

1969　Nuclear Installations Act 1969 made various amendments to the Nuclear Installations Act 1965.

1970　The National Radiological Protection Board was created by the Radiological Protection Act 1970.

1971　The Atomic Energy Authority (AEA) was divided into three parts under the Atomic Energy Authority Act 1971.

1973　The Weapons Group of the Atomic Energy Authority was transferred to the Secretary of State for Defence by the Atomic Energy Authority (Weapons Group) Act 1973.

1976　Congenital Disabilities (Civil Liability) Act 1976 made provision as to injuries to unborn children.

1986　The Atomic Energy Authority was put onto a trading fund basis by the Atomic Energy Act 1986.

1990　Environmental Protection Act 1990 made various amendments to the Radioactive Substances Act 1960.

1991　Atomic Weapons Establishment Act 1991 made provision for contractorisation of the Establishment.

Radioactive Material (Road Transport) Act 1991 provided controls and regulation-making powers in relation to the transportation by road of radioactive material.

1993　Radioactive Substances Act 1993 consolidated and replaced the Radioactive Substances Act 1960.

1995　Atomic Energy Authority Act 1995 made provision for transfer schemes to facilitate privatisation of the Authority.

Environment Act 1995 created new agencies, the Environment Agency and the Scottish Environment Protection Agency (SEPA), with responsibility for enforcing the Radioactive Substances Act 1993, and made consequential amendments to that Act.

CHAPTER TWO

THE LICENSING OF NUCLEAR INSTALLATIONS

Applicable legislation

2–01

The requirements for the licensing of nuclear reactors and other installations are contained in the Nuclear Installations Act 1965 ("the 1965 Act"), which consolidate and replace the requirements of the Nuclear Installations (Licensing and Insurance) Act 1959, and the Nuclear Installations (Amendment) Act 1965. The background to the development of this legislation is briefly described in Chapter 1.

The licensing system was instituted in the context of a growth in civil nuclear power. The Minister of Power (Lord Mills) pointed out on the Second Reading of the Bill which became the 1959 Act:

> "As is well known to your lordships, several nuclear power stations are under construction for the Electricity Boards and others are at the planning stage. One research reactor in private industry—that of Associated Electrical Industries Limited, at Aldermaston—has just begun to operate. Though the Atomic Energy Authority are acting as the consultants for each of these projects and are advising the owners about the requirements of safety, it is nevertheless in the interests of all that the ultimate liabilities and obligations of the owners should be defined without delay."[1]

Part of this process was the provision of a licensing system:

> "At the present time, any person can operate a nuclear reactor without a licence, apart from the ordinary patent licence for construction, from the Atomic Energy Authority. The whole purpose of this Bill is to institute licences for safety reasons, and to enable the Minister to set up an organisation to ensure that proper conditions are laid down and followed."[2]

Moving the Second Reading of the Bill in the Commons, the Paymaster-General (Mr Reginald Maudling) also stressed the lack of adequate controls; neither the town and country planning system nor the theoretical possibility of control under existing atomic energy legislation were "... adequate to give the kind of regular and definite assurance needed that reactors are constructed and operate in accordance with the

---

[1] *Hansard*, H.L. Vol. 212, col. 503.
[2] *Hansard*, H.L. Vol. 212, col. 1004.

highest standards of safety."[3] The existing legislation, the Atomic Energy Act 1946, was passed at a time when no one anticipated a nuclear programme for the generation of electricity, and was described by Sir Ian Horobin, Parliamentary Secretary to the Minister of Power, as "... merely an Act to regularise things which had been going on under great secrecy as part of the war effort".[4] It contained, for example, no power to vary licences or to exercise licensing powers where an installation was closed down.

The 1959 Act rectified that lack of control by instituting a requirement for the licensing of sites of nuclear installations; this system remains essentially unchanged under the 1965 Act.

2–02 Nuclear reactors

By subsection 1(1)(a) of the 1965 Act, no person shall use any site for the purpose of installing or operating any nuclear reactor (other than one comprised in a means of transport, whether by land, water or air) unless a licence to do so (a nuclear site licence) has been granted by the Health and Safety Executive and is for the time being in force.

"Nuclear reactor" is defined by subsection 26(1) to mean any plant (including fixed or movable machinery, equipment or appliances) designed or adapted for the production of atomic energy by a fission process in which a controlled chain reaction can be maintained without an additional source of neutrons. The original form of words in the Bill referred to plant "designed or adapted for the production of atomic energy by a maintained and controlled fission process". This wording originated from the Atomic Energy Authority and was designed to cover nuclear reactors.[5] However, at the House of Lords stage of the Bill, Lord Shackleton proposed an amendment to exclude sub-critical reactors (*i.e.* those where the output of neutrons cannot be large enough to produce a super-critical self-sustaining reaction) such as those used in universities for research purposes.[6] The Government undertook to consider this issue, and brought forward an amendment introducing the requirement that the chain reaction be capable of being maintained without an additional source of neutrons.[7] Subject to this qualification, it is clear that the Government intended all reactors to be licensed, no matter how small, whether in a university department or anywhere else.[8]

Reactors comprised within any means of transport are excluded, *e.g.* nuclear ships or submarines. Part of the reason behind this appears to have been that the Government did not want to legislate for ships of foreign flags outside British territorial waters, as this would be contrary to

[3] *Hansard*, H.L. Vol. 599, col. 862.
[4] *Hansard*, H.C. Vol. 599, col. 929.
[5] *Hansard*, H.L. Vol. 213, col. 334.
[6] *Hansard*, H.L. Vol. 213, col. 331.
[7] *Hansard*, H.L. Vol. 217, col. 924.
[8] *Hansard*, H.L. Vol. 599, col. 933.

the view of international law as embodied in Article 6(1) of the Convention on the High Seas, adopted at Geneva in April 1958. This states that a ship on the high seas should be subject to the exclusive jurisdiction of the State of its flag; and that "... it shall not be within the unilateral power of one Government to legislate for ships upon the high seas".[9] It is also clear that the use of the word "site" effectively excludes both ships and aircraft (*ibid.*). This view appears to be reflected in the context of liability by the distinction drawn at subsection 13(2) of the 1965 Act between injury and damage incurred on vessels registered in the United Kingdom and other vessels.

Other installations   2–03

As well as for the use of a site for installing or operating a nuclear reactor, a nuclear site licence is required by subsection 1(1)(b) of the 1965 Act for other installations of such a class or description as may be prescribed, being an installation designed or adapted for:

  (i) the production or use of atomic energy; or

 (ii) the carrying out of any process which is preparatory or ancillary to the production or use of atomic energy and which involves or is capable of causing the emission of ionising radiations; or

(iii) the storage, processing, or disposal of nuclear fuel or of bulk quantities of other radioactive matter, being matter which has been produced or irradiated in the course of production or use of nuclear fuel.

Together with nuclear reactors, such installations are termed "nuclear installations" (subs.26(1)).

The ability to prescribe such classes of installation provides a means by which ancillary processes may be controlled, in particular the storage and processing of nuclear fuel or of bulk quantities of associated radioactive matter. The term "bulk quantities" is not defined. The purpose of this provision:

> "... is to enable the [Act] to be applied, as the progress of the nuclear industry may require, to any installation other than a reactor which presents a comparable public hazard. The provision as originally drafted would have enabled the [Act] to be applied to some installations which, although used for processes ancillary to the production of atomic energy, nevertheless present no hazard; for example, installations where graphite blocks for reactor moderators or beryllium cans for containing fuel elements are made. The same considerations

[9] *Hansard*, H.C. Vol. 599, col. 930.

apply to places for the treatment, storage or disposal of radioisotopes in small quantities or of the less radioactive types. This Amendment is intended to define more precisely the types of potentially dangerous installation to which the Bill may be extended by regulations."[10]

2–04 Prescribed installations

The regulations which prescribe such classes of installations may also exempt, or make provision for exempting, certain installations (subs. 1(2)). This power may not be used, however, to exempt "relevant installations", *i.e.* those subject to an international agreement to which the United Kingdom is a party dealing with third-party lliability in the field of nuclear energy (subss.1(2) and 26(1)).

The current regulations are the Nuclear Installations Regulations 1971 (S.I. 1971 No. 381). Regulation 3 prescribes for licensing purposes, installations of any of the following descriptions:

(1) any installation designed or adapted for the carrying out of any process involved in the manufacture from—
  (a) enriched uranium (meaning uranium enriched so as to contain more than 0.72 per cent of isotope 235),
  (b) plutonium,
  (c) any alloy, chemical compound, mixture or combination containing enriched uranium,
  (d) any alloy, chemical compound, mixture or combination containing plutonium,
  of fuel elements to be used for the production of atomic energy;
(2) any installation designed or adapted for the carrying out of any process (not being a process carried out solely for the purposes of chemical or isotopic assay or metallographic investigation) involved in—
  (a) the production from:
    (i) enriched uranium,
    (ii) any alloy, chemical compound, mixture or combination containing enriched uranium,
    of any alloy, chemical compound, mixture or combination containing enriched uranium;
  (b) the production from:
    (i) plutonium,
    (ii) any alloy, chemical compound, mixture or combination containing plutonium,
    of any alloy, chemical compound, mixture or combination containing plutonium,

[10] *Hansard*, H.L. Vol. 217, col. 925.

    (c) the production, from any alloy, chemical compound, mixture or combination containing enriched uranium, of enriched uranium,

    (d) the production, from any alloy, chemical compound, mixture or combination containing plutonium, of plutonium;

(3) any installation designed or adapted for the incorporation of—

    (a) enriched uranium,

    (b) any alloy, chemical compound, mixture or combination containing enriched uranium,

    (c) plutonium,

    (d) any alloy, chemical compound, mixture or combination containing plutonium,

in any device designed to form part of a nuclear assembly or designed for irradiation in a nuclear reactor other than a device designed solely for the purpose of measuring neutron flux;

(4) any installation comprising a nuclear assembly designed or adapted for the production of neutrons and containing—

    (a) enriched uranium,

    (b) any alloy, chemical compound, mixture or combination containing enriched uranium,

    (c) plutonium,

    (d) any alloy, chemical compound, mixture or combination containing plutonium,

and in which a controlled chain reaction can be maintained with an additional source of neutrons;

(5) any installation designed or adapted for the processing of irradiated nuclear fuel other than processing carried out solely for the purpose of chemical or isotopic assay or metallographic investigation of such nuclear fuel;

(6) any installation designed or adapted for storage of—

    (a) fuel elements referred to in paragraph (1) of this Regulation,

    (b) irradiated nuclear fuel,

    (c) bulk quantities of any other radioactive matter which has been produced or irradiated in the course of the production or use of nuclear fuel,

other than storage incidental to carriage and in the case of irradiated nuclear fuel other than storage incidental to any of the excepted purposes referred to in paragraph (5);

(7) any installation designed or adapted for—

    (a) any treatment of irradiated matter which involves the extraction therefrom of plutonium or uranium,

    (b) any treatment of uranium whether enriched or not such as to increase the proportion of isotope 235 contained therein;

59

(8) any installation designed or adapted for the carrying on of any process involved in the production from nuclear matter, not being excepted matter, of isotopes prepared for use for industrial, chemical, agricultural, medical or scientific purposes.

Government policy is that the nuclear site licensing regime should apply to any deep repository for nuclear waste constructed by NIREX (see further para. 5–20 below). This will require minor amendment to the 1971 Regulations, which has been agreed with the Department of Trade and Industry, and awaits a suitable legislative opportunity.[11]

HSC also considers it desirable to clarify that both decommissioning and the "handling" of radiocactive matter are licensable activities, by amendments to the Regulations.[12]

2–05 Licensing: the position of the Atomic Energy Authority

Originally, the United Kingdom Atomic Energy Authority was exempted from the licensing requirement under section 1 of the 1965 Act. The reason was that most of the knowledge of experimental reactors which existed in 1959 was located within the Authority. On that basis, it seemed to the Government "... rather pointless to subject them to this statutory licensing system".[13] Another justification for this initial exemption was that the work of the Authority was mainly experimental, and that to subject it to licensing would entail enormous demands on the inspectorate in terms of familiarity with the experimental work in progress:

> "The inspectors would ... have to be as expert in advanced research as the scientists of the Authority itself—a requirement which the Government considers to be unnecessary, if not impracticable."[14]

However, the exemption was removed as from October 31, 1990 by the Nuclear Installations Act 1965 (Repeal and Modifications) Regulations 1990 (S.I. 1990 No. 1918), which repealed the words in section 1, "other than the Authority".

2–06 Offence of installation or operation without licence

Contravention of the prohibition on using a site for the purpose of installing or operating a nuclear installation without a nuclear site licence is an offence (subs.1(3)). The issue of offences is discussed below at para. 2–49 below.

[11] see: HSC, *Submission to the Nuclear Review* (1994), para. 60(a).
[12] see: HSC, Submission to the Nuclear Review (1994), para. 60(b) and (c).
[13] *Hansard*, H.C. Vol. 599, col. 964.
[14] *Hansard*, H.L. Vol. 212, col. 508.

Operations requiring a permit

2–07

Additional to the requirements for a nuclear site licence as described above, section 2 of the 1965 Act applies to prohibit certain activities except under, and in accordance with, a permit in writing granted by the United Kingdom Atomic Energy Authority, or by a government department. The section applies to the use of a site by any person other than the Authority for:

(a) any treatment of irradiated matter involving the extraction of plutonium or uranium; or

(b) any treatment of uranium so as to increase the proportion of isotope 235 contained therein.

Plutonium production reactors were designed for nuclear weapons purposes; those in the United Kingdom were built at the site of a disused ordnance factory on the Cumbria coast, re-named Windscale. Increasing the proportion of isotope 235 is known as "enrichment"; this is a complex and difficult process which was carried out at the Capenhurst gaseous diffusion plant in Cheshire, now shut down. The provision was inserted to "... ensure Government control over the extraction of plutonium and other dangerous fissile materials which might be used as components for nuclear weapons".[15]

Unless the permit under section 2 is granted by the Minister, it will not authorise the use of a site other than for the purposes of research and development (subs.(1A)). By subsection (1B), where a permit granted by the Minister to a body corporate authorises use of a site for purposes other than research and development, the Minister may, by order, direct that the provisions of Schedule 1 to the Act shall have effect in relation to the body corporate. Schedule 1 contains various provisions relating to security, for example restricting rights of entry and requiring compliance with directions given by the Minister for the purpose of safeguarding information. The only order made applying Schedule 1 to date applies to British Nuclear Fuels Ltd.[16] This refers to the grant of permits under section 2, authorising the use of the Capenhurst and Windscale sites for purposes other than, or not limited to, research and development.

Any permit given under section 2, whether by the Authority, the Minister, or by any other government department, may at any time be revoked by that body, or may be surrendered by the person to whom it was granted (subs.2(1D)).

Contravention of section 2 is an offence (subs.2(2)).

---

[15] *Hansard*, H.L. Vol. 217, col. 926.
[16] Nuclear Installations (Application of Security Provisions) Order 1971 (S.I. 1971 No. 569).

2–08 Ministerial responsibility

The 1965 Act refers in various sections to "the Minister" as having functions and responsibilities. Subsection 26(1) defines this term to mean:

(a) in relation to England and Wales, the Minister of Power; and
(b) in relation to Scotland, the Secretary of State for Scotland.

In 1969 the Ministry of Power was dissolved, and the relevant functions transferred to the Minister of Technology.[17] The functions were in turn passed to the Secretary of State for Trade and Industry, where they currently remain.[18]

2–09 The role of the Health and Safety Executive in licensing

Nuclear site licences are granted by the Health and Safety Executive (s.1(1)). The references to the Health and Safety Executive were inserted into the Act by the Nuclear Installations Act 1965 (Repeals and Modifications) Regulations 1974 (S.I. 1974 No. 2056), reg. 2(1)(b) and Sched. 2. The Health and Safety Executive is constituted and governed by the Health and Safety at Work, etc. Act 1974, and is an independent public corporate body with statutory duties to enforce health and safety legislation. It is the executive arm of the Health and Safety Commission, to which it reports. In relation to general health and safety matters, the HSE advises the Secretary of State for Employment; however, for nuclear safety matters, it advises the President of the Board of Trade as head of the Department of Trade and Industry, and in Scotland, the Secretary of State for Scotland.

Section 24, as substituted for England, Wales and Scotland by the Nuclear Installations Act 1965 (Repeals and Modifications) Regulations 1974 (subsequently amended by the Energy Act 1989, s.6) makes provision for the appointment of inspectors by the Secretary of State. Under that section, the Secretary of State may appoint inspectors for the purpose of assisting him in the execution of the provisions of the Act, other than those mentioned in Schedule 1 to the Health and Safety at Work, etc. Act 1974. The provisions of the 1965 Act which are mentioned in Schedule 1 are sections 1, 3–6, 22, 24A and Schedule 2 (*i.e.* the main licensing provisions). It follows that appointment of inspectors for licensing purposes is under the Health and Safety at Work, etc. Act, rather than the 1965 Act.

Section 24 is therefore of only limited importance, but any inspectors appointed under it may, for the purpose of assisting the Secretary of State, exercise the powers set out in subsection 20(2) of the Health and Safety at

[17] The Minister of Technology Order 1969 (S.I. 1969 No. 1498).
[18] The Secretary of State for Trade and Industry Order 1970 (S.I. 1970 No. 1537).

Work, etc. Act 1974 as are specified in their instrument of appointment (subs.(2)). They will have the same range of powers available as those appointed under the 1974 Act. By subsection 24(2), various provisions of the Health and Safety at Work, etc. Act 1974 dealing with restrictions on disclosure of information (s.28), offences (s.33), and prosecutions by inspectors (s.39), apply to inspectors appointed under section 24 in the same way as they apply to inspectors appointed under section 19 of the Health and Safety at Work, etc. Act.

Subsection 24(3) provides a power of cost recovery, whereby the Secretary of State may, with the agreement of the Treasury, require a licensee to repay to the Secretary of State such part of the sums paid by the Secretary of State for remuneration of inspectors and other expenses as appears to be attributable to the nuclear installations in respect of which the site licences have been granted. This can include costs in respect of pensions (subs.24(4)). The HSE also enjoys powers of cost recovery under section 24A (see para. 2–47 below).

The Advisory Committee on the Safety of Nuclear Installations (ACSNI)   2–10

ACSNI was established by the Health and Safety Commission in 1977, in succession to the Nuclear Safety Advisory Committee, to provide the HSC and ministers with independent advice on matters relating to the safety of nuclear installations. Its terms of reference are:

> "To advise the [Health and Safety] Commission, and when appropriate, the Secretaries of State, on major issues affecting the safety of nuclear installations including design, siting, operation and maintenance which are referred to them or which they consider require attention."

Committee members are appointed by the Commission from a wide field of specialisms and experience; members are also appointed by the CBI and TUC, and assessors nominated by the nuclear industry provide technical input. The Committee operates various sub-groups on particular subjects, and has published a diverse series of reports on various topics, as well as a biennial report. One of the most important recent pieces of work by ACSNI is its joint study group formed with the RWMAC to examine the approach for site selection and health criteria for radioactive waste facilities (see para. 5–57 below).

The Nuclear Safety Division (NSD)   2–11

The NSD is one of 10 divisions within the HSE, headed by the Director of Nuclear Safety. Information about the Division's composition and activities can be found in the HSE publication, *The Work of HSE's Nuclear*

Installations Inspectorate (February 1995) and in the *Nuclear Safety Newsletter,* which is issued three times a year by the HSE.

As at February 1995, some 280 staff were employed by the NSD, about 50 working from London and most of the remainder from the NSD's offices in Bootle, Merseyside. Some 60 per cent of NSD staff are scientists, a high proportion of which will have 10 or more years' experience in industry, many of them in the nuclear industry. Most of the remaining 40 per cent are professional civil servants. The NSD will also make use of independent consultants and can call on specialists in other parts of the HSE.

The NSD is organised into six branches: Branch A (Policy: Administration and International); Branch B (Assessment: Scientific); Branch C (Assessment: Engineering); Branch D (Chemical Plant: Inspection); Branch E (Inspection: Power Reactors and Emergency Arrangements); and Branch F (Policy: Radiation, Regulation and Research). It is Branches D and E which carry out the site inspections and regulatory control functions.

2–12 The Nuclear Installations Inspectorate (NII)

The NII was formed in April 1960 under the Nuclear Installations (Licensing and Insurance) Act 1959. It was originally accountable to the Ministry of Power, but in January 1975, was incorporated within the newly formed HSE, where it is now part of the Nuclear Safety Division (see para. 2–11). An inspector may be a member of an assessment, policy or inspection branch of the NSD.

H.M. Chief Inspector of Nuclear Installations is also the Director of the NSD. Each branch of the NSD will be headed by a Deputy Chief Inspector or equivalent grade, and units in each branch will be led by a Superintending Inspector or equivalent. The detailed licensing, regulation, and overseas work, is carried out by Principal Inspectors and Inspectors. As mentioned in para. 2–11, inspectors are technically qualified to a high level and have relevant industry experience.

2–13 Appointment and powers of inspectors

As mentioned above (see para. 2–09), the inspectors carrying out nuclear site licensing functions are technically appointed under the Health and Safety at Work, etc. Act 1974, rather than the 1965 Act itself. The licensing provisions of the 1965 Act are among the "existing statutory provisions", forming in turn part of the "relevant statutory provisions" mentioned in sections 1, 53, and Schedule 1 of the 1974 Act. Accordingly, the HSE must, by subsection 18(1), make adequate arrangements for the enforcement of those provisions, and may appoint inspectors to do so under section 19.

Those inspectors may be given the wide-ranging powers specified at section 20, and may serve improvement notices under section 21, and prohibition notices under section 22. It is important to note that inspectors will not only enforce the licensing requirements of the 1965 Act in relation to the installation, but also the general duties applying under the 1974 Act itself (see para. 2–28 below).

Nuclear site licences: generally   2–14

A nuclear site licence may only be granted to a body corporate and is not transferable (subs.3(1)). Subsection 3(2) allows the Health and Safety Executive, if it thinks fit, to treat two or more installations in the vicinity of one another as being in the same site for the purposes of site licensing, *i.e.* they may be licensed together. Provision is also made for the situation where part of the licensed site is no longer required by the licensee for any use needing a licence; by subsection 3(5), provided the HSE is satisfied that there is no danger from ionising radiations from anything on that part of the site, the site licence may be varied so as to exclude it.

The fact that a licence can only be granted to a company or other corporate body, and is not transferable, is both unusual and important. A prospective licensee must show not only that the plant to be used will be safe, but also that the corporate body, as licensee, is viable in relation to management of the site and dealing with post-closure liabilities.[19] A prospective change in the operator of a licensed nuclear site will therefore involve the grant of a replacement licence to the incoming corporate body. Before granting such a licence, the NII will apply the same evaluation criteria as it would for an initial licensee. The relicensing of nuclear sites as part of the restructuring of the nuclear industry for privatisation is explained at paragraph 2–52.

Where more than one company is working on the same site, the NII will wish to grant only one licence, to the company in charge of the day to day operation of the nuclear plant and whose staff manage its operation.[20] The need for the licensee to be the user of the site does not rule out the possibility of using contractors for certain functions. The licensee will need to demonstrate sufficient supervision and oversight to demonstrate that its chain of command and ability to control activities on the site have not been compromised.[21]

The HSE will generally impose a standard condition preventing the licensee from transferring ownership or possession of all, or part of, the site to any other person without the HSE's consent (see para. 2–27 below).

---

[19] see: HSE, *The Work of the HSE's Nuclear Installations Inspectorate* (1995), para. 23.
[20] see: HSE, *The Work of the HSE's Nuclear Installations Inspectorate* (1995), para. 28.
[21] *Nuclear Site Licences: Notes for Applicants* (HS(G) 120, 1994), paras. 42–43.

2-15 Non-prescriptive nature of the licensing system

The regime of nuclear site licensing is essentially non-prescriptive. As the Health and Safety Commission stated in its Submission to the Government's Nuclear Review:

> "The licence conditions set goals but do not specify the means of achieving them. A licence applicant must provide a comprehensive demonstration—a safety case—that safety will be controlled through all stages of the plant's life. Licensees, being responsible for safety, have to propose their own solutions to safety issues; these are not imposed by NII. The format for safety cases is not prescribed."[22]

On the grant of the licence, the conditions will make the licensee responsible for the application of detailed safety standards and safe procedures; the role of the NII is to review the licensee's arrangements to ensure that they are clear and unambiguous and that they address the main safety issues adequately.[23]

2-16 Nuclear site licences: acceptable risk

As analysed in Chapter 3 dealing with liability and insurance, subsection 7(1) of the Act contains a duty on the licensee of a nuclear site to secure that injury or damage is not caused by specified occurrences, or by the emission of ionising radiations. In the case of *Re Friends of the Earth*,[24] the issue arose as to the relationship between that duty and the grant of site licences. Friends of the Earth made application for judicial review of the Secretary of State's decision relating to the Sizewell B pressurised water reactor nuclear power station. The application of leave for judicial review in fact failed due to lack of promptitude; but the Court of Appeal in upholding the decision to refuse leave commented on one substantive argument raised by the applicants. This was that subsection 7(1) was to be interpreted as requiring an assurance of absolute safety before a site licence could be granted. Gibson L.J. found this proposition untenable:

> "If [the applicant's] contention were right Parliament had to have intended to prohibit the licensing of a nuclear power station unless it could be shown that no such occurrence could possibly occur, even from an unforeseen natural event or from some combination, however wildly improbable, of equipment failure or human error or malice on the part of the operators. That would in effect mean that no licence could be granted."

---

[22] see: HSC, *Submission to the Nuclear Review* (1994), para. 20.
[23] see: HSC, *Submission to the Nuclear Review* (1994), para. 32.
[24] [1988] J.P.L. 93.

Gibson L.J. did not feel that section 7 had been intended by Parliament to impose upon the NII the duty of requiring proof from an applicant that the proposed nuclear power station was so designed, and would be so operated, that no such harmful occurrence could occur.

This conclusion, it has been pointed out, is consistent with statements in the course of Parliamentary debates, and with the history of the Nuclear Installations (Licensing and Insurance) Act 1959 as a consequence of the serious fire at the Windscale Pile; "there is no evidence here of an absolutist conception of safety".[25]

Tolerability of risk   2–17

Following the inquiry into the proposal of Sizewell B nuclear power station (see para. 2–23 below), the inspector, Sir Frank Layfield, recommended that the HSE should "formulate and publish guidelines on the tolerable levels of individual and social risk to workers and the public from nuclear power stations". The response to that recommendation was the HSE document, *The Tolerability of Risk from Nuclear Power Stations* (revised 1992). The document received close public examination and general endorsement at the public inquiry chaired by Michael Barnes Q.C. into the subsequently proposed nuclear power station at Hinkley Point (1990). The concept of tolerability of risk implies considerations of public opinion, as well as purely expert assessment; the approach and philosophy of the HSE document has been increasingly applied to the regulation of other major industrial risks in the United Kingdom. The document[26] distinguishes "tolerability" from "acceptability". Acceptability implies a willingness to take a risk as it is; to tolerate a risk involves keeping it under review and reducing it further when this is possible.

The starting point of the HSE's approach involves three tests[27]:

(a) whether a risk is so great or the outcome so unacceptable that it must be refused altogether; or

(b) whether the risk is, or has been made, so small that no further precaution is necessary; or

(c) if a risk falls between these two states, that it has been reduced to the lowest level practicable, bearing in mind the benefits following from its acceptance and taking into account the costs of any further reduction. The injunction laid down in safety law is that any risk must be reduced so far as reasonably practicable or to a level which is "as low as reasonably practicable" (ALARP principle.)

The relevant working methods in the case of nuclear risk are complex,

---

[25] Dr C. Miller, "Radiological Risks and Civil Liability" [1989] J.Env.L. (Vol. 1), No. 1, p. 10 at p. 11.
[26] HSE, *The Tolerability of Risk from Nuclear Power Stations* (rev. 1992), para. 10.
[27] HSE, *The Tolerability of Risk from Nuclear Power Stations* (rev. 1992), para. 25.

but ultimately involve applying one of these three main tests. The document considers risk in the context of normal operation of nuclear installations (which effectively means risks to workers), and the risk of nuclear accidents, involving the calculation of the risk of plant failure, natural catastrophes and human error. The levels of risk which are tolerable for members of the public are much lower than for radiation workers; unlike such workers, the public have no choice as to whether they are exposed or not, and include vulnerable groups such as children and pregnant women.

The HSE considers a risk of death of around one in 1,000 per annum as the maximum tolerable for nuclear workers; this roughly equates to the maximum levels of risk acceptable to workers in relatively high-risk occupations such as offshore oil workers, or roofing contractors in the construction industry. For the general public, the risk could not be less than 10 times below this figure, *i.e.* one in 10,000 per annum. This is about the average annual risk of dying in a traffic accident, and can be compared with everyone's general chance of contracting fatal cancer, which is an average of one in 300 per annum.

At the Hinkley Point inquiry, Michael Barnes Q.C. recommended a figure for the general public of one in 100,000 per annum. Whilst maintaining its general approach that one in 10,000 is the maximum tolerable level for large industrial plant (which is likely to require further reduction under the ALARP principle), the HSE proposed to adopt a risk of one in 100,000 per annum as the benchmark for new nuclear power stations in the United Kingdom. In suggesting this proposal, the HSE recognised that this standard is, broadly speaking, achievable and measurable in the case of a new station.[28]

In practice, HSE considered that the measures taken for nuclear installation safety means that the risk borne on average by members of the public from *normal* nuclear installation activities (*i.e.* disregarding outside accidents) is usually no more than one in 1 million per annum.

At the other end of the scale to risk regarded as intolerable, there are levels of risk which can be regarded as broadly acceptable, *i.e.* below which no further improvement should be required if these entail cost. The HSE suggests that in the light of the ordinary risks of life, this level might be taken as one in 1 million per annum,[29] *i.e.* about the same level of risk as being electrocuted in the home, and about 100 times less than the average annual risk of dying in a traffic accident.

[28] HSE, *The Tolerability of Risk from Nuclear Power Stations* (rev. 1992), para. 173.
[29] HSE, *The Tolerability of Risk from Nuclear Power Stations* (rev. 1992), para. 175.

Applications for licences   2–18

Detailed guidance is provided for prospective licensees in HSE (1994) *Nuclear Licences under the Nuclear Installations Act 1965 (as amended)—Notes for Applicants* (HS(G) 120). This makes it clear that an application will be assessed under three broad areas:

1. *Organisation*

That the applicant is the user of the installation and has an adequate management structure and resources to discharge the obligations and liabilities connected with the licence. This will usually involve submission of a "management prospectus" dealing with management commitment and structure, clear lines of authority, staff resources, definition and documentation of duties, training and experience of staff, and provision of a high level of health and safety expertise. Effectively it is that part of the safety case dealing with management issues and should cover company policy, audit and independent reviews of safety, preparation of safety cases, and quality assurance (see para. 2–52 below as to the possible implications of privatisation of the nuclear industry on this aspect of licensing).

2. *Location*

For new plants, details will be expected of present and predicted population around the site, including information on schools, hospitals, institutions and other places where people may congregate (*e.g.* shopping and entertainment complexes); the licensed site boundaries must be clearly defined; the licensee must demonstrate ownership or sufficient security of tenure for the anticipated lifetime of the site, and for the period of decommissioning.

3. *Activities*

The approach here will depend on the type of installation, and whether the plant is of a new or established type. Safety guidelines for design must be acceptable before a safety case can be considered. Submissions to the NII will be made in a staged approach, as aspects of design reach the point where their safety can be assessed. There will thus be an ongoing dialogue with the NII during development of the safety case. Approaches to safety and risk are discussed below (see para. 2–21 below). The NII identifies a number of submission stages in this process:

  (a) reference design (initial statement of design and safety criteria);
  (b) preliminary safety report (the means, in principle, by which the reference design can meet the safety criteria);
  (c) pre-construction safety report (a more comprehensive statement on safety analysis);
  (d) proposed research and development work in support of the safety case;

(e) proposals for quality assurance; the means for ensuring that design, manufacture, inspection and construction are carried out to the required standard; and

(f) a contract design: the design intended for instruction.

2–19 Licensing future plant

In the course of the Government's Nuclear Review, the HSE considered the issue of licensing possible new nuclear power plant which might be of foreign design or manufacture.[30] In considering the issue, the HSE divided the options into those applicable in the near, medium and long-terms. Whilst in principle the HSE found all the options considered to be licensable, the long-term options require considerably more assessment on a site-specific safety case basis. Replication of an earlier plant already built in the United Kingdom (*e.g.* Sizewell B) is regarded as both feasible and practicable, and could lead to significant reductions in the time required for assessment.

Off-the-shelf designs, for which a full standardised safety case is available, are still likely to require some additional safety justification, if not modification. It may be possible for the NII to give credit for the regulatory assessment received as part of an overseas licensing process, provided that the HSE can gain a full and detailed understanding of those regulatory processes; bilateral international exchange agreements between the NII and its overseas counterparts may be helpful in this respect.[31] The proposed licensee cannot delegate its own safety responsibilities to a vendor of plant, although participation of the vendor in discussions between the NII and proposed licensee may be helpful.

2–20 Nuclear site licences: consultation

In all cases before granting a nuclear site licence in respect of a site in Great Britain, the HSE must consult the Environment Agency, or if the site is in Scotland, the Scottish Environment Protection Agency (subs.3(1A) of the 1965 Act, inserted by the Environment Act 1995, Sched. 22, para. 7(1)).

By subsection 3(3), where it appears to the HSE appropriate to do so, the applicant for a nuclear site licence may be directed by the HSE to serve notice on any of the following bodies:

(a) any local authority;

(b) any water undertaker or local fisheries committee;

(c) in Scotland, any river purification board, local water authority, salmon fisheries district board, or the Tweed Fisheries Board of Commissioners; and

(d) any other public authority.

---

[30] see: HSC, *Submission to the Nuclear Review* (1994), para. 99.
[31] see: HSC, *Submission to the Nuclear Review* (1994), paras. 118–119.

The notice must give such particulars as may be specified with regard to the use to be made of the site, and state that representations may be made to the HSE within three months of the date of service. Where such a direction is given, the HSE shall not grant the licence unless satisfied that three months have elapsed since service of the last of such notices, and only after considering any representations made in accordance with the notices.

Subsection 3(3) does not, however, apply in relation to an application in respect of a site for a generating station where a consent under section 36 of the Electricity Act 1989 is required for its operation (subs.(4)).

Safety assessment principles   2–21

The granting of a nuclear site licence involves the submission of a safety case and supporting reports for assessment by the NII. In order to achieve a consistent and uniform approach to this assessment process, the NII works within a framework of safety assessment principles (SAPs). These principles are explained in the HSE document, *Safety Assessment Principles for Nuclear Plants* (1992).

SAPs are used primarily for assessing the safety of proposed new nuclear installations, but assessment continues throughout the life of a plant. Once the licence has been granted, the SAPs will be augmented and implemented by licence conditions (para. 9). Four fundamental principles have continued to be applied from previous publications of SAPs and remain the basis of the current approach. These are:

(1) no person shall receive doses of ionising radiations in excess of statutory dose limits as a result of normal operation (see the Ionising Radiations Regulations 1985: para. 2–29 below);
(2) the exposure of any person to radiation shall be kept as low as reasonably practicable (ALARP);
(3) the collective effective dose to operators and to the general public as a result of operation of the nuclear installation shall be kept as low as is reasonably practicable;
(4) all reasonably practicable steps shall be taken to minimise the radiological consequences of any accident.

These principles are related to the HSE's approach to tolerable levels of risk (see para. 2–17 above). The concept of a level of tolerability is translated into basic safety limits (BSLs) for risks from normal operation and accident conditions. Any proposed plant must satisfy those limits to be considered for licensing. This approach involves decisions on a case-by-case basis, with no generally applicable numerical guide being available. To avoid disproportionate use of NII resources in assessment, BSLs are complemented by basic safety objectives (BSOs); these objectives define the point beyond which the NII need not seek further safety improvements. However, the overriding determinant is the ALARP principle, in that if, by applying this principle, the operator can provide a higher standard of safety than the BSO, the onus is on him to do so (para. 29). BSLs and BSOs are related to individual and societal risk and cover:

(a) radiation doses likely to be received by workers and members of the public in normal operation; and

(b) the chances of accidents leading to radiation doses or the release of radioactive materials.

They therefore provide measures against which the NII assessors of safety cases can make judgments. Inevitably, this is a process of uncertainty rather than numerical exactitude, particularly in relation to the risks from accidents, which are difficult to predict accurately.

The current principles comprise some 333 in number, and effectively provide a guide to all aspects of assessment of safety, such as construction and design, fault analysis, probabilistic safety assessment, principles of siting, engineering and equipment, ventilation and containment, human factors, external hazards such as earthquake and aircraft impact, accident management, waste disposal, and decommissioning. Some are general in nature; *e.g.* Principle P82 on Codes and Standards, which states that:

"The design should be conservative and follow appropriate national or international codes and standards and the plant should satisfy the requirements of the best practicable standards of manufacture, construction, inspection, maintenance and operation, commensurate both with the safety and with any relevant reliability requirements of its component parts."

Others, such as P238, are more detailed in scope:

"Movement of material into and out of glove boxes should be by means of an engineered transfer system."

In some cases the principle provides a yardstick as to what is acceptable, *e.g.* P56(a) on siting states that, allowing for some national growth in the population of the area over the life of the plant, it should be possible to evacuate all persons from an affected area of up to 1 km around the site within two hours from the time a decision to evacuate is taken.

Taken together, the principles provide a comprehensive and painstaking checklist of matters which need to be addressed, and minimum standards which need to be achieved, in any safety case put forward for a new plant. As mentioned above, many of these principles will be secured in the case of actual installations through licence conditions.

Another tool used in the licensing process by the NII is that of Probabilistic Safety Analysis (PSA). This is part of a methodical accident analysis process which produces numerical estimates of the risk from the plant. By this means, weaknesses in the design can be identified, anticipated and remedied; it should allow the "balance" of the plant to be confirmed, *i.e.* that no particular class of accident or feature makes a disproportionate contribution to the overall risk.

Nuclear site licences: relationship with Electricity Act consents and planning permission   2–22

As well as satisfying the NII that nuclear site licensing requirements can be met, licence applicants must also comply with planning requirements or provisions in the Electricity Act 1989. There has, in the past, been a measure of doubt as to some aspects of these planning requirements in relation to the special problems presented by major nuclear projects.[32] Consent to construct any form of power station with a capacity greater than 50 megawatts will require consent under section 36 of the Electricity Act 1989. Having obtained such consent for the construction process and for use of the site, a licence under the 1965 Act will be needed to install and operate the reactor. The NII will wish to be satisfied that consent requirements have been satisfied before granting a licence.

The European Council Directive[33] on the assessment of the environmental effects of certain public and private projects includes, as a Schedule 1 type of project (requiring environmental assessment in all cases), a nuclear power station or other nuclear reactor, excluding research installations for the production and conversion of fissionable and fissile materials, the maximum power of which does not exceed 1 kilowatt continuous thermal load. Such installations fall within Schedule 1 to the Town and Country Planning (Assessment of Environmental Effects) Regulations 1988 (S.I. 1988 No. 1199), and the Electricity and Pipe-line Works (Assessment of Environmental Effects) Regulations 1990 (S.I. 1990 No. 442). Originally, the 1988 Regulations simply referred to a nuclear power station or other nuclear reactor—the words corresponding to the directive's exception for small research reactors were added by the Town and Country Planning (Assessment of Environmental Effects) (Amendment) Regulations 1990 (S.I. 1990 No. 367). The Electricity and Pipe-line Works Regulations 1990 refer simply to environmental assessment being required for all section 36 Electricity Act consents (reg. 3(1)).

The Sizewell B Inquiry   2–23

The relationship between nuclear licensing procedures, and consent to construct a nuclear power station, was an important component of the public inquiry into the proposal by the Central Electricity Generating

---

[32] R. Macrory, *Planning Procedures in the Nuclear Age* [1980] J.P.L. 148.
[33] Dir 85/337/EEC: [1985] O.J. L175/40.

Board to seek ministerial consent under section 2 of the Electricity Act 1989 for Britain's first pressurised water reactor (PWR): the Sizewell B project. This immensely long and complex inquiry, held under the Electricity Generating Stations and Overhead Lines (Inquiries Procedure) Rules 1981, ran from January 1983, to March 1985.[34]

Whilst expressing strong confidence in the safety standards applied by the NII, it is clear that for the purpose of giving consent, the Government was not satisfied simply with the endorsement of the NII as licensor:

> "... the Government appeared to want more: it looked to the Inquiry to reinforce the NII position. The safety case had to be legitimised by the Inquiry, over and above the statutory licensing process."[35]

The ministers, inspector, and CEGB, were all well aware of public sensitivity as to safety of PWR design following the Three Mile Island Incident in the USA, which involved a PWR but of a different design to that proposed at Sizewell B (see further para. 1–33 above). However, it was also clear that the inquiry could not duplicate the nuclear site licensing process, which it was neither designed to perform, nor technically capable of performing. The pre-construction review stage of the CEGB's safety case alone ran to 26 volumes and over 300 supporting documents, of which the CEGB presented "highlights".[36]

The approach of the inquiry was therefore to look into the entire philosophy of nuclear safety:

> "It demanded justification of the principles and procedures adopted by various responsible and advisory bodies. It analysed the fundamentals of safety assessment calculations for certain key components and possible accident sequences where there was historical experience of danger and expressed public concern. And it explored the competence and efficiency of organisational relationships between the various parties involved in the licensing process. In sum, the Inquiry sought to pass judgment on the reliability of the total safety process— safety assessment reactor design, component manufacture, project management, operational performance and the handling of spent fuel and the decommissioned reactor, so that the whole 'safety phenomenon' need not be subject to a similar degree of public scrutiny within a future inquiry."[37]

---

[34] see: O'Riordan, Kemp and Purdue, *Sizewell B—An Anatomy of the Inquiry* (MacMillan Press, 1988).

[35] O'Riordan, Kemp and Purdue, *Sizewell B—An Anatomy of the Inquiry* (MacMillan Press, 1988), p. 22.

[36] O'Riordan, Kemp and Purdue, *Sizewell B—An Anatomy of the Inquiry* (MacMillan Press, 1988), p. 182.

[37] O'Riordan, Kemp and Purdue, *Sizewell B—An Anatomy of the Inquiry* (MacMillan Press, 1988), pp. 182–183.

This involved a probing consideration of the internal arrangements of the HSE, the safety criteria applied by the HSE, and the approach adopted to licensing; the experience was a salutary one for HSE.[38] One specific aspect of design which was discussed in detail related to the integrity of the reactor pressure vessel (RPV), through which pressurised and super-heated water would cascade to cool and moderate the fuel rods. This crucial component of the design had to be shown to have a probability of failure so low as to be "incredible".

Nuclear site licences: provisions as to insurance   2–24

Section 19 of the 1965 Act, discussed elsewhere, requires the licensee of a nuclear site to make financial provision against possible claims under the 1965 Act. This issue may, with the Secretary of State's consent, be dealt with in the licence, by including provisions as to the time from which section 19 is to apply. Where such provision is included in the licence, the requirements of section 19 do not apply until that time, or the first occasion after the grant of the licence on which the site is used for the operation of a nuclear installation, whichever is the earlier. The consequence of not making such special provision is that the requirements of section 19 operate from the time of the grant of the licence (subss.19(1) and 5(3)).

The provision of insurance is not enforced by HSE under the Health and Safety at Work, etc. Act 1974, but by the Department of Trade and Industry. HSE will consult with that Department when granting a licence.[39]

Form of licence   2–25

Whilst each licence will be unique to its site, the NII adopts a form and structure which is common to all nuclear site licences. Schedules attached to the licence will provide:

(1) a brief definition of the site (with reference to a site map) and a description of the plant or definition of the processes; and
(2) a series of licence conditions (these will often be accompanied by brief explanatory notes).[40]

---

[38] O'Riordan, Kemp and Purdue, *Sizewell B—An Anatomy of the Inquiry* (MacMillan Press, 1988), pp. 228, 231.
[39] see: HSE, *The Work of the HSE's Nuclear Installations Inspectorate* (1995), para. 29.
[40] *Nuclear Site Licences: Notes for Applicants* (HS(G) 120, 1994).

2–26 Attachment of conditions to licences

By subsection 4(1), the Health and Safety Executive must, on granting the licence, attach to it, by instrument in writing, such conditions as appear to them to be necessary or desirable in the interests of safety, whether in normal circumstances, or in the event of accident or other emergency. After the grant of the licence, the HSE has the power to attach such conditions from time to time. There is no statutory right of appeal against conditions (see para. 2–33 below), although, as the HSE is accountable to the Health and Safety Commission, that may offer a route of complaint for an aggrieved licensee.

Some guidance is provided by subsection 4(1) as to the type of matters which may in particular be covered by conditions. These are:

(a) securing the maintenance of an efficient system for detecting and recording the presence and intensity of any ionising radiations from time to time emitted from anything on the site or from anything discharged on or from the site;

(b) with respect to the design, construction, installation, operation, modification, and maintenance of any plant or installation on the site, or to be installed on the site;

(c) with respect to preparations for dealing with, and measures to be taken on the happening of any accident or other emergency on the site; and

(d) without prejudice to control under the Radioactive Substances Act 1993, with respect to the discharge of any substance on or from the site.

2–27 Standard conditions

The HSE has evolved a standardised set of conditions "with the aim of producing consistent safety requirements which are non-prescriptive and flexible".[41] These are set out at Appendix 4 of the *Notes for Applicants*, and comprise the following:

1. Interpretation
2. Marking of the site boundary
3. Restriction on transferring or parting with possession of all or part of the site without the consent of the HSE
4. Restrictions on nuclear matter on site except in accordance with adequate arrangements
5. Consignment of nuclear matter
6. Documentation and records
7. Notification, recording, investigation and reporting of incidents
8. Warning notices

[41] *Nuclear Site Licences: Notes for Applicants* (HS(G) 120, 1994).

9. Instructions to persons on the site
10. Training
11. Emergency arrangements
12. Duly authorised, qualified and experienced personnel
13. Establishment by licensee of a nuclear safety committee or committees
14. Production and assessment of safety case documentation
15. Periodic review and systematic assessment of safety case
16. Submission to the HSE of site plan, design and specifications
17. Implementation of quality assurance arrangements
18. Arrangements for radiological protection
19. Construction and installation of new plant
20. Modifications to design of plant under construction
21. Arrangements for commissioning
22. Modification or experiment on existing plant
23. Operating rules – *i.e.* a safety case demonstrating the safety of operation and identifying the limits and conditions necessary in the interests of safety
24. Operating instructions
25. Operational records
26. Control and supervision of operations
27. Safety mechanisms, devices and circuits
28. Examination, inspection, maintenance and testing
29. Duty to carry out tests and inspections
30. Periodic shutdown
31. Shutdown of specific operations
32. Arrangements for minimising the rate of production and accumulation of radioactive wastes so far as is reasonably practicable
33. Disposal of radioactive waste
34. Prevention of leakage or escape of radioactive material and radioactive waste
35. Arrangements for decommissioning—including the submission of arrangements and programmes to the HSE for approval

Two general points may be made on these conditions. First, the conditions in the main require the licensee to make appropriate arrangements to be reviewed by the NII; those arrangements will form the basis of the licensee's safety management system. Secondly, the conditions provide the basis for control by NII, but do not relieve the licensee of its general statutory responsibilities for safety under the Health and Safety at Work, etc. Act 1974. These are discussed in the next paragraph.

General duties as to safety   2–28

It is important to recognise that regulation of licensed nuclear sites under the 1965 Act co-exists with the general duties placed on the operator of the site under the Health and Safety at Work, etc. Act 1974, and enforced by the HSE. It is not appropriate to provide a detailed analysis of those duties here. However, the key aspects are:

1. The duty of the licensee as an employer to ensure, so far as is reasonably practicable, the health, safety and welfare of all his employees (1974 Act, s.2).

2. The duty of the licensee to conduct his undertaking in such a way as to ensure, so far as is reasonably practicable, that persons not in his employment who may be affected by it are not thereby exposed to risks to their health and safety (1974 Act, s.3).

3. The duty of the licensee as the person having control of the site to take reasonably practicable measures to ensure the safety of non-employees working on the site (1974 Act, s.4).

The duty under section 3 (point (2) above), which is owed to the general public, is obviously of significance in the case of a nuclear site. The duty will be broken by a risk of exposure to harm, even if the actual exposure or harm did not ultimately materialise.[42] The concept of a licensee conducting his undertaking has been interpreted widely by the courts.[43] It is therefore likely to extend to cases where a nuclear installation has been shut down, as well as where it is in operation.

As a general principle, it may be argued that a company is not conducting an undertaking where it is properly employing an independent contractor to do the work.[44] However, in the context of nuclear installations licensing, a very high degree of control is expected of the licensee as the person using the site (see para. 2–52 below); arguments that the licensee was not the person carrying on the undertaking are therefore likely to be scrutinised with scepticism.

As major industrial installations, nuclear sites will, of course, also raise many issues of health and safety of a non-nuclear nature; these are dealt with by officers from the HSE's Field Operations Division.

2–29 The Ionising Radiations Regulations 1985

The Nuclear Safety Division of the HSE is also responsible for enforcing the Ionising Radiations Regulations 1985 (see para. 2–11 above) at licensed sites. These Regulations provide strict limits for the exposure of workers and members of the public from activities involving work with radiation. The policy Branch of NSD, which is based in London, develops policies to ensure that such risks are properly controlled.

---

[42] *R. v. Board of Trustees of the Science Museum* [1993] 1 W.L.R. 1171, C.A.

[43] see: *R. v. Mara* [1987] 1 W.L.R. 87, C.A.; and *R. v. Associated Octel Ltd* [1994] 4 All E.R. 1051, C.A.; [1996] 1 W.L.R. 1543; [1996] 4 All E.R. 846, H.L.

[44] *R.M.C. Roadstone Products Ltd v. Jester* [1994] I.C.R. 456, D.C.

One example of a prosecution under the 1985 Regulations involved an incident at the Amersham International plc radiography source manufacturing plant at Harwell, in September 1993. A radiation source was manually lifted part of the way out of a container, resulting in an unplanned radiation dose to an employee. In February 1995, Didcot magistrates found the company guilty on two charges. The company was fined £3,000 on a charge of failing to make written local rules under reg. 11(1), and £2,500 on a charge of failing to restrict as far as reasonably practicable the radiation exposure of an employee under reg. 6(1). Costs of £12,800 were awarded to the HSE.[45]

Conditions on handling, treatment and disposal of nuclear matter 2–30

Special provision is made by subsection 4(2) allowing the Health and Safety Executive at any time by instrument in writing, to attach to a site licence such conditions as they think fit with regard to the handling, treatment and disposal of nuclear matter. "Nuclear matter" is defined by subsection 26(1) to cover (subject to any exceptions which may be prescribed):

(a) any fissile material in the form of uranium metal, alloy or chemical compound (including natural uranium), or of plutonium metal, alloy or chemical compound, and any other fissile material which may be prescribed; and

(b) any radioactive material produced in, or made radioactive by exposure to the radiation incidental to, the process of producing or using such fissile material.

Various of the standard conditions listed above at para. 2–27 relate to the handling, treatment and disposal of nuclear matter.

Periodic safety reviews 2–31

The earliest Magnox reactors have been in operation for between 25–40 years, as against an original conservative expectation of 20–25 years. Since the late 1970s, operators have been required to carry out a phased programme of Long Term Safety Reviews (LTSRs). The standard form of nuclear site licence now requires periodic and systematic reviews and reassessment of safety cases; these are known as PSRs (Periodic Safety Reviews) and replace the earlier LTSRs. The licence condition is not prescriptive as to the frequency of PSRs, but a period of 10 years has been adopted in accordance with good practice internationally. Essentially, the

---

[45] *HSE Nuclear Safety Newsletter—Issue 7* (June 1995).

objective of PSRs is to compare the safety case of the reactor with its original design intent and with modern standards, thus providing reassurance that the reactor will continue to be safe for a further period of operation.

A programme of PSRs for Magnox reactors and AGRs is under way, and PWRs will be brought within the programme in due course. Any modifications identified as desirable or necessary in the course of PSRs are controlled through the site licensing process.

2–32 Variation and revocation of conditions

As well as attaching additional conditions to a site licence under subsections 4(1) and 4(2), the Health and Safety Executive may at any time by instrument in writing vary or revoke any condition (subs.4(3)). In practice, this is important as it makes the licence a flexible regulatory tool, and provides scope for it to be tailored to specific circumstances applying at the relevant phase of the life of the installation.[46]

2–33 Lack of appeal against licensing decisions

The Health and Safety at Work, etc. Act 1974, s.44, provides a general right of appeal to the Secretary of State against decisions of licensing authorities for refusal to issue licences, or relating to conditions, revocation, variation or refusal to vary. However, that general provision does not apply to nuclear site licensing decisions; a nuclear site licence for that purpose being defined by subsections 44(7) and 44(8), consistently with the Nuclear Installations Act definition.

A person aggrieved by a decision of the NII would therefore have to seek redress either by complaint to the Health and Safety Commission (to which the Inspectorate is accountable), or if suitable grounds exist (which may well not be the case), by way of judicial review.

2–34 Representations by Trades Unions, etc.

Subsection 4(4) requires the Health and Safety Executive to consider any representations made by any organisation representing persons having duties upon the licensed nuclear site, though not necessarily employed there, with a view to the exercise of the powers relating to the attachment, variation and revocation of conditions.

---

[46] *Nuclear Site Licences: Notes for Applicants* (HS(G) 120, 1994).

Relationship of conditions to controls under the Radioactive Substances Act 1993

2–35

Radioactive discharges from a licensed nuclear site require authorisation under the Radioactive Substances Act 1993 (see Chapter 5), from the Environment Agency or SEPA. The distinction in theory is therefore between the nuclear processes operated on the site (which fall to the NII to control), and the emissions in gas or liquid form leaving the site. The relationship between the two systems of regulation is governed by a memorandum of understanding between the Agencies and HSE designed to minimise duplication and conflict.

The duties of the two authorities are obviously complementary, and require co-ordination. It has been noted above (see para. 2–27) that standard site licence conditions deal with the accumulation and disposal of radioactive waste. In particular, standard condition 33 requires the licensee, if so directed by the HSE, to ensure that radioactive waste is disposed of as the Executive may specify, and in accordance with an authorisation granted under the Radioactive Substances Acts 1960 or 1993. The purpose of this condition is to give discretionary powers to the HSE to direct that waste be disposed of in a specific manner.

In relation to the requirements of the Radioactive Substances Act 1993 as to registration for keeping and use of radioactive material (s.6), an exemption applies in favour of the licensee of a nuclear licensed site (subs.8(1)). The exemption applies in respect of any premises situated on the site, and in respect of keeping and use on those premises of radioactive material of every description. It applies until such time as the licence is revoked or surrendered, and thereafter until the licensee's period of responsibility comes to an end (see para. 2–42 below).

Posting of conditions

2–36

Whilst the site licence remains in force, the licensee must cause copies of the current conditions to be kept posted upon the site, in such characters and in such positions as to be conveniently read by persons having duties upon the site which are, or may be, affected by the conditions (subs.4(5)). In particular, the inspector may direct part or parts of the site where such notices are to be posted. Contravention of this requirement is an offence, as is removing, injuring or defacing such a notice without reasonable cause (subs.4(6)).

Construction and commissioning

2–37

A site licence will not be granted until the NII is satisfied with issues of design, organisational arrangements and safety concepts. This means that the risk of major changes being required for safety reasons is minimised once construction has commenced. The general approach of the NII to the construction and commissioning of nuclear installations is described in: *The Work of HSE's Nuclear Installations Inspectorate* (1995), paras. 42–49.

The NII's practice is to impose various "hold points" on the course of construction by way of licence conditions. Consent is required to progress beyond each hold point to the next stage. The work of the NII at this stage will extend to monitoring the construction of important items of plant, witnessing tests and quality assurance procedures.

Commissioning is also carried out in a phased way, with test results and consent required at each stage. Detailed case histories must be kept by the licensee relating to the construction of important components such as pressure vessels; these must be retained for the lifetime of the site.

With regards to possible new structures for projects under a privatised regime, and the implications for licensing, see para. 2–52 below.

2–38 Contravention of conditions

Contravention of conditions attached to a site licence is an offence under subsection 4(6). "Contravention" is expressly stated to include failure to comply with the condition by subsection 26(1), presumably to avoid any argument that a positive act is required. The offence may be committed by the licensee and by any person having duties on the site who committed the contravention.

Examples of prosecutions for contravention of site licence conditions are relatively rare. In 1988, BNFL was prosecuted under subsection 4(6) of the 1965 Act (and under subsection 13(1) of the Radioactive Substances Act 1960) in respect of the discharge of a large volume of slightly radioactive liquid from its Sellafield plant into the Irish Sea.[47] One of the six counts related to contravention of a condition requiring the licensee to take "all reasonable steps to minimise the exposure of persons to radiation". BNFL were found guilty on that count and fined £2,500.

More recently, both BNFL and Nuclear Electric were prosecuted during 1995. In July 1995, BNFL pleaded guilty to five contraventions of site licence conditions, and were fined £3,000 for each of the five offences by Whitehaven Magistrates' Court. The charges related to non-compliance with operating rules and instructions in transporting a fuel flask and inadequate operational records associated with flash operations. The magistrates indicated that the incident showed complacency on the part of the plant operator; they regarded breaches of the site licence as a serious matter, since the basis of public confidence lay in the operator's compliance.[48]

The incident at Nuclear Electric was more serious, involving failure to

---

[47] Dr C. Miller "Radiological Risks and Civil Liability" [1989] J.E.L. (Vol. 1), No. 1, p. 10 at p. 15.
[48] *HSE Nuclear Safety Newsletter—Issue 8* (October 1995).

shut down a reactor at its Wylfa power station for nine hours following the failure of refuelling equipment which might have restricted the flow of coolant to the reactor core. Charges were brought under the Health and Safety at Work, etc. Act 1974 and for three breaches of site licences. The 1974 Act charge was under section 2 in relation to safety of employees. A charge under section 3 of the 1974 Act relating to endangering the public, and a fourth breach of condition charge were left on the file. Having pleaded guilty at Mold Crown Court, Nuclear Electric was fined £250,000 and ordered to pay £138,000 prosecution costs. The incident was regarded by the trial judge, Morland J., as a very serious one, although he rejected suggestions that a "meltdown disaster" could have taken place. Imposing sentence he said:

> "The total fine must be commensurate with the criminal breaches. It should be exemplary in amount to underline the public's insistence that the nuclear industry is conducted with absolute safety."[49]

Statutory enforcement powers   2–39

The 1965 Act does not itself contain enforcement powers; these are to be found in the Health and Safety at Work, etc. Act 1974. The two main powers are improvement notices and prohibition notices. Improvement notices may be served under section 21 of the 1974 Act where an inspector is of the opinion that a person:

(a) is contravening one or more of the relevant statutory provisions, or

(b) has contravened one or more of those provisions in circumstances that make it likely that the contravention will continue to be repeated.

Since the "relevant provisions" include subsection 4(6) of the 1965 Act (contravention of site licence conditions), an improvement notice is a means of securing compliance with such conditions. The notice must specify the alleged contravention by reference to the relevant provision, and require it to be remedied. Failure to comply is an offence under subsection 33(1)(g) of the 1974 Act, and is subject to maximum penalties of up to a £20,000 fine and six months' imprisonment on summary conviction, and an unlimited fine and up to two years' imprisonment on conviction on indictment (subs.33(2A)).

A prohibition notice may be served under section 22 in relation to activities carried on, or to be carried on, under the control of any person, being activities to which the "relevant provisions" apply (see above). The

---

[49] *The Times*, September 15, 1995.

inspectorate must be of the opinion that the activities involve, or will involve, a risk of serious personal injury (subs.33(2)). The notice will direct that the activities shall not be carried on unless the matters giving rise to the risk, and any associated statutory contraventions involved, have been remedied. Failure to comply with the notice is an offence carrying the same penalties as for improvement notices (see above).

Such powers are sometimes used in relation to nuclear sites; for example, in 1993 an improvement notice was issued to BNFL following an investigation by the NII into a breach of operating rules at Calder Hall power station.[50] In March 1995, an improvement notice was served on Scottish Nuclear following inadvertent isolation of emergency feed and back-up cooling systems; deficiencies in maintenance procedures were identified and required to be rectified by the notice.

2–40 Powers arising from site licences

In practice, one of the most important means of control exercised by the NII derives from the terms of licence conditions rather than express statutory powers. The three main mechanisms are set down below.[51]

1. CONSENTS — required before the licensee can carry out an activity specifically defined in the licence. For example, a consent will be required before a reactor may be started up again following shutdown; this is one of the most widely used forms of control.

2. APPROVALS — used to "freeze" a licensee's arrangements; the licensee is required to submit its arrangements for approval by the NII and, once approved, these cannot be changed without further approval. This mechanism will be used, for example, in relation to operating rules.

3. DIRECTIONS — issued by the NII when it requires the licensee to take a specific action. For example, various conditions give the NII power to direct a licensee to shut down a plant, operation or process. Such directions will relate to matters of major or immediate safety importance and have been used only rarely.

In addition to these three mechanisms, in some cases conditions will require the licensee to obtain the *agreement* of the NII to a particular course of action, or will allow the NII to *notify* the licensee that a course of action is required, or to *specify* a requirement. Effectively, these constitute

---

[50] *HSE Nuclear Safety Newsletter—Issue 1* (June 1993).

[51] see also *Nuclear Site Licences: Notes for Applicants* (HS(G) 120, 1994), para. 16 and Appendix 2; and HSE, *The Work of the HSE's Nuclear Installations Inspectorate* (1995), paras. 25–27.

communications between the NII and the licensee which have a legal effect. In order to administer such arrangements, the NII makes use of a standard form of letter known as a "licence instrument".

Revocation and surrender of licences   2–41

Subsection 5(1) of the 1965 Act states simply that "... a nuclear site licence may at any time be revoked by the HSE or surrendered by the licensee". Whilst no conditions or restrictions are placed upon the ability to revoke a licence, it is submitted that some justification for such action would be required by the courts in practice. Revocation of a licence would be the ultimate sanction where the NII was concerned about the capacity of the licensee to operate safely, or about the licensee's commitment to safety. As at February 1997, no licence has yet been revoked in the United Kingdom.

Following revocation or surrender of the licence, the licensee shall, if so required by the HSE, deliver up, or account for, the licence to such person as the HSE may direct.

Responsibility following revocation or surrender   2–42

The licensee's responsibility does not end following surrender or revocation of a licence. There is then a "period of responsibility" during which the licensee must keep posted on the site notices indicating the site's limits. The licensee must also comply with any directions given by the HSE for the purpose of preventing, or giving warning of, any risk of injury to any person, or damage to any property by ionising radiations from anything remaining on the site (subs.5(2)). Contravention of any such direction, or pulling down, injuring or defacing a notice, are offences under subsection 5(4).

What constitutes the "period of responsibility" is defined by subsection 5(3). It means the period beginning with the grant of the licence and ending with the earlier of the following dates:

(a) when the Health and Safety Executive gives written notice that in its opinion there has ceased to be any danger from ionising radiations from anything on the site or, as the case may be, any part thereof;

(b) the date when a new nuclear site licence in respect of a site comprising the whole, or part of the site, is granted either to the same licensee or to some other person.

The language of the subsection contemplates the possibility of different periods of responsibility for different parts of a site; for example, if the HSE is able to give notice as to cessation of danger as to part of a site, or if part of a site in respect of which the licence has been surrendered is subject to the grant of a new licence. The issue of decommissioning is considered in the next paragraph.

The HSC has pointed out that the criterion of "no danger" is an onerous one, and has been the subject of internal review with the HSE; "There is no need to determine whether the existing legislation should be amended or whether guidance should be issued on the HSE's interpretation of the criterion."[52]

2–43 Decommissioning

The HSE will not give notice terminating the period of responsibility under subsection 5(3) of the 1965 Act (see para. 2–42 above) until satisfied that the site has been decommissioned and decontaminated to a high standard.[53]

Whilst sites being decommissioned are currently subject to nuclear licensing by virtue of their storage of nuclear matter, the HSE considers that it would be prudent, in view of the likely wider variety of such sites in the future, to clarify in the Nuclear Installations Regulations 1971 (see para. 2–04 above) that decommissioning as such is a licensable activity.[54] Any such amendment would be the responsibility of the DTI.

Licence conditions will require the licensee to make and implement adequate arrangements for decommissioning and for the production of decommissioning programmes, to be submitted to the HSE for approval. As with construction of nuclear plant, decommissioning is likely to be a strictly phased and staged process, requiring formal approval at each stage.[55] The relevant conditions will also give the HSE power to direct that decommissioning be halted at any stage.

Whilst some of the safeguards required while the plant was operational may no longer be needed as decommissioning proceeds, the full licence will still remain in force until such time as the HSE is satisfied there is no more danger from ionising radiations.

The issue of financing decommissioning costs and liabilities is potentially a difficult one. The NII does not have power under the site licensing process to require financial provision to be made for decommissioning. The HSE has made known its concern that a system should be established to ensure that adequate funds exist; this might take the form of requiring funds to be set aside, or making regular payments into a segregated decommissioning fund.[56] The DTI, to whom issues of financial provision fall, has instituted a study on the management of such liabilities.

The issue of decommissioning also has implications for radioactive waste management which is discussed elsewhere (see para. 5–12 below).

---

[52] see: HSC, *Submission to the Nuclear Review* (1994), para. 60(d).
[53] *Nuclear Site Licences: Notes for Applicants* (HS(G) 120, 1994), para. 29.
[54] see: HSC, *Submission to the Nuclear Review* (1994), para. 60(b).
[55] HSE, *The Work of the HSE's Nuclear Installations Inspectorate* (1995), para. 18.
[56] see: HSC, *Submission to the Nuclear Review* (1994), para. 130.

Maintenance of list of licensed sites   2–44

The Secretary of State is required by subsection 6(1) of the 1965 Act to maintain a list of all sites in respect of which a nuclear site licence has been granted, including maps showing the position of each site; the list is to be made available for inspection by the public. This obligation is qualified by subsection 6(2) in that the list is not required to show any site, or part of a site, in the case of which no nuclear site licence is in force and 30 years have elapsed since the expiration of the last licensee's period of responsibility (as to which, see para. 2–42 above).

Siting and emergency procedures   2–45

The Government took a cautious approach to the siting of the first Magnox stations, locating them in comparatively remote or rural areas to minimise the numbers of people at risk in the event of an escape of radioactivity.[57] This safety and siting policy was reviewed in 1968, as a result of which AGRs with pre-stressed concrete pressure vessels were allowed to be built in semi-urban environments such as Hartlepool and Heysham.

Standard licence conditions will require the licensee to make and rehearse arrangements for dealing with emergencies, including consultation with any other body whose co-operation is necessary.

Requirements are also imposed by the Public Information for Radiation Emergencies Regulations 1992 (S.I. 1992 No. 2997), implementing Council Directive 89/618/EURATOM on informing the general public about health protection measures to be applied, and steps to be taken in the event of a radiological emergency.[58] Under these Regulations, it is the duty of every employer who conducts an undertaking from which a radiation emergency is reasonably foreseeable, to ensure that members of the public in the area likely to be affected are supplied (without having to request it) with at least the information specified in Schedule 2 to the Regulations, and that such information is made publicly available (reg. 3). The information specified at Schedule 2 includes: basic facts about radioactivity and its effects; the various types of emergency and their consequences; measures envisaged in the event of an emergency; appropriate information on action to be taken by the public in that event; and the authority or authorities responsible for implementing those emergency measures. In preparing that information there must be consultation with "first-tier local authorities" (*i.e.* county council or metropolitan county fire and civil defence authorities) and "second-tier local authorities" (*i.e.* district councils). Without prejudice to this duty, the employer must endeavour to enter into an agreement with each

---

[57] HSE, *The Work of the HSE's Nuclear Installations Inspectorate* (1995), para. 36.
[58] [1989] O.J. L357.

second-tier local authority in order to disseminate the information to the relevant members of the public.

By regulation 4, first-tier local authorities are under a duty to prepare and maintain arrangements to supply, in the event of any emergency, information and advice on the facts, the steps to be taken and the health protection measures applicable. The information must include that specified in Schedule 3, for example, recommendations to stay indoors, evacuation arrangements and basic rules on hygiene and decontamination. These duties are subject to the enforcement and offence provisions of the Health and Safety at Work, etc. Act 1974 as if made under that Act (reg. 6).

2–46 Dangerous occurrences and nuclear incidents

Section 22 of the 1965 Act makes provision for the reporting to the HSE of occurrences of a prescribed class on a licensed site, or in the course of carriage of nuclear matter. Failure to report such an occurrence is an offence (subs.22(2)) and the Secretary of State may direct an inspector to make a special report into such an occurrence, or indeed cause an inquiry to be held (subss.22(4) and (5)). The Nuclear Installations (Dangerous Occurrences) Regulations 1965 (S.I. 1965 No. 1824), in prescribing occurrences for this purpose, refer to:

(a) occurrences causing or likely to cause death or serious injury to health;
(b) occurrences involving the breaking open of any outside container in which nuclear matter is being carried;
(c) explosion or outbreaks of fire on a licensed site, affecting or likely to affect the safe working or safe condition of the installation; and
(d) any "uncontrolled criticality excursion".

Apart from these statutory provisions, the HSE operates a system of issuing a statement of nuclear incidents at nuclear installations on a quarterly basis under arrangements announced to Parliament by the Parliamentary Under-Secretary of State for Energy on April 30, 1987,[59] and slightly modified in June 1993.[60] The arrangements derive from the HSC's general powers under section 11 of the Health and Safety at Work, etc. Act 1974. These quarterly incident statements are published free of charge by the HSE and comprise a brief description of the incident and the required remedial action. Normally, each incident will already have been made public by the licensee or site operator before inclusion in the statement.

[59] *Hansard*, H.C. Vol. 115, cols. 203–204.
[60] *HSE Press Notice—E108:93.*

These incidents vary considerably in seriousness, and include anticipated, as well as actual problems; for example, Incident 91/4/1 in the report for the fourth quarter of 1991 relates to the discovery of high alumina cement present in roof beams of Building B466 at Harwell (UKAEA), causing concern as to the possibility of structural failure. A recent incident reported in this way which led to some public concern was the location of 10 radioactive hotspots at the UKAEA's Dounreay site.[61] The contamination was thought to date from the 1960s or 1970s, and led to further investigation and remedial measures by the AEA.

Incidents are classified on a seven-point international nuclear event scale (INES) depending upon their seriousness and consequences. The scale ranges from 0 (no significance) to 7 (major accident). The incident at Chernobyl rated 7 on this scale and that at Three Mile Island (see para. 1–33 above) rated 5 (accident with off-site risks). Between 1984–1993, the United Kingdom has always had a number of site incidents equivalent to 0 on the scale (numbers ranging between 425 in 1984, to 176 in 1991). Incidents on scale 1–2 are much rarer, and there has been only one incident on scale 3 (in 1992). Fortunately, since the system was initiated there have been no accidents above scale 3. The Dounreay "hotspots" referred to above, by way of example, rated 0 on the INES scale, *i.e.* of no significance.

The HSE will also occasionally publish reports of investigations into nuclear incidents; an example is the report published in 1994 into the *Leakage of Plutonium Nitrate into the Plutonium Evaporator Cell of the B205 Chemical Separation Plant Sellafield on 8 September 1992.*

Cost recovery by Health and Safety Executive   2–47

Section 24A of the 1965 Act (added by the Atomic Energy Act 1989) provides the HSE with the power to recover expenses in relation to site licensing functions (subss.(3) and (4)), as well as the carrying out of research into nuclear safety at the direction of the HSC (subs.(1)(b)). The expenses recoverable include sums paid by way of remuneration, allowances, or other payments to inspectors (subs.(2)), including pensions (subs.(6)). The HSE may require advance payments on account of anticipated liability by a licensee or licence applicant (subs.(7)). By subsection (8), provision is made for repayment of any overpayments made on this basis.

The basis on which charges are made is explained in HSE's *Notes for Applicants.*[62] Licensees are charged for the direct cost of the NII's regulatory activities, all expenses including administrative support staff within the NSD and a proportionate share of the HSE's central services. Such costs include extramural studies and research commissioned to

---

[61] Incident 92/2/1; reported in the 2nd-quarter report for 1995.
[62] (HS(G) 120, 1994) paras. 22–25.

assist the NII, these costs being allocated according to the nature of the work done under each research contract.

Total costs are distributed between licensees on the basis of the percentage of inspector time allocated against the site, so that each licensee should pay the true economic costs of licensing; as an example, the charges for a power station site with two reactors were approximately £750,000 in 1993.

The HSE also recovers the cost of a programme of research co-ordinated by the HSC under arrangements with the Department of Trade and Industry. This cost is shared between the current major licensees (Nuclear Electric, Scottish Nuclear and BNFL) proportionately, as their activities relate to the subject-matter of the research.

2–48 Offences: directors and similar officers

Subsection 25(1) of the 1965 Act contains the normal provision whereby, when a body corporate is guilty of an offence and the offence is proved to have been committed with the consent or connivance of, or be attributable to any neglect on the part of, any director, manager, secretary or similar officer, or any person who was purporting to act in any such capacity, he shall also be guilty of the offence. As referred to above (para. 2–14), any holder of a nuclear site licence must be a body corporate.

The term "manager" will be confined to those in positions of real authority within the corporate structure, not simply those with managerial functions.[63] In practice, the most likely situation where the section might be used is where an offence results from neglect on the part of a director to carry out duties with which he was charged in relation to safety; one of the requirements of licensing will be the establishment of a nuclear safety committee with an advisory role, and the formulation of a management structure for safety.

Two special aspects of this section deserve mention. Some of the offences under the 1965 Act relate to the capacity of the licensee; in such cases, the director or other officer will be liable as if he, as well as the body corporate, were the licensee. Secondly, the expression "director" is extended to a member of any body corporate established for the purpose of carrying on industrial activity under national ownership, where the affairs of the body corporate are managed by its members.

2–49 Institution of proceedings

The ability to bring proceedings for alleged offences under the 1965 Act are restricted in only two cases: the offence of a failure to have a permit under section 2, and failure to make financial provision under section 19.

---

[63] *R. v. Boal (Francis)* [1992] 1 Q.B. 591.

In these cases, proceedings may not be instituted in England and Wales except by the Minister, or with the consent of the Director of Public Prosecutions (subs.25(3)).

Penalties 2–50

As originally drafted, the sections covering offences under the 1965 Act provided relatively low penalties; for example, a maximum fine of £100 on summary conviction, or a maximum fine of £500 (or up to 5 years' imprisonment) for operation of a nuclear installation without a site licence. These maximum figures were, however, disapplied by subsection 25(2) in the case of any body corporate convicted on indictment, leaving the body corporate subject to a fine of such amount as the court thought just. These upper limits on penalties were removed by the Nuclear Installations, etc. (Repeals and Modifications) Regulations 1974 in relation to most offences. Maximum penalties were retained however in relation to subsections 2(2) and 19(5). However, subsection 25(2) as referred to above remains in operation, so that in the case of a body corporate convicted on indictment, the fine is at the discretion of the court.

In serious cases, charges are likely to be brought under the Health and Safety at Work, etc. Act 1974 as well as, or instead of, the 1965 Act. The penalties for such offences are discussed in the next paragraph.

Offences under the Health and Safety at Work, etc. Act 1974 2–51

Failure to discharge the statutory duties as to safety (referred to at para. 2–28) is an offence under subsection 33(1)(a) of the 1974 Act. The maximum penalties are a fine not exceeding £20,000 on summary conviction, and an unlimited fine or conviction or both on indictment.

Proceedings may only be instituted by an inspector or by or with consent of the Director of Public Prosecutions (s.38). Directors, secretaries, managers and similar officers may, in certain circumstances, be subject to prosecution under subsection 37(1) (see para. 2–48 above).

Licensing a privatised nuclear industry 2–52

On May 19, 1994, the Government announced the terms of its review of the role of nuclear power within the United Kingdom energy supply industry. As part of that process, the HSC made a detailed submission to the Government.[64] Important aspects of this submission are set out below.

---

[64] see: HSC, *Submission to the Nuclear Review* (1994).

1. *Adequacy of existing regime*

In considering the adequacy of the current licensing regime the conclusion of the HSC was that it represented, on balance, the best mechanism for maintaining nuclear safety and should be retained (para. 2.12).

2. *Use of contractors*

The HSC points out that when the 1965 Act was drafted, the operators of nuclear installations were primarily Government-owned organisations, which had longevity, financial security and adequate resources. The NII expects that any new organisation wishing to become a licensee will display the same characteristics, particularly in terms of technical capability within its staff; in this respect the NII will look closely at the use of "bought-in expertise" and at any proposed reductions in staff. The HSC views with concern the possibility that greater commercialisation and fragmentation within the industry "... may result in applications for licensing under the 1965 Act from companies which have little knowledge or experience of nuclear safety matters and may not appreciate the regulatory requirements" (para. 67). With this in mind, the HSE has revised its *Notes for Applicants* to stress the requirement of a management prospectus (see para. 2–18 above) in order to deal with possible issues relating to management and safety organisation. The HSE draws a distinction between the "ordinary" use of contractors for specific activities on the one hand, and management-related "contractorisation" on the other. The former situation is relatively easily dealt with by arrangement under licence conditions to ensure proper selection and monitoring of contractors. Contractorisation is less simple; the holder of the licence will be expected to have the competence to oversee and take responsibility for the activities of management contractors, and their use should not be allowed to compromise either the licensee's chain of command or its ability to control activities on site. Eligibility for a licence will depend upon an applicant being able to demonstrate a sufficient degree of control over the licensable activity. This may be particularly important where a different company employs the staff.

3. *"Turnkey contracts"*

The touchstone of control means that in some circumstances the owner of an installation may not be its licensee. Where a contractor is employed to act autonomously, without direct control and supervision by the owner (as under a turnkey contract), it will probably be necessary for the contractor to be the licensee. The essence of such a contract is that the contractor manages the relevant construction project autonomously to completion, before handing over the completed plant to the operator, normally before fuel is loaded. The HSE's view is that on completion there is no reason why the site could not be re-licensed to the ultimate operator, provided it can meet the appropriate criteria, including a thorough understanding of the plant design and operation. If a suitable operator to hold a new licence does not exist, the contractor will retain the legal liability (para. 73).

4. *Dual licensing*

One issue which has been tentatively considered is that of "dual licensing" whereby, in relation to decommissioning, both the owner of the plant and the decommissioning contractor would be licensed; the suggestion apparently emanated from the Atomic Energy Authority and Nuclear Electric Plc. However, the HSE regards it as fundamental that at any one time a single body should have responsibility under the licensing system, and that there should be no doubt as to the identity of the responsible body. Accordingly, the HSE would wish to avoid dual licensing.

5. *Retention of skills and expertise*

The HSE regards licensees as having an obligation to ensure that sufficient skills, knowledge and expertise are available to them to meet safety contingencies; this includes the commissioning of relevant research, either by the licensee or through the HSC Co-ordinated Research Programme. The HSE will press for the retention of adequate research despite any pressures to reduce costs in that respect (para. 82). The HSE also notes it is possible that, if no further nuclear power stations are ordered, there might be a tendency for a general run-down in the availability of skills and expertise. Licence conditions relating to the training and appointment of suitably qualified persons will provide the NII with a regulatory basis for challenging any such loss of expertise, a loss which could result in safety inadequacies (para. 84).

6. *Financing of decommissioning liabilities*

The HSC has flagged its view that adequate funds must be ensured to meet decommissioning obligations in a manner, and to a timetable, which keeps risks to workers and members of the public as low as are reasonably practicable. This may involve a change to the legislative and regulatory structure, which at present does not allow a specific requirement of financial provision to be imposed (para. 130).

The Government is committed to the view that changes of ownership within the nuclear industry will not affect standards of safety.[65] Existing principles and arrangements will therefore continue, whatever the future structure of the industry.[66]

As a consequence of the structural changes to the industry connected with privatisation, the HSE determined that the changes were sufficient

---

[65] *The Prospects for Nuclear Power in the U.K.*, Cm. 2860, para. 10.12; endorsing the "White Paper", *Privatising Electricity*, Cm. 322 (1988).
[66] *The Prospects for Nuclear Power in the U.K.*, Cm. 2860, paras. 10–19.

to require re-licensing of all Nuclear Electric and Scottish Nuclear sites.[67] This re-licensing process was completed on schedule, with new licences for the 16 sites coming into force on March 31, 1996.[68] The process is described in detail in the HSE Report, *Restructuring and Privatisation of the Nuclear Industry: Report on the Work of the Health and Safety Executive to Grant Replacement Site Licences.*[69]

2–53 Security requirements

The Nuclear Generating Stations (Security) Regulations 1996 (S.I. 1996 No. 665) which came into force on April 1, 1996, impose various requirements as to security which apply to generating stations and laboratories, as defined in the Regulations. The generating stations covered are nuclear installations which are designed or used, currently or previously, for generating electricity for supply, and at which nuclear fuel is kept, or is proposed to be kept. The laboratories covered are those which are nuclear installations in this sense, and which are or have been used for the purpose of examining radiated fuel on behalf of the operator of a generating station. Regulation 3 requires the operation of a generating station or laboratory to submit a security plan to the Secretary of State for approval, *i.e.* a description in writing of the security regime (standards, procedures and arrangements) which apply thereto. At all times when nuclear fuel is on site, the operator must ensure that the site is subject to a security regime which conforms to an approved security plan (reg. 4). Any directions given by the Secretary of State with regard to security must be complied with (reg. 5) and regular security assessments must be made (reg. 6). Further requirements relate to the security regimes applicable to any structural alterations or extensions carried out at the generating station or laboratory (reg. 7) and to the unloading, storage and transportation of any nuclear fuel in transit to or from the site (reg. 8).

The Regulations are made under the Health and Safety at Work, etc. Act 1974 and are enforceable as such, the responsible authority being the Secretary of State for Trade and Industry.

2–54 European regulation

The issue of regulating nuclear safety has been considered by two Advisory Groups of the European Commission: the Nuclear Regulations Working Group, and the Regulatory Assistance Management Group. These Groups have published a report, *Establishing an Effective Nuclear*

---

[67] *Memorandum of Evidence* submitted by HSE to the House of Commons Trade and Industry Committee, November 1995, H.C. 43.
[68] *H.S.E. News Release—E50:96* (April 1, 1996).
[69] *H.S.E. News Release—E100:96* (May 31, 1996).

Safety Regulatory Regime.[70] Part I of the Report considers the objectives and requirements of an effective regulatory regime, drawing on internationally recognised good practices in nuclear safety, experience in Western European countries, and the Safety Fundamentals identified by the Nuclear Safety Standards Advisory Group (NUSSAG) of the International Atomic Energy Agency.[71]

The Report is essentially an advisory document aimed at those in senior governmental and regulatory positions. It mirrors many of the fundamental concepts of safety applied by the NII in the United Kingdom. This is not surprising since the NII was represented on the Task Force which produced the Report, and indeed the Report was prepared by a former member of the NII. The Report distinguishes between nuclear and radiological safety. Its General Nuclear Safety Objective is:

"To protect individuals, society and the environment from harm by establishing and maintaining in nuclear installations effective defences against radiological hazards."

Radiological safety (or radiological protection) is regarded as a separate but inherently linked discipline, the Radiological Safety Objective being:

"To ensure that in all operation states radiation exposure within the installation or due to any planned release of radioactive material from the installation is kept below prescribed limits and as low as reasonably achievable, and to ensure mitigation of the radiological consequences due to accidents."

Issues covered in the Report include organisation of regulatory bodies, licensing assessment, inspection, compliance and enforcement, regulatory liaison, and maintaining regulatory effectiveness through adequate staffing and training.

As in the United Kingdom, the fundamental starting points are that there is no such thing as absolute safety, and that responsibility for safety rests firmly with the operator of the installation.

[70] EUR 15397 E.N., 1994.
[71] IAEA, *The Safety of Nuclear Installations—Safety Fundamentals* (1992).

CHAPTER THREE

LIABILITY AND INSURANCE

Introduction: problems of liability and insurance   3–01

With the post-war expansion of work in developing modern sources of energy came the problem of the potential damage which might flow from a nuclear accident. This was particularly so given the possibility of power stations being constructed and run by private operators. In the United Kingdom, Calder Hall nuclear power station was commissioned in 1956, and was the world's first commercial power station to use nuclear energy; there followed the nuclear power station at Chapelcross, Ayrshire in 1958, and the rapid programme of constructing Magnox reactors during the 1960s. Progress was, however, constrained by the reluctance of private companies to build and operate commercial nuclear plants because of the risk of heavy financial liabilities in the event of an accident, and the reluctance of insurance companies to provide cover. A 1957 report by the United States Atomic Energy Commission estimated that in a "worst case" accident, 3,400 fatalities and 43,000 injuries would arise along with some $7 billion in property damage; such predictions heightened existing concerns of electricity utility companies and equipment suppliers.[1]

Before the international instruments mentioned below were formulated, a number of countries, Britain included, had adopted their own legislation on the issue of liability. These included the U.S. Price-Anderson Act 1957, the German Atomic Energy Act 1959, and the Swiss Federal Law on the Exploitation of Nuclear Energy for Peaceful Purposes and Protection from Irradiation, 1959. All followed the principle of channelling legal liability on a strict liability basis to the owner and operator of a nuclear installation, coupled with the limitation of liability both in amount and in time. This balancing of liabilities against protection for operators was a reflection of "the desire of the U.K. Government, and indeed the Governments of Western Europe in general, not to risk stifling this new industry at birth by imposing impossibly heavy financial responsibility on the operators and others who would be concerned in the industrial and commercial development of nuclear energy".[2]

[1] Campbell-Mohn, Breen and Futrell (eds.), *Sustainable Environmental Law* (St. Paul, Minn. 1993), p. 891.
[2] James C. Dow, *Nuclear Energy and Insurance* (1st ed., London, 1989), p. 83.

3–02 Early United Kingdom legislation

In the United Kingdom, the Nuclear Installations (Licensing and Insurance) Act 1959 was the outcome of this strategy. As well as making provision for the licensing of nuclear installations (ss.1–3), the 1959 Act provided a general framework of liability which has remained largely unchanged in subsequent legislation. The licensee of a nuclear installation was placed under a strict duty to secure that no ionising radiations emitted from the site, or from waste discharged on or from the site, caused hurt to any person or damage to any property, whether that person or property was on the site or elsewhere (subs.4(1)(a)). This duty also applied to ionising radiations from irradiated nuclear fuel in the course of carriage on behalf of the licensee within the United Kingdom (subs.4(1)(b)). The only exception to such liability related to emissions or damage attributable to hostile action in the course of armed conflict.

CHANNELLING LIABILITY — Channelling liability was accomplished by subsection 4(2), which provided that no person other than the licensee should be under any liability in respect of hurt to any person or damage to property to which subsection 4(1) applied. By subsection 4(5) the licensee's liability under subsection 4(1) was provided to be in substitution for any other liability of the licensee apart from that subsection, thus achieving channelling in terms of the cause of action as well as the defendant.

LIMITATION OF LIABILITY: TIME — Limitation of liability in terms of time was provided by subsection 4(4), by which no action to establish a claim by virtue of subsection 4(1) could be commenced after the expiration of 30 years from the "relevant date", *i.e.* the date of the occurrence on, or in connection with the use of, the site in question which gave rise to the claim. The corollary of this limitation was that the licensee was also protected in relation to claims made more than 10 years after the relevant date, in that the licensee was not required to make any payment in satisfaction of such claims unless and until Parliament had made provision to secure that the amount required to satisfy the claim would be reimbursed to the licensee (subs.4(4) proviso).

LIMITATION OF LIABILITY: AMOUNT — The other aspect of limitation related to the amount of liability, and was dealt with as part of the issue of insurance cover or other financial provision. By subsection 5(1), the licensee was required to make provision, by insurance or other means, for sufficient funds to be available to ensure that duly established claims were satisfied up to an aggregate amount of £5 million, in respect of certain stated "cover periods". Where a claim was established successfully, but funds created under subsection 5(1) were not required to be made available for the full satisfaction of that claim, again, the licensee would not be required to meet the claim until Parliament had provided a means by which the excess over the insured amount would be reimbursed to the licensee (subs.4(4) proviso).

INSURANCE — The question of insurance cover is dealt with in more detail below, but it is important to stress that the scheme of insurance and limitation was designed in close consultation with insurers, and was largely based upon what insurance capacity was available.[3] When introducing the Bill (which became the 1959 Act) for second reading in the House of Commons, the Paymaster-General (Mr Reginald Maudling) referred to the need to clarify the liability of operators to those hurt or damaged by an accident. He went on to say that in the case of private operators it was the responsibility of such operators, as with every type of commercial risk, to obtain adequate insurance against such liability:

> "It is, however, not possible to provide insurance cover of an unlimited amount or over a period so long as 30 years. This is not only the experience in this country. It is the experience in some neighbouring countries as well ... Equally, in the case of the length of claim, it is, as a practical matter, impossible to ensure [sic] against claims arising more than about ten years ahead. Should claims arise after ten years up to a period of thirty years, these, again, will devolve upon Parliament for settlement."[4]

Thus the United Kingdom developed its own framework of liability in advance of international action in the field. At the same time, the Organisation for European Economic Co-operation (now the OECD) was studying the issue, and there were soon to be Conventions agreed. The subsequent evolution of the United Kingdom legislation was largely a response to the requirements of those Conventions, and before considering the United Kingdom legislation further, it is useful to refer to the Conventions themselves.

International provisions: generally   3–03

At international, as at national level, early appreciation of the potential benefits of the peaceful applications of nuclear power was rapidly tempered by the realisation of the catastrophic consequences that could follow from failure to maintain adequate levels of safety:

> "In atomic energy, because of its spectacularly military origins and the potential risk involved in its utilisation, the notion of calamities has taken high precedence over other considerations

[3] *ibid.*
[4] *Hansard*, H.C. Vol. 599, cols. 866–867; and see also *Hansard*, H.L. Vol. 212, cols. 1035–1036.

both in the public mind and in the minds of lawyers and administrators concerned with problems of regulation."[5]

This realisation led to efforts on the part of relevant international organisations, specifically the Organisation for Economic Co-operation and Development (OECD) and International Atomic Energy Agency (IAEA), not only to devise means of seeking to prevent nuclear accidents, but also to secure adequate mechanisms for compensation in the event that such disasters did occur:

> "The principal national and international legal problems posed by the development of the pacific uses of atomic energy have been on the one hand to keep calamities from happening and on the other to devise appropriate remedies so that damage to health and property may be compensated in the most humane, equitable and expeditious way possible."[6]

Essentially, there are two main Conventions on third party liability in the field of nuclear energy. The first was the Paris Convention of 1960, concluded under the aegis of the OECD, which came into force on April 1, 1968, and to which the United Kingdom is a party. The Paris Convention is supplemented by the Brussels Convention of 1963, to which the United Kingdom is also a party.

The other main Convention is that negotiated by the IAEA and signed in Vienna in 1963. However, it did not come into force until 1977, when the 10 necessary ratifications were achieved. The Convention has not been successful in terms of the number of state parties, which currently (1997) do not include any Western European countries, or any leading countries in the field of nuclear energy; indeed, of the first 10 ratifying states only two (Argentina and the former Yugoslavia) had nuclear power stations in operation at that time.[7]

An attempt to regulate the relationship between the Paris and Vienna Conventions (no state being a party to both) was provided by the Joint Protocol relating to the application of the Vienna Convention and the Paris Convention; this Protocol entered into force in April 1992, although the United Kingdom is not a party. It provides that either the Vienna or Paris Convention shall apply to a nuclear incident to the exclusion of the other; the determining factor will be whether the nuclear installation involved is situated in the territory of a party to the Vienna Convention, or the Paris Convention.

---

[5] Jerry L. Weinstein, *Progress in Nuclear Energy, Service X, Law and Administration* (1966), p. ix.
[6] *ibid.*
[7] James C. Dow, *Nuclear Energy and Insurance* (1st ed., London, 1989), p. 102.

The Paris Convention: generally

3–04

Together with its Supplementary Brussels Convention, the Paris Convention represents an impressive feat of international diplomacy. The Convention was negotiated and agreed under the auspices of the Organisation for European Economic Co-operation (later to become the OECD). The specific arm of the OEEC involved was the European Nuclear Energy Agency, which was later re-named the Nuclear Energy Agency. The impetus for negotiation sprang in part from Article 98 of the newly-signed Treaty of Rome, which committed the six members of the European Atomic Energy Community to "take all necessary measures to facilitate the conclusion of insurance contracts covering nuclear risks", and in part from a commitment in an Agreement for Co-operation between EURATOM and the USA to develop suitable measures providing protection against third-party liability.

The preamble to the Convention refers to the desire of the parties to ensure "adequate and equitable compensation for persons who suffer damage caused by nuclear incidents whilst taking the necessary steps to ensure that the development of the production and uses of nuclear energy for peaceful purposes is not thereby hindered." It refers also to the conviction of the parties of the need for unification of the basic liability rules applying nationally, whilst leaving parties free to take additional measures on a national basis where they deem it appropriate.

The Convention was signed on July 29, 1960, by the 16 OEEC countries involved in its negotiation. Ratification was required by five signatories before the Convention could come into force (Art. 19(b)). With the need to pass national legislation to ratify the Convention, it did not come into force until 1968, the United Kingdom being among the initial group of signatories to ratify; all the major European countries, with the exception of Austria, Luxembourg and Switzerland, have now ratified. Before it came into force, the Convention was amended by an Additional Protocol of January 28, 1964, with the object of making it compatible with the Vienna Convention, thus solving problems which could arise for States wishing to participate in both the Paris and Brussels Conventions.

The Paris Convention has been said to rest on four fundamental principles, namely:

(a) channelling of liability;
(b) limitation of liability;
(c) compulsory cover for liability; and
(d) single jurisdiction.[8]

Each of these principles is considered separately in the following paragraphs.

---

[8] See Rafaello Fornassier in Weinstein, *Progress in Nuclear Energy, Service X, Law and Administration* (1966), p. 24.

3–05 The Paris Convention: channelling of liability

Article 3(a) of the Convention provides that the operator of a nuclear installation shall be liable for:

    (i) damage to or loss of life of any person; and

   (ii) damage to or loss of any property (other than the nuclear reactor itself, any on-site property used in connection with the installation, and the means of transport in the case of incidents in the course of carriage);

upon proof that such damage or loss was caused by a nuclear incident involving either nuclear fuel, radioactive products or waste (as defined) in the installation, or nuclear substances coming from the installation.

The term "damage" is not defined, a problem which is discussed below. Therefore, the requirement is that first there be a "nuclear incident", *i.e.* an occurrence or succession of occurrences causing damage, provided that the occurrence or the damage arises out of, or results from, the radioactive properties of nuclear fuel, radioactive products or waste, or a combination of radioactive properties with toxic, explosive or other hazardous properties. Secondly, it must be proven that the incident involved either nuclear fuel, radioactive products or waste in the installation, or nuclear substances coming from the installation.

Article 3(c) allows any Contracting Party by legislation to provide that the liability of the operator of a nuclear installation situated in its territory shall include liability for damage which arises out of, or results from, ionising radiations emitted by any source of radiation inside the installation, and not just those referred to in Article 3(a).

Article 3(b) of the Convention deals with mixed damage, *i.e.* that which is caused only in part by a nuclear incident. If the "nuclear" and "non-nuclear" damage are not reasonably separable, then the "non-nuclear" damage shall be considered to be damage caused by the nuclear incident.

Article 4 deals with the carriage of nuclear substances, including storage incidental thereto. The provisions are complex, but Article 4(a) essentially deals with incidents involving nuclear substances in the course of carriage from a nuclear installation, and Article 4(b) deals with carriage to the installation. In relation to carriage from the installation, the operator shall be liable only if the incident occurs:

    (i) before liability has been assumed by the operator of another nuclear installation, expressly and in writing; or

   (ii) in the absence of such express terms, before the operator of another nuclear installation has taken charge of the nuclear substances; or

   (iii) where the nuclear substances are intended to be used in a reactor comprised in a means of transport, before the person duly authorised to operate the reactor has taken charge of the substances; or

(iv) where the nuclear substances have been sent to a person within the territory of a non-Contracting State, before they have been unloaded from the means of transport by which they have arrived in the territory of that non-Contracting State.

By Article 4(b), the operator of an installation shall be liable in respect of nuclear incidents involving nuclear substances in the course of carriage to the installation only if the incident occurs:

(i) after liability has been assumed by him, pursuant to the express terms of a contract in writing from the operator of another nuclear installation;

(ii) in the absence of such express terms, after he has taken charge of the nuclear substances; or

(iii) after he has taken charge of the nuclear substances from a person operating a reactor in a means of transport; or

(iv) where the nuclear substances have, with the written consent of the operator, been sent from a person within the territory of a non-Contracting State, after they have been loaded onto the means of transport by which they are to be carried.

The provisions of Articles 4(a) and (b) therefore mirror each other to provide what should be a clear scheme of liability. Article 4 is expressed to be without prejudice to Article 2, which deals with the territorial scope of the Convention. As a regional agreement, the Convention does not apply to nuclear incidents occurring in the territory of a non-Contracting State or to damage suffered in such territory. The two exceptions to this principle are where the legislation of the operator's Contracting State provides otherwise, and in relation to the rights under Article 6(e) of those who have their principal place of business in the territory of a Contracting State.

The channelling of liability in accordance with these provisions is achieved by Article 6(g), which provides that the right to compensation for damage caused by a nuclear incident may be exercised only against an operator liable for the damage in accordance with the Convention. Direct rights of action against the insurer or other financial guarantor of the operator, are preserved, if national law permits such action. By Article 6(f), the operator has a right of recourse against third-parties in respect of nuclear incidents only in very limited circumstances, where the damage results from an act or omission done with intent to cause damage, and if, and to the extent that, it is so provided expressly by contract.

3–06 The Paris Convention: limitation of liability in amount

By Article 7, the aggregate of compensation required to be paid in respect of damage caused by a nuclear incident shall not exceed the maximum liability established in accordance with the article. The scheme of the Convention was to fix a maximum amount of liability, but to provide that any contracting party can, taking into account the possibilities for the operator of obtaining the necessary insurance or other financial security, establish by national legislation a greater or lesser amount, but in no event to be less than a minimum figure. These maximum and minimum amounts were originally set at 15 million and 5 million European Monetary Agreement Units (effectively equivalent to 15 million and 5 million U.S. dollars at that time). Subsequently, the units of account were changed by the Protocol of November 16, 1982, to adopt as the currency unit the Special Drawing Rights of the International Monetary Fund.

The main weakness of the Paris Convention was the inadequacy of these amounts; this problem was addressed by the Brussels Supplementary Convention, discussed below. Article 15 allows any contracting party to take such a measure as it deems necessary to provide for an increase in the amount of compensation specified in the Convention and indeed a number of countries have done so, increasing the sums payable by significant amounts.

3–07 The Paris Convention: limitation of liability in time

Article 8 of the Convention provides that the right of compensation shall be extinguished if an action is not brought within 10 years from the date of the incident. This limitation was a response to representations made on behalf of insurers, who pointed out the impracticability of requiring them to keep their books open indefinitely after the occurrence of an incident which might, or might not, give rise to claims many years hence.[9] The article does, however, allow national legislation to establish a longer period if measures have been taken by the contracting party, in whose territory the nuclear installation is situated, to cover the liability. Special provision is made for damage caused by nuclear matter which is stolen, lost, jettisoned or abandoned. The period of 10 years from the incident still applies, but is subject to a further limit of 20 years from the date of the theft, loss, jettison or abandonment (Art. 8(b)).

3–08 The Paris Convention: compulsory cover for liability

Article 10 provides that to cover liability under the Convention, the operator shall be required to have and maintain insurance or other financial security of the amount established pursuant to Article 7 and of

---

[9] James C. Dow, *Nuclear Energy and Insurance* (1st ed., London, 1989), pp. 91–92.

such type and terms as the competent public authority shall specify. By Article 10(b), contracting parties must not allow the insurer or other financial guarantor to suspend or cancel the insurance or other financial security without giving notice in writing of at least two months or, in the case of the carriage of nuclear substances, during the period of carriage. By Article 12, the compensation payable, insurance and reinsurance premiums and sums, must be freely transferable between the monetary areas of the contracting parties.

The Paris Convention: single jurisdiction   3–09

Article 13(a) provides for jurisdiction for actions under the compensation provisions of the Convention to lie only with the courts of the contracting party in whose territory the nuclear incident occurred.[10] Where the incident occurs outside the territory of any of the contracting parties, or where its location cannot be determined with certainty, jurisdiction lies with the courts of the contracting party in whose territory the nuclear installation of the relevant operator is sited (Art. 13(b)). Provision is made by Article 13(c) for determining jurisdiction in case of potential conflicts. Article 13(d) provides for the enforcement of final judgments to be entered by the competent court under this article throughout all contracting parties; it is specifically provided that the merits of the case may not be subject to further proceedings. Another aspect of jurisdiction is that by Article 13(e), if action is brought against a contracting party itself, that party may not invoke any jurisdictional immunities before the competent court, save in matters of execution.

The Brussels Supplementary Convention   3–10

As mentioned above, the amounts of compensation provided by the Paris Convention were relatively low. Article 15 of that Convention, however, provided that any contracting party may take such measures as it deems necessary to provide for an increase in the amount of compensation specified in the Convention. Considerable progress was made in negotiations between France, Germany, Italy, Belgium, The Netherlands, Luxembourg (as the original six members of the European Community) and Britain towards agreeing a convention within the framework of the European Atomic Community (EURATOM). This Convention would have involved agreement to increase the Paris limits, but in the event the proposal was superseded by the Convention Supplementary to the Paris

---

[10] A detailed discussion of the problems of jurisdiction which might arise is provided by Paolo Galizzi, *Questions of Jurisdiction in the Event of a Nuclear Accident in a Member State of the European Union* [1996] J.Env.L. (Vol. 8), No. 1, p. 7.

Convention on Third Party Liability in the Field of Nuclear Energy, formulated by the OECD; this Convention (the Brussels Convention) was signed at Brussels on January 31, 1963.

By Article 3 of the Brussels Convention the contracting parties undertake that compensation in respect of damage caused by nuclear incidents should be provided up to the amount of 120 million Special Drawing Rights per incident. As with the Paris Convention, the amounts were originally expressed as European Monetary Agreement units of account, but the units were changed to International Monetary Fund Special Drawing Rights (SDR) by the Protocol of November 16, 1982. By Article 3(b), such compensation is to be provided:

(a) up to an amount of at least 5 million SDR, out of funds provided by insurance or other financial security, such amount to be established by the legislation of the contracting party where the nuclear installation is located;

(b) between this amount and 70 million SDR, out of public funds to be made available by the contracting party where the nuclear installation is located;

(c) between 70 million and 120 million SDR, out of public funds to be made available by the contracting parties according to the formula for contributions specified in article 12.

Article 3(c) gives contracting parties a choice on how to implement this requirement. They may either establish the maximum liability of the operator pursuant to the Paris Convention of 120 million SDR, or may establish such maximum liability at the amount of at least 5 million SDR which they have fixed pursuant to (a) above, and provide that the excess of that amount up to 120 million units shall be available from public funds. In any event, by Article 3(d) the obligation of the operator to pay compensation, interest or costs out of public funds to be made available under the Convention shall only be enforceable against the operator as and when such funds are in fact made available.

The formula under Article 12 for contributions between contracting parties is based on two ratios: gross national product, and the thermal power of the reactors located in each state at the time of the nuclear incident.

With the scale of commitment of public funds involved, it was necessary to define accurately the ambit of the Convention. Accordingly, it applies only to nuclear installations used for peaceful purposes, situated in the territory of a contracting party, and appearing on the list of such installations produced and maintained by contracting parties under Article 13 (Art. 2(a)(I)). The Convention does not apply to incidents occurring entirely in the territory of a state which is not a contracting party, i.e. it covers incidents occurring within the territory of a contracting party, or on the high seas.

Following the requisite number of ratifications, the Convention came into force in December 1974. As with the Paris Convention, the Brussels Supplementary Convention was amended by an Additional Protocol of January 28, 1964, to minimise inconsistencies with the Vienna Convention. It is easy to overlook the magnitude of the advance marked by the Brussels Supplementary Convention, which has been described as "... an unprecedented step in international collaboration and mutual confidence by agreeing to pay victims of nuclear incidents out of joint Government funds."[11] Effectively, the Supplementary Convention creates a system of mutual assistance on the principle of financial solidarity; this may "... result in a Contracting Party being obliged to make available funds ... where the incident occurs in a foreign country and where the operator liable and all the victims are foreigners."[12] When the Nuclear Installations (Amendment) Bill was introduced in 1965 to implement the Supplementary Convention, this risk-sharing arrangement was described to Parliament by Lord Stonham (Joint Parliamentary Under-Secretary of State at the Home Office) in terms of collective arrangements by members of a "club", which would provide reassurance to the public, suppliers and carriers, without imposing an unacceptable burden on the United Kingdom.[13]

The Vienna Convention   3–11

The Vienna Convention on Civil Liability for Nuclear Damage of May 1963 offers an alternative international regime to the Paris and Brussels Supplementary Conventions. Whilst the United Kingdom is not a party to the Vienna Convention, the Convention has been referred to judicially as an aid to construing United Kingdom legislation (see para. 3–26 below). It is also clear that when passing the Nuclear Installations (Amendment) Act 1965, the Government had the Vienna Convention in mind as well as the Paris and Brussels Conventions, and regarded all three as "fully compatible".[14] Similarly, in relation to the Nuclear Installations Act 1969, three conventions (Paris, Brussels and Vienna) were regarded as lying behind the legislation, which was to be consistent with all three.[15] The long title of that Act is: "An Act to make in the Nuclear Installations Act 1965 certain amendments necessary to bring that Act into conformity with international agreements".

The intention of the International Atomic Energy Agency in formulating the Vienna Convention was that, unlike its OECD counterpart, it would have universal, as opposed to regional, application. The Convention was not, however, ratified by the 10 states required for it to

---

[11] Jerry L. Weinstein, *Progress in Nuclear Energy, Service X, Law and Administration* (1966), p. x.
[12] See Rafaello Fornassier in Weinstein, p. 29.
[13] *Hansard*, H.L. Vol. 263, col. 1280.
[14] *Hansard*, H.L. Vol. 263, col. 1275; and *Hansard*, H.C. Vol. 702, cols. 48–50.
[15] *Hansard*, H.L. Vol. 301, cols. 329–330.

come into effect for some years, and only entered into force in February 1978. As mentioned above, those ratifying do not include the leading states in developing nuclear power.

The schemes of the Vienna and Paris Conventions are basically similar: the operator of a nuclear installation is strictly liable for damage caused by a nuclear incident, that liability is limited in time and amount, and the operator is required to provide financial security. However, there are some significant differences in the drafting of the two Conventions. The Vienna Convention, for example, provides a useful definition of "nuclear damage" which is lacking in the Paris Convention; it is also provided more explicitly in the Vienna Convention that liability is absolute (Art. V.1). The minimum amount of liability per incident which can be limited by national legislation is 5 million U.S. dollars; a limit which is now inadequate. The $5 million figure was the subject of much debate in Conference. It was recognised as insufficient for the proper protection of victims, inadequate in comparison with the figures in the Brussels Supplementary Convention of 70 million and 120 million SDR, and with the figures used in some national legislation (as high as $500 million). Running counter to this view was the argument that a high figure for minimum liability would prevent many financially weak countries from signing the Convention; this view eventually prevailed.

3–12 Scope of the Paris, Brussels and Vienna Conventions —non-peaceful uses of nuclear energy

The preamble to the Vienna Convention refers to the desirability of establishing minimum standards to provide financial protection against damage resulting from "certain peaceful uses of nuclear energy". No such reference appears in the body of the Convention. The Paris Convention contains no reference at all to peaceful uses, though its published *Exposé des Motifs* is written in such terms as to indicate that this is what was in mind. The Brussels Supplementary Convention is quite clearly confined to installations used for peaceful purposes (Art. 2(a)(i)).

It is, therefore, clear that the mutual funding regime for compensation under the Brussels Convention would not apply to installations used for military purposes. The position is unclear in relation to the rules on liability in the Vienna and Paris Conventions, though it has been suggested that the guiding principle should be that "the [Vienna] Convention should in no way be interpreted as intending to facilitate or further the non-peaceful uses of nuclear energy".[16] Whilst the application of the liability rules of both Conventions in some respects favours the operator, and in others the victim of damage, the main benefits almost

---

[16] see Wolff in Weinstein, pp. 5–6.

certainly lie with the victim, so that non-application to military installations might, in fact, favour or facilitate non-peaceful uses. Ultimately, the issue would be one for the relevant national courts to decide.

Another problem is that whilst some installations may clearly be non-peaceful in nature, *e.g.* weapons manufacture or nuclear submarine refuelling, it is quite possible that some installations might be used only partly for the production of fissile material destined for non-peaceful purposes. The status of such installations may be obscure, in that the nuclear incident giving rise to damage or injury may or may not involve material intended for military use; if it does not involve such material then it might be argued that the "peaceful" status of the installation is compromised by its occasional military-related use.

As explained below, the United Kingdom legislation draws no distinction between peaceful and non-peaceful uses of nuclear matter in the context of liability.

The Convention on the Liability of Operators of Nuclear Ships   3–13

This Convention was adopted in Brussels in May 1962, as the result of a Diplomatic Conference on Maritime Law. It is not in force and the United Kingdom is not a signatory. The definition of "nuclear ship" is wide enough to cover both naval and merchant vessels, although by Article X.3, warships or other State-owned or State-operated ships are not liable to arrest, attachment or seizure, or to the jurisdiction of the courts of any foreign state. The Convention primarily had in mind the potential growth in nuclear-powered merchant vessels; the simple fact that very few nuclear ships, other than naval vessels and submarines, have been produced accounts for the general lack of interest in the Convention or national legislation on the subject.

Article II of the Convention channels absolute liability to the operator of a nuclear ship for nuclear damage caused by a nuclear incident involving the nuclear fuel of the ship, or radioactive products or wastes produced in the ship. Such liability is limited to 1,500 million francs (a unit of account related to gold) per nuclear ship in respect of any one nuclear incident (Art. III). An operator is required to maintain insurance or other financial security to cover this liability to such an amount as the licensing state may specify; the licensing state must ensure the payments of claims for compensation up to the 1,500 million franc limit to the extent that the insurance or other financial security is inadequate. By Article V, a limitation period of 10 years is applied, running from the date of the nuclear incident.

At the stage when the Nuclear Installations (Amendment) Act 1965 was under consideration, it was pointed out that anomalies could arise if nuclear matter in transit were treated differently to that related to the propulsion of a ship. The response of the Government was that in the absence of special legislation dealing with nuclear ships, the ordinary law would apply once an operator governed by the U.K. nuclear installations legislation had ceased to have liability.[17] At that time, apart from warships (which were, and are, subject to special intergovernmental arrangements) there were only two nuclear ships operating in the world; the then recent visit of one of these, the "Savannah", to Southampton was the subject of a special agreement with the U.S. Government. Even under the Convention, contracting states may deny access to their harbours and waters to nuclear ships licensed by other contracting states, even if they have fully complied with the Convention (Art. XVII).

3–14 The Convention relating to Civil Liability in the Field of Maritime Carriage of Nuclear Material

This Convention, which has been signed but not ratified by the United Kingdom, was the result of concerns, following the entry into force of the Paris Convention, as to how the liability provisions of that Convention would relate to the large existing body of law on the liability of carriers of goods by sea. The IAEA and the Nuclear Energy Agency of the OECD organised a joint symposium, including the relevant maritime organisations, to consider these issues. A diplomatic conference followed, and led to the signing of the Convention by the United Kingdom and others. The Convention entered into force on July 15, 1975, and has been ratified by Denmark, France, Norway, Spain, Sweden, Argentina, Germany, Italy, Liberia and Yemen, but not by the United Kingdom.

The key point of the Convention is that a person who, by virtue of an international Convention or national law applicable in the field of maritime transport, might be held liable for damage caused by a nuclear incident, is exonerated from such liability if the operator of a nuclear installation is liable for such damage either under the Paris or Vienna Convention, or by virtue of a national law which is in all respects as favourable to the victims of damage as those Conventions (Art. 1). This exoneration extends by Article 2 to damage to a nuclear installation, property on the site of the installation, or the means of transport of nuclear material for which the operator of the nuclear installation is not liable because his liability is excluded under the Conventions or under the relevant national law. By Article 3, the Convention does not affect the liability of the operator of a nuclear ship in respect of damage caused by a nuclear incident involving the nuclear fuel or nuclear waste of the ship (see para. 3–13 above).

---

[17] *Hansard*, H.C. Vol. 702, col. 63; see also *Hansard*, H.C. Vol. 706, col. 694 (February 11, 1965).

The Nuclear Installations Act 1965: generally 3–15

The Nuclear Installations Act 1965 ("the 1965 Act") consolidates the Nuclear Installations (Licensing and Insurance) Act 1959, and the Nuclear Installations (Amendment) Act 1965. The latter Act was passed to implement the United Kingdom's international obligations under the Paris and Brussels Supplementary Conventions (see above) and also took into account the Vienna Convention. Given the 1965 Act's status as consolidating legislation, it is helpful to have regard to the Parliamentary debates on the 1959 and 1965 Acts.

Sections 1–5 of the 1965 Act deal with the licensing of nuclear installations and are discussed in Chapter 2. Sections 7–21 deal with liability and insurance issues and are analysed in this chapter.

Section 7: duty of licensee of licensed site 3–16

Subsection 7(1) of the 1965 Act imposes a duty on the licensee of a nuclear site to secure that specified matters do not cause injury to any person, or damage to any property of any person other than the licensee. Those matters are as follows:

(a) occurrences of the type mentioned in subsection (2) involving nuclear matter, the injury or damage arising out of, or resulting from, the radioactive properties of the nuclear matter, or a combination of those and any toxic, explosive or other hazardous properties of the nuclear matter;

(b) the emission of ionising radiations emitted during the period of the licensee's responsibility from anything caused or suffered by the licensee to be on the site which is not nuclear matter, or from any waste discharged (in whatever form) on or from the site.

Category (a) corresponds broadly to the definition of "nuclear incident" contained in the Paris Convention and which must be covered under that Convention. Category (b) in fact corresponds to the wording used in subsection 4(1)(a) of the Nuclear Installations (Licensing and Insurance) Act 1959. Article 3(c) of the Paris Convention allows the United Kingdom as a contracting party to make an operator liable for damage arising out of, or resulting from, ionising radiations emitted by any source other than nuclear fuel, radioactive products or waste in the installation, or nuclear substances coming from the installation.

3–17 Occurrences involving nuclear matter

Subject to any exceptions which may be prescribed, "nuclear matter" is defined by section 26 of the 1965 Act to mean:

(a) any fissile material in the form of uranium metal, alloy or chemical compound (including natural uranium), or of plutonium metal, alloy or chemical compound, and any other fissile material which may be prescribed; and

(b) any radioactive material produced in or made radioactive by exposure to the radiation incidental to the process of producing or utilising any such fissile material.

This definition represents a paraphrase of the definitions of "nuclear fuel" and "radioactive products or waste" as contained in the Paris Convention (Art. 1(a)(iii) and (iv)).

The occurrences referred to in subsection 7(1) are specified in subsection 7(2), and fall into three categories covering both occurrences on and off the nuclear site. Each of the three categories is discussed separately below. The definition of "occurrence" contained in subsection 26(1) is not relevant to the term as used in section 7, though as appears from subsection 15(1) the occurrence may be either a single event, continuing event, or succession of events.

Whilst it might be possible to argue that, in the absence of a statutory definition, the term implies some specific accident or incident rather than an ongoing release of radiation in the normal course of operations, it seems unlikely that a court would be attracted to such an argument. The Paris and Vienna Conventions both use the term "nuclear incident" to mean any occurrence or succession of occurrences that have the same origin (Arts. 1(a)(I) and I(1) respectively).

3–18 Occurrences on the licensed site

Subsection 7(2)(a) refers to any occurrence on a licensed site. The question is therefore whether the occurrence causing the injury or damage involved nuclear matter (see above), whether it took place on the licensed site, and whether it took place during the period of the licensee's responsibility. Whether the occurrence took place on the licensed site will be a question of fact; the extent of the site will be clear from the licence and it should be noted that under subsection 3(2) of the 1965 Act, the Health and Safety Executive as licensing authority may treat two or more installations in the vicinity of one another as a single site for licensing purposes. The 1965 Act does not use the term "nuclear incident" which occurs in the Paris Convention (Art. 1(a)(I)).

The licensee's period of responsibility is defined by subsection 5(3) to cover the period beginning with the grant of the licence, and ending with the earlier of the following two dates:

(a) the date when the HSE gives written notice that in its opinion there has ceased to be any danger from ionising radiations from anything on the site; or

(b) the date when a new nuclear site licence in respect of the site is granted, whether to the same licensee or to some other person.

The licensee may therefore be responsible for occurrences that take place after the installation has ceased to be operational.

Occurrences involving nuclear matter being carried   3–19

Subsection 7(2)(b) refers to two types of occurrence elsewhere than on the licensed site involving nuclear matter which is not excepted matter. Excepted matter is that prescribed by the Nuclear Installations (Excepted Matter) Regulations 1978 (S.I. 1978 No. 1779) (explained below). The forms of occurrence are those involving nuclear matter which at the time of the occurrence is either:

(a) in the course of carriage on behalf of the licensee as licensee of the site; or

(b) in the course of carriage to the site with the agreement of the licensee from a place outside the "relevant territories" (*i.e.* a country for the time being bound by an international agreement to which the United Kingdom is a party relating to third-party liability in the field of nuclear energy—currently the Paris and Brussels Supplementary Conventions).

In either case, the nuclear matter must not at the time of the incident be on any other "relevant site" in the United Kingdom (*i.e.* a licensed site during the period of the licensee's responsibility, premises occupied by the United Kingdom Atomic Energy Authority, or sites occupied for nuclear purposes by Government departments). The nuclear matter must then be carried on behalf of the licensee, or with the licensee's agreement to the site in order to render the licensee liable. Unlike occurrences on the licensed site, the occurrence need not take place during the period of the licensee's responsibility.

The 1978 Regulations defining excepted matter are technically complex, but essentially cover:

(a) substances where the content of uranium 235 does not exceed one per cent of the total mass of all the uranium isotopes present and which do not exceed certain stated limits of radioactivity; and

(b) nuclear matter (other than waste discharged on or from a relevant site or consigned therefor) which has been consigned from the site, is not at the time on a relevant site and when it left the site was duly packaged and labelled and did not exceed certain stated limits of radioactivity.

The parameters of the 1978 Regulations correspond to those established in 1977 in relation to the Paris Convention (Decision on the Exclusion of Small Quantities of Nuclear Substances, adopted by the OECD/NEA Steering Committee on October 27, 1977) and, in relation to the Vienna Convention, in 1978 (Resolution of the Board of Governors Concerning the Establishment of Maximum Limits for the Exclusion of Small Quantities of Nuclear Material, adopted on September 14, 1978).

3–20 Occurrences involving matter which has been on the licensed site or in the course of carriage

The third main type of occurrence is mentioned at subsection 7(2)(c). This is any occurrence elsewhere than on the licensed site involving nuclear matter which is not excepted matter (as to which, see para. 3–19 above) and which:

(a) has been on the licensed site at any time during the licensee's period of responsibility; or
(b) has been in the course of carriage on behalf of the licensee as licensee of the site.

In either case, the nuclear matter must not have subsequently been:

(a) on any "relevant site"; or
(b) in the course of any "relevant carriage"; or
(c) within the territorial limits of any country which is not a "relevant territory" (except in the course of relevant carriage).

The various terms "relevant site", "relevant carriage" and "relevant territory" are defined at section 26. The repeated use of the horrible phrase, "relevant territories" was described as using the word "relevant" ... "to the point of nausea" during the Second Reading of the Nuclear Installations (Amendment) Bill.[18] "Relevant site" and "relevant territory" are referred to in para. 3–19 above; "relevant carriage" means carriage on behalf of the licensee of a licensed site, the United Kingdom Atomic Energy Authority, a Government department for the purposes of using a nuclear site, a relevant foreign operator (*i.e.* the operator of an installation in a relevant territory outside the United Kingdom), or a person authorised to operate a nuclear reactor comprised in a means of transport and in which the nuclear matter in question is intended to be used. For example, therefore, the operator of a nuclear reactor who sends

[18] *Hansard*, H.C. Vol. 702, col. 55 (Mr John Peyton, Yeovil).

spent fuel for reprocessing would remain liable until the nuclear matter reaches the licensed reprocessing plant; unless the matter is carried on behalf of the licensee of the reprocessing plant, in which case the liability of the operator of the reactor ceases—and that of the reprocessor commences—when carriage begins.

The general concept is therefore that the licensee assumes liability for nuclear matter which has been on its site or which has been carried on its behalf, until such time as responsibility passes to another nuclear operator or analogous person. Apart from this, there is no time limit on such liability relating to when the material left its site, other than the general limitation periods under section 15, which relate to the occurrence giving rise to the claim. This accords with the principle at Article 5(c) of the Paris Convention, that where nuclear matter has been in a number of nuclear installations before a nuclear incident occurs, it is the operator of the last installation, or an operator who has subsequently taken the nuclear matter in charge, who would be liable.

The meaning of "occurrence"   3–21

The concept of an "occurrence" is central to the liability scheme of section 7. Whilst the term is defined by section 26, that definition is only expressed to be for the purposes of sections 16(1), (1A), 17(3) and 18.

Ionising radiations   3–22

Apart from the liability for nuclear occurrences referred to above, section 7 also imposes a duty on the licensee to secure that no ionising radiations emitted during his period of responsibility:

(a) from anything caused or suffered by the licensee to be on the site which is not nuclear matter, or

(b) from any waste discharged (in whatever form) on or from the site,

cause injury to any person, or damage to any property of any person other than the licensee. The duty is wider than that based on occurrences in that the source of the radiation need not be nuclear matter but may be any article or substance, or waste in any form. As mentioned previously, this corresponds to the wording originally used in subsection 4(1)(a) of the Nuclear Installations (Licensing and Insurance) Act 1959. As it was put by Mr Reginald Maudling (Paymaster-General) in relation to the 1959 Act:

"The absolute liability will apply to everything that happens on the site of the nuclear installation and anything that emerges

117

from it whether by reason of the operation of the reactor or the discharge of waste ..."[19]

The duty extends to material which has left the licensed site (*e.g.* waste discharged from the site), but only applies to ionising radiations emitted during the period of responsibility; liability does not therefore extend indefinitely in respect of waste which has left the site.

One question is the potentially difficult relationship between subsections 7(1)(a) and 7(1)(b) of the Act. In relation to waste consigned from a site to another licensed site for storage or treatment, subsection 7(2)(c) would, as mentioned above, terminate the consignor's responsibility at the latest when it reaches the licensed site of the consignee, and possibly earlier. Subsection 7(1)(b) contains no such restriction, and the possibility of the consignor remaining liable under subsection 7(1)(b) after the consignee has become liable under subsections 7(2)(a) or (b), would run counter to the basic principle of channelling liability.

3–23 Injury to persons

The Paris and Vienna Conventions include within their respective definitions of nuclear damage, "damage to or loss of life of any person", and "loss of life and personal injury". The wording originally used in the 1959 Act was "hurt to any person". This terminology was the subject of lengthy debate in the context of the Nuclear Installations Amendment Bill, in which it was changed to "injury" in preference to "physical hurt", which had been proposed. The intention was to avoid the use of the "inelegant" word "hurt", to substitute the more familiar term "injury" and by omitting the word "physical", to extend the benefits conferred by the Act.[20]

Subsection 26(1) defines "injury" to mean personal injury and to include loss of life. The issue of whether the duty extends to the avoidance of risks to health was considered in *Merlin v. British Nuclear Fuels PLC.*[21] The case related to contamination of the plaintiffs' home by radioactive matter emanating from the defendant's plant at Sellafield, Cumbria. It was claimed that as a result of the discharge of waste from the site into the Irish Sea, radioactive matter had found its way back onto the coastline, where it had become deposited in the mud of the Ravenglass Estuary and thence, by the action of wind and by the carriage of the sediment on the feet of the plaintiffs, their family and pets, into their house.

The main arguments in the case related to whether the contamination constituted damage to the property (this is discussed at para. 3–26 below). In rejecting the argument that contamination of the plaintiff's house *per*

[19] *Hansard*, H.C. Vol. 599, col. 865.
[20] *Hansard*, H.C. Vol. 706, col. 668.
[21] [1990] 2 Q.B. 557; [1991] J.Env.L. (Vol. 3), No. 1, p. 122.

se amounted to damage to property, Gatehouse J. referred to the issue of risk to health in the following terms:

"The Act of 1965 compensates for proved personal injury, not the risk of future personal injury. If the Act were concerned with risk a number of very difficult questions would arise. For instance risk to whom? Is it the plaintiffs' health risk that has to be evaluated, or, (and this was their concern) that of their children, or is it that of potential purchasers of the house? The degree of risk depends, among other factors, on the length of time over which the individual is exposed to radioactivity. Is the court to attempt to forecast how many years each individual concerned is likely to live in the house?"[22]

In closing, Gatehouse J. also expressed the view that:

"The presence of alpha-emitting radionuclides in the human airways or digestive tracts or even in the bloodstream merely increases the risk of cancer to which everyone is exposed from both natural and artificial radioactive sources. They do not *per se* amount to injury."[23]

As to the impairment of the ability to have normal children, or effects on pregnant women as injury, see the Congenital Disabilities (Civil Liability) Act 1976, s.3(2), outlined in the following paragraph.

Unborn children   3–24

The potential hazards of radiation to unborn children, or indeed children not yet conceived, was appreciated at the time of the Nuclear Installations (Licensing and Insurance) Bill in 1958. For example, Lord Taylor raised the issue in the context of concerns that the 10-year limitation period originally proposed (see para. 3–48 below) was too short:

"... radioactive caesium gives out gamma rays which can have the effect of irradiating the gonads or sex glands from a distance. The pioneers of radiology showed on autopsy an atrophy of the sex glands. But before this occurs there are other changes in the cellular structure of the sperm or ova. This is what has created so much discussion, the question of the congenital malformation of infants as a result of this long-continued irradiation of the sexual cells. This risk continues throughout the entire procreative life of the individual and I

[22] *ibid.*, p. 130.
[23] *ibid.*, p. 131.

think that this again points to the need to abolish or at least to modify drastically the ten-year period."[24]

Even after the Government extended the limitation period to 30 years in the course of the Bill, this left considerable concern as to the position of persons unborn at the time of the occurrence, or even future generations affected.[25]

Such concerns were still being raised in 1965, when the Nuclear Installations (Amendment) Bill was under debate. Mr Nicholas Ridley (Cirencester and Tewkesbury) sought clarification from the Minister as to the position under the Bill of the unborn child and the un-conceived child, quoting from the Ballad of Chevy Chase:

"The child may rue that is unborn
The hunting of that day."

The Minister Without Portfolio (Sir Eric Fletcher) responded by saying he was advised that no addition to the Bill was necessary to enable "an unborn child to make a claim in the unhappy result of that child, when born, suffering injury as a result of a nuclear incident occurring before his birth".[26] In relation to a child not conceived at the time of the incident he thought it "very doubtful" that such a person could substantiate a claim, but felt that this was a matter best left to the judiciary to decide.[27] Neither the Paris or Vienna Conventions give any help in this respect.

The position was not fully clarified until the passage of the Congenital Disabilities (Civil Liability) Act 1976, which was passed to give effect to recommendations in the Law Commission's Report on Injuries to Unborn Children.[28] The general provisions in section 1 of that Act on civil liability to children born disabled do not affect the operation of the 1965 Act as to liability and compensation in respect of injury or damage caused by occurrences involving nuclear matter, or the emission of ionising radiation (subs.3(1)). However, subsection 3(2) provides, for the avoidance of doubt, that anything which affects a man in his ability to have a normal, healthy child, or which affects a woman in that ability, or affects her while pregnant so that her child is born with disabilities, is an injury for the purposes of the 1965 Act.

By subsection 3(3) of the 1976 Act, if a child is born disabled as a result of injury to either of its parents caused in breach of a duty imposed by sections 7–11 of the 1965 Act, the child's disabilities are to be regarded for the purpose of compensation and related matters under the 1965 Act as injuries caused on the same occasion, and by the same breach of duty, as was the injury to the parent. The combined effect of subsections 3(2) and

---

[24] *Hansard*, H.L. Vol. 212, col. 1031.
[25] *Hansard*, H.L. Vol. 213, cols. 344–348; and *Hansard*, H.C. Vol. 599, col. 891.
[26] *Hansard*, H.C. Vol. 706, col. 669.
[27] *ibid.*
[28] *Law Commission—No. 60*, Cmnd. 5709 (1974).

(3) is therefore that the disability of the child is equated with the injury to its parent, either in the sense of impairment of healthy reproductive function, or effects on the unborn child. The child will be regarded as having suffered injury in its own right on the same occasion as the parent. The statutory limitation period will therefore begin to run before the child is born, or even conceived in some cases; this can be contrasted with the position for other types of injury under the 1976 Act where the liability is treated as relating to personal injury sustained by the child immediately after its birth (subs.4(3)).

The ability to obtain compensation under the 1976 Act is qualified in three ways:

1. subsection 13(6) of the 1965 Act, dealing with contributory fault (see para. 3–46 below), is applied in relation to the child as if the reference in that section to fault were that of the parent, *i.e.* the parents' contributory fault may apply to limit the award to the child (subs.3(4)).

2. By subsection 3(5), compensation is not payable to the child at all if the relevant injury to the parent preceded the time of conception, and at the time of conception either or both parents knew the risk of the child being born disabled, *i.e.* the particular risk created by their own injury. Effectively, the principle of *volenti* on the part of a parent is applied against the child. The would-be parents who may thus be prevented from trying to have children will have their own cause of action if they are affected in their ability to have a normal, healthy child. There is, however, a potential problem in that a person may be at risk of giving birth to a disabled child though it may not be certain that this will be the case. If the person takes that risk, the child, if born disabled, will be penalised. If the person chooses not to take the risk, it may be said that the adverse effect on their ability to have a healthy child has not been proven and thus there is no injury. It is to be hoped, however, that the risk of producing a disabled child will be regarded by the courts as an injury.

3. Compensation for loss of expectation of life is not recoverable unless the child lives for at least 48 hours (subs.4(4)).

Personal injury: the Sellafield litigation

3–25

The issue of personal injury arising from parental exposure to radiation was litigated at great length in the cases of *Reay v. British Nuclear Fuels* and *Hope v. British Nuclear Fuels*.[29] The first plaintiff was the mother of Dorothy Reay, who was born in October 1961, and died from early acute lymphatic

---

[29] [1994] 5 Med.L.R. 1; [1994] P.I.Q.R. P171; [1994] Env.L.R. at 320.

leukaemia in September 1962. The second plaintiff was Vivien Hope, who was born in 1965 and was diagnosed as having non-Hodgkins lymphoma (NHL) in 1988 (from which she recovered). It was alleged that the plaintiffs' conditions were caused by paternal pre-conception irradiation (PPI), causing mutation in the sperm of their fathers who worked at Windscale (later Sellafield). The cause of action was not in fact the 1965 Act, but rather the duty applying to the defendants' predecessor, the Atomic Energy Authority, under subsection 5(3) of the Atomic Energy Act 1954 to secure that no ionising radiations from anything on their premises, or from waste discharged from those premises, caused any hurt to any person or damage to property.

The case turned on expert evidence and on the results of various studies, in particular, epidemiological research by Professor Martin Gardner associating leukaemia cases in Seascale, West Cumbria with PPI.[30] The judge concluded that the observation of an excess number of cases of cancer in the area was not the result of PPI, but could most reasonably be explained by a combination of chance, socio-demographic and statistical factors. The plaintiffs put forward an argument that, if causality through PPI was not the explanation, then the fact that those excess cases were children of fathers who had been subject to high doses of radiation must be put down simply to chance. This argument gave the judge "cause for pause and reflection",[31] but he concluded that, considering the evidence, the scales tilted decisively in favour of the defendants; the plaintiffs had therefore failed to satisfy him on the balance of probabilities that PPI was a material cause of this excess, or of their own injuries. In particular, there was no evidence of excess leukaemia among children born to fathers who were victims of the atom bombs in Japan. Whilst there were factors which might have gone some way towards explaining the differences in the Japanese data and the Gardner thesis, the fact remained that far from being "in the same ballpark", the Gardner thesis would be "way out in Australia or whatever".[32] The "synergy theory", put forward by the plaintiffs as a possible explanation for this difference, was found to have several flaws.

The Gardner Report, on which the plaintiffs relied heavily, was found to be virtually unsupported by other studies, and had shortcomings which reduced confidence in it; in particular, it could not explain excesses of leukaemia in sites where PPI could not be the explanation. The Gardner thesis of PPI could not be excluded on "mechanistic" grounds on the state of current knowledge on genetics. However, on the other hand, the mechanisms proposed to explain PPI were speculative and did not carry forward the case against BNFL. Finally, it was not established on the evidence that leukaemia and NHL were a single disease; this greatly weakened the case for PPI as a cause of NHL.

[30] B.M.J. (1990), 300, pp. 423–429.
[31] [1994] Env.L.R. at 369.
[32] [1994] Env.L.R. at 365

Shortly after the judgment, the HSE published a report following up the work of Professor Gardner. This concluded that, for West Cumbria as a whole, there was little evidence to support any link between PPI and leukaemia/NHL. However, a strong statistical association of these factors was acknowledged in children born to Sellafield mothers resident in Seascale who had started work at Sellafield before about 1965. Research into possible workplace-factors other than radiation yielded no clear explanations. The association might be explained by a combination of causes; no single factor seemed capable of explaining the findings beyond all doubt.[33]

Damage to property   3–26

The duties created by section 7 also apply to "damage to any property of any person other than the licensee". The question of what constitutes damage to property was considered in *Merlin v. British Nuclear Fuels PLC*,[34] the facts of which are set out above in para. 3–23 above. Having discovered the extent of the contamination of their property, the plaintiffs in that case decided to move. They acquired another property with the aid of a bridging-loan, but experienced difficulty in selling their original house following the broadcast of a Yorkshire Television documentary entitled *"Sellafield—the Nuclear Dustbin"*, which featured the family's problems (and in which the plaintiffs voluntarily co-operated). Under pressure from the bank, the property was ultimately sold at auction (to a Sellafield employee) at a low value. The contention of BNFL was that the 1965 Act provided for compensation in respect of proven personal injury or damage to property and did not compensate for mere economic loss, which was the essence of the plaintiffs' claim. Despite an initially unfavourable reaction to this argument—which would give no remedy for a case which might be the typical result of an accidental emission of radioactive material from a nuclear site—Gatehouse J. was ultimately convinced that the defendants' approach was correct. The reasons for this were as follows:

  (1) the Vienna Convention, Art. I(k), defines "nuclear damage" to include (i) loss of life, personal injury or loss or damage to property, together with "(ii) any other loss or damage ... if and to the extent that the law of the competent court so provides." Thus, a contracting party would have to provide redress for damage to property but could choose whether or not to give redress for other forms of loss, *e.g.* economic loss;

---

[33] HSE, *Investigation of Leukaemia and other Cancers on the Children of Male Workers at Sellafield* (1993).
[34] [1990] 2 Q.B. 557; [1991] J.Env.L. (Vol. 3), No. 1, p. 122. See also: *Blue Circle Industries plc v. Ministry of Defence, The Times*, November 26, 1996, noted in the Preface to this text.

(2) the 1965 Act went as far as the Vienna Convention required, but did not provide compensation for "any other loss or damage", as it could have done in accordance with Article I(k)(ii);

(3) "personal injury or damage to property is a familiar enough phrase and in my judgment it means, as it does in other contexts, physical (or mental) injury or physical damage to tangible property. The word 'property' may well have a wider meaning in some contexts (*e.g.* in testamentary dispositions or in the field of company law ...), but where used in the Vienna Convention and in the Act of 1965, it does not in my judgment extend to incorporeal property or property rights. The plaintiff's argument that property including the airspace within the walls, ceilings and floors of Mountain Ash, that this has been damaged by the presence of radionuclides and the house rendered less valuable as the family's home, seems to me to be too far-fetched";

(4) the 1965 Act contains various compromises, one of which is a restriction on the nature of the harm which qualifies for compensation;

(5) whilst enormous doses of radioactivity would be required to produce any detectable damage to the molecular structure of building materials and other inanimate objects, the phrase "damage to property" could well apply, for example, to injury to livestock, and thus have a sensible function;

(6) it was not the case that "the jurisprudence of the English court" would fill the gap intentionally left by Parliament. Gatehouse J. could see no reason why compensation under the 1965 Act should extend to pure economic loss when such loss would not be recoverable at common law. No "special relationship" existed between the plaintiffs and the defendants to sustain any such claim;

(7) whilst bearing in mind the dangers of accepting arguments based upon the "floodgates" principle, the judge inclined away from "a construction of the Act of 1965 which would result in the operator being in continued breach of the statutory duty to a possibly very large number of people". The judge thought it was in the very nature of nuclear installations that there would be some additional radionucludes present in the houses of local people; if the mere presence of such radiation sources was enough to constitute damage, the result would be to confer a claim for compensation on "possibly thousands of citizens";

(8) the use of the word "cause" in section 7 implied to the judge the necessity for cause and effect between the relevant incident or emission and the damage. The mere presence of ionising radiations was not enough without some consequential damage.

This reasoning is, however, open to question in certain respects:

(a) it is curious that the judgment relied so heavily upon the Vienna Convention (to which the United Kingdom is not a party) whilst making no reference to the Paris Convention. The Vienna Convention does indeed draw a distinction in defining nuclear damage between loss of life, personal injury and damage to property on the one hand, and "any loss or damage" on the other. The Paris Convention does not contain the same explicit distinction, but simply refers to damage to, or loss of, any property. However, whilst the relevant Parliamentary Debates were not referred to in Gatehouse J.'s judgment, it is clear that Parliament had the Vienna Convention, as much as the Paris Convention, in mind when framing the 1965 Act and earlier legislation;

(b) it has been suggested by Professor Richard Macrory[35] that Gatehouse J. was "obviously influenced by the terms of the Vienna Convention and could have adopted a rather more flexible approach". However, Macrory also acknowledges that the judge might well have reached a similar conclusion in any event, "given the current reluctance of British courts to extend tortious liability for pure economic damage beyond certain confined categories such as negligent misstatements".

Consequential loss 3–27

The 1965 Act makes no express reference to the issue of consequential loss. At the Commons stage of the Nuclear Installations (Amendment) Bill, an amendment was proposed to make it clear that damage to property included loss of profits or other earnings suffered by the victim. In the case of an individual suffering injury, such loss could include loss of earnings and medical expenses. In the case of a corporate plaintiff, there might be loss of profits or other purely monetary loss.[36] In view of the arguments raised on non-material damage in *Merlin v. BNFL* (see para. 3–26 above), it is interesting that Article I(k)(ii) of the Vienna Convention was referred to in support of the amendment.[37]

The Minister Without Portfolio, in rejecting the amendment, denied that it was originally the intention of the 1959 Act to cover such consequential loss; it was neither the function nor the intention of Parliament to legislate on issues of remoteness of damage.[38] Loss of earning capacity would, on normal principles, be a possible head of damages for individuals.[39] Sir Eric Fletcher stated the Government's intention so far as corporate plaintiffs were concerned as follows:

---

[35] Macrory; [1991] J.Env.L. (Vol. 3), No. 1, p. 132.
[36] *Hansard,* H.C. Vol. 706, col. 683 (February 11, 1965).
[37] *ibid.,* col. 685.
[38] *ibid.,* col. 687.
[39] *ibid.,* col. 688.

"It is the Government's intention to place a corporate plaintiff in an action under the Bill in the same position as he would be in an action for negligence at common law in which liability were either admitted or proved. It is my belief that the words of the Bill adequately give effect to that intention."[40]

The Government reconsidered the issue before the Lords stage and remained confident that the Bill would leave a plaintiff, including a corporate plaintiff, in an action under the Bill, in the same position as he would be in an action for negligence at common law in which liability was admitted or proved:

"... the Bill as drafted leaves it to the court to decide, as at common law, the extent and quantum of damages and, in particular, any question as to remoteness of damage."[41]

In other words, the normal rules of remoteness of damage will govern whether loss of profit or other consequential loss is recoverable.[41a]

3–28 The licensee's property and that of third parties

The duties under section 7 apply to property of any person other than the licensee. As originally drafted in the 1959 Act, the duty applied expressly to property on the site of the installation or elsewhere. However, Article 3(a)(ii) of the Paris Convention excludes damage to the installation itself and any property on the site of the installation which is used, or is to be used, in connection with the installation; the 1965 Act now reflects that exclusion. In determining whether there is liability in respect of an occurrence, any property which at the time of the occurrence is on the licensed site may, by subsection 7(3), be deemed to be the property of the licensee (and so not protected by the section 7 duties), even though it is in fact the property of some other person. The subsection applies to property which is:

(a) a nuclear installation; or
(b) is other property on the site for use in connection with the operation of the nuclear installation by the licensee, or the cessation of such operation.

Effectively therefore, a third party will have no remedy under subsection 7(1)(a) for occurrences affecting their property if, for example, it is on-site plant or equipment used for operation or decommissioning of the reactor.

---

[40] *ibid.*, col. 690.
[41] *Hansard*, H.L. Vol. 263, col. 1287.
[41a] see also: *Blue Circle Industries plc v. Ministry of Defence, The Times,* November 26, 1996, noted in the Preface to this text.

However, subsection 7(3) applies only to determine liability "in respect of any occurrence". This leaves open the possibility that the deeming provision would not apply to the detriment of a person relying on breach of the duty under subsection 7(1)(b), since this relates to damage caused by the emission of ionising radiations, rather than an "occurrence".

Liability of third parties in respect of licensee's property   3–29

As explained above, subsection 7(1) excludes damage to the licensee's property. Where that property is damaged by a nuclear occurrence caused by a third party, section 12 (discussed below) would not prevent that third party incurring liability since the section applies to damage caused in breach of a duty under section 7, and, if it is the licensee's own property which has been damaged, there is no breach of such duty by the licensee. However, subsection 12(3A) (added by the Nuclear Installations Act 1969, s.1) allows a third party to incur such liability only:

(a) in pursuance of an agreement to incur liability in respect of such damage entered into in writing before the occurrence of the damage; or

(b) where the damage was caused by an act or omission of that person done with intent to cause injury or damage.

This reflects the underlying intention of Article 6(f) of the Paris Convention, which states that the operator shall have a right of recourse only:

(a) if the damage caused by a nuclear incident results from an act or omission [of a third party] done with intent to cause damage, against the third party acting or omitting to act with intent;

(b) if and to the extent that it is so provided expressly by the contract.

The reasoning of subsection 12(3A) would appear to apply equally whether it is actually the licensee's property which has been damaged, or property of a third party which is deemed to belong to the licensee by subsection 7(3). In debates on the 1969 Act, the purpose of the amendment was said to be to make the United Kingdom legislation match the Conventions, and to restrict the possibility of an operator seeking to establish liability on the basis of negligence against the supplier of a faulty component which resulted in damage to the reactor.[42]

---

[42] *Hansard*, H.L. Vol. 301, col. 332; and *Hansard*, H.C. Vol. 779, col. 624.

3–30 Duty of United Kingdom Atomic Energy Authority

Subsection 7(1) is expressly made subject to subsection 7(4), which provides that section 8 shall apply to sites occupied by the Authority; this wording was added by the Nuclear Installations Act 1965 (Repeal and Modifications) Regulations 1990.

Section 8 provides that the duties of section 7 apply to the Authority whether or not a nuclear site licence is in force in relation to premises occupied by the Authority. If no such licence is in force, then the reference to the "period of responsibility" in section 7 cannot of course be established by reference to licensing, as is provided by subsection 5(3); the period is thus simply that of occupation by the Authority (subs.8(b)). It should be noted that section 7 is applied to the Authority in relation to *all* sites which are currently occupied by the Authority or which have been so occupied. By contrast, section 9 applies to the Crown only in relation to sites used for purposes within section 1.

The application of the section 7 duties to the Authority in 1990 did not represent a new burden on the Authority, since the Authority was subject to a corresponding and more onerous duty under subsection 5(3) of the Atomic Energy Authority Act 1954. The subsection was repealed by the Energy Act 1983, ss.34, 36, Pt. 2, Sched. 2. This duty was to secure that no ionising radiations from anything on any premises occupied by the Authority, or from any waste discharged (in whatever form) on or from such premises, caused hurt to any person or damage to any property, whether he or it was on the premises or elsewhere. This duty was described as making the Authority "completely and absolutely liable to an unlimited extent for the result of any of its operations and for any period of time"—a position which could not be expected of the private sector.[43]

For a case brought on the basis of the subsection 5(3) duty against BNFL as a successor to the Authority see *Merlin v. British Nuclear Fuels Plc.*[44]

3–31 Duty of Crown

A Government department which uses a site for a purpose described in section 1 of the 1965 Act is subject to section 7 as if the Crown were the licensee under a nuclear site licence (s.9). As with the United Kingdom Atomic Energy Authority, the "period of responsibility" corresponds to the period of occupation by the Department.

3–32 Duty of foreign operators

Section 10 applies a limited version of the section 7 duty to certain foreign operators. A "relevant foreign operator" is defined by section 26 as a person who operates an installation to which a relevant international

---

[43] *Hansard*, H.L. Vol. 212, col. 1059.
[44] [1990] 2 Q.B. 557; [1990] 3 W.L.R. 383.

agreement applies in a territory for the time being bound by such agreement, other than the United Kingdom. An obvious example would be the operator of a nuclear reactor in a country which is a party to the Paris Convention.

Where nuclear matter (which is not excepted matter) either:

(a) is in the course of carriage on behalf of a relevant foreign operator or is in the course of carriage to that operator's site with their agreement (by subsection 13(3), the agreement must be in writing) from a place outside the relevant territories, and in neither case is for the time being on any relevant site in the United Kingdom; or

(b) having been on the foreign operator's site or in the course of carriage on behalf of the operator, has not subsequently been on any relevant site or in the course of any relevant carriage or (except in the course of relevant carriage) within the territorial limits of a country which is not a relevant country;

then it is the duty of the operator to secure that no occurrence mentioned in subsection (2) causes injury or damage resulting from radioactive properties of the material, or a combination of radioactive, toxic or other hazardous properties. The occurrences referred to in subsection (2) are:

(a) an occurrence taking place wholly or partly within the territorial limits of the United Kingdom; or

(b) an occurrence outside such territorial limits which also involves nuclear matter in respect of which a duty is imposed on any person by section 7, 8 or 9.

The first type of occurrence might relate to material entering the United Kingdom for reprocessing (see para. 3–33 below). The second type might apply where materials of different origin are being carried together outside the United Kingdom.

Movements of nuclear matter to and from the United Kingdom   3–33

It may be helpful to give examples of how these complex sections 7–10 might operate. 1, 2 and 3 below set out some possibilities:

1. Spent nuclear fuel from a foreign operator subject to the Paris Convention is being transported to a site in the United Kingdom for reprocessing. The foreign operator would be subject to the section 10 duty if, (a) it was being carried on his behalf and was not yet on a relevant site in the United Kingdom, or (b) (since it was material that had been on the foreign operator's site) it had not subsequently been

at another installation subject to the Convention, or entered the territory of a non-party except in the course of carriage on behalf of the foreign operator. The foreign operator would be liable in respect of an occurrence taking place wholly or partly within the United Kingdom (including territorial waters). If the occurrence also involves nuclear matter in respect of which a duty is imposed by section 7, 8 or 9 (*e.g.* because it is in the course of carriage on behalf of the licensee of the United Kingdom site) then the duty extends to occurrences outside United Kingdom territorial limits.

It should be noted that in the circumstances described above, the United Kingdom licensee would be under the section 7 duty if—

(a) the material was being carried on his behalf rather than on behalf of the foreign operator (if this were the case then it would not be in the course of carriage on behalf of the foreign operator), or

(b) the material was being carried with his agreement from a place outside the relevant territories. If it was being carried straight from the foreign operator's site then this would not be the case. If it had entered the territory of a non-party (other than simply in the course of carriage) then the United Kingdom licensee would be liable under subsection 7(2)(b), but the foreign operator would not.

There is, therefore, a degree of symmetry inherent in sections 7 and 10 which should prevent both a United Kingdom licensee and a foreign operator being liable.

2. Nuclear fuel produced or reprocessed in the United Kingdom is transported to a relevant foreign operator for use in the foreign operator's reactor. If it is carried on behalf of the foreign operator, that operator will be liable under subsection 10(1)(a)(i) once it leaves the licensed site in the United Kingdom. The United Kingdom licensee will not be liable since it is not being carried on his behalf, nor will he be liable under subsection 7(2)(c) since it will have entered "relevant carriage" on behalf of the foreign operator. If it is carried on behalf of the United Kingdom licensee, then he will be liable: the foreign operator will not be liable under subsection 10(1)(a)(ii) since it is not being carried to the operator's site from a place outside the relevant territories.

3. Waste arising from the reprocessing in the United Kingdom of spent nuclear fuel from a relevant foreign operator is returned to the foreign operator for disposal. The position is essentially the same as for 2 above, but with one added complication. It could be said that the United Kingdom licensee is not only subject to the duty of subsection 7(1)(a) in relation to "occurrences", but is also subject to the duty under subsection 7(1)(b) in relation to waste discharged (in

whatever form) from the site. However, if the United Kingdom licensee continued to be liable under subsection 7(1)(b) after the foreign operator had taken charge of the material or assumed responsibility for it in writing, this would be contrary to the principles of channelling of liability as expressed in relation to the carriage of nuclear substances in Article 4(a)(i) and (ii) of the Paris Convention. The legislation ought to be construed, if possible, to avoid a situation where both the United Kingdom licensee and foreign operator are both liable in respect of the same occurrence.

Duty of carriers   3–34

Section 11 of the 1965 Act deals with nuclear matter which is in the course of carriage within the United Kingdom (including territorial waters) where the carriage is not "relevant carriage", and the nuclear matter is not for the time being on any "relevant site". In such cases it is the duty of the person on whose behalf the material is being carried, to secure that no occurrence involving the nuclear matter causes injury or damage (other than damage to their own property) to be incurred within the territorial limits of the United Kingdom, resulting from the radioactive properties of the matter.

In most foreseeable cases, it is likely in practice that carriage within the United Kingdom will be relevant carriage, *i.e.* on behalf of either a site licensee, the Atomic Energy Authority, the government, or a foreign operator. In that case, section 11 will not apply. At the time of introducing clause 3 of the 1965 Nuclear Installations (Amendment) Bill (which in due course became section 11), it was pointed out that by administrative means, steps would normally be taken to ensure that nuclear matter entering the United Kingdom would engage the liability of an operator from a Convention party. However, there might be cases of traffic between non-Convention states crossing United Kingdom territorial waters, and vessels carrying nuclear matter might need to dock in the United Kingdom in an emergency.[45] The section imposes absolute liability, unlimited in time or amount; since no Convention party is involved, there was no basis or reason for setting any such limit.[46]

Under the 1976 Convention on Limitation of Liability for Maritime Claims, which has the force of law in the United Kingdom by virtue of section 17 of, and Part 1, Schedule 4 to, the Merchant Shipping Act 1979, it is generally possible for shipowners and salvors to limit their liability for claims in respect of injury and damage. However, certain types of claim are excepted from limitation under the Convention, and these include claims subject to any international Convention or national law governing or prohibiting liability for nuclear damage and claims against the

---

[45] *Hansard*, H.C. Vol. 702, col. 51; and *Hansard*, H.L. Vol. 263, col. 1278.
[46] *ibid.*

shipowner of a nuclear ship for nuclear damage (Art. 3(c) and (d), Sched. 4).

3–35 Right to compensation: section 12

Section 12 of the 1965 Act is the key section which gives rise to a right to compensation in respect of injury or damage caused in breach of section 7, 8, 9 or 10. The section is qualified in various respects by following sections and needs to be read in conjunction with them. To be precise, the section does two separate things, in accordance with the principle established by the international Conventions of channelling liability—

(1) it provides that compensation is payable in accordance with section 16 (which limits liability financially) wherever the injury or damage was incurred (*i.e.* whether inside or outside the territory of the United Kingdom),
(2) it precludes any liability for such injury or damage being incurred by any other person than the person subject to the duty under section 7, 8, 9 or 10.

The Health and Safety at Work, etc. Act 1974 contains various provisions on safety and risk which may be applicable to nuclear sites; however, subsection 47(1)(c) of the 1974 Act provides that nothing contained in Part I of the 1974 Act is to be construed as affecting the operation of section 12 of the 1965 Act.

3–36 Damage, injury or loss outside section 12

Section 12 applies only to injury or damage caused in breach of a duty under sections 7–10 of the 1965 Act. If the damage or injury results entirely from a release of radiation which does not constitute such a breach, section 12 does not apply, and liability will be determined on common law principles. Similarly, if there is no "injury or damage" as required by the Act,[47] common law principles will also apply. A release of radioactive material might cause economic loss without physical damage; for example, loss of the use of property, diminution in value, or the inability to sell crops or produce. Whether a claim in negligence, nuisance or *Rylands v. Fletcher*[48] would succeed would depend largely upon the approach taken to the recovery of economic loss.

It is conceivable that damage may result from mixed causes, and this is indeed contemplated by section 12. Injury or damage which is not caused by breach of such a duty, but is not reasonably separable from that which is, shall be deemed for the purposes of compensation to be caused in

---

[47] *Merlin v. British Nuclear Fuels PLC* [1990] 2 Q.B. 557; [1991] J.Env.L. (Vol. 3), No. 1, p. 122 (see para. 3–26 above).
[48] (1868) L.R. 3; H.L. 330.

breach of duty (subs.12(2)). This principle is, however, subject to subsection (3), in that the application of section 12 to any emission of radiation which is not a breach of duty does not affect any liability of any person arising outside the Act. The two causes of action may therefore co-exist in this particular situation, but subsection (3) expressly provides that a claimant may not recover compensation both under and outside the Act for the same injury or damage.

Exclusion of other claims   3–37

Section 12 provides the only remedy for injury or damage falling under it; by subsection 12(1)(b), no other liability shall be incurred by any person in respect of that injury or damage. The expressed policy of this provision is the avoidance of a multiplicity of claims, for example against, or between, contractors.[49]

The principle, therefore, will exclude claims at common law or under other statutory provisions against the person in breach of the duty. It would not of course exclude criminal liability, which is not liability in respect of the injury or damage. Importantly, it also excludes any possible claims against third parties, for example, the liability of a negligent contractor or subcontractor whose work at a nuclear reactor gives rise to an occurrence, or the liability of a negligent driver who causes an accident involving nuclear matter which is being carried.

The principle is subject to some exceptions:

(1) in respect of injury or damage caused partly by a breach of duty under the 1965 Act and partly by an emission of ionising radiations which does not constitute such a breach, liability outside the Act is not excluded in relation to the emissions (subs.(3));

(2) the exclusion does not affect the operation of the Carriage by Air Act 1931, the Carriage by Air Act 1961 or the Carriage by Air (Supplementary Provisions) Act 1962 in relation to international carriage covered by one of the relevant conventions for those purposes (subs.(4)(b));

(3) nor does it affect the operation of any Act giving effect to the 1956 Geneva Convention on the Contract for the International Carriage of Goods by Road (subs.(4)(c));

(4) in cases where an occurrence involving nuclear matter in the course of carriage results in a claim in respect of damage to the means of transport being made against a foreign operator under section 10, the foreign operator may be protected by subsection 16(2)(a) if he would not have been liable under the

---

[49] *Hansard*, H.L. Vol. 212, col. 504.

law of his own territory had the occurrence taken place there. In such cases subsection 12(1)(b) does not apply to prevent a claim outside the Act being made in respect of that damage.

The reference to the specific Conventions at paragraphs (2) and (3) above is explicable by reference to Article 6(b) of the Paris Convention, which provides that the rule as to channelling of liability shall not affect the application of any international agreement in force or open for signature, ratification or accession as at the date of the Convention (July 29, 1960). Originally, subsection 12(4)(a) of the 1965 Act provided that the exclusion in subsection 12(1)(b) was not to affect the operation of the Carriage of Goods by Sea Act 1924, but that provision was repealed by the Carriage of Goods by Sea Act 1971, s.6(3)(b).

3–38 Limitation for maritime claims

The general ability to limit liability for maritime claims provided for in Part 1, Schedule 7, to the Merchant Shipping Act 1995 does not apply to claims made by virtue of sections 7–11 of the 1965 Act (Pt. II, Sched. 7, para. 4(3)). This is the effect of paragraph 3(c) of Part 1, Schedule 7, and the same applies to claims against a shipowner of a nuclear ship for nuclear damage (para. 3(d)).

3–39 Compensation for damage to licensee's property

The principle that a third party cannot be liable for injury or damage under the 1965 Act requires careful application to cases where it is the property of the licensee or foreign operator itself which is damaged. Damage to the licensee or operator's property does not fall within the duties imposed by sections 7 and 10. That being the case, section 12 would not apply to prevent a claim being made in contract or tort by the licensee/operator or their insurers against the person who caused the damage.

Subsection 12(3A) restricts cases where any such liability can be incurred to:

(1) where there was a written agreement to incur such liability entered into before the damage occurred; or
(2) where the damage was caused by an act or omission of that person done with intent to cause injury or damage.

The subsection was inserted by the Nuclear Installations Act 1969, s.1, after it was discovered that the 1965 Act might allow the operator to make a claim in negligence against the supplier of a faulty component for damage to property on the site. This would have been contrary to the

Paris Convention.[50] The relevant provision of the Paris Convention states that the operator shall have a right of recourse only:

    (i) if the damage results from an act or omission done with intent to cause damage, and against the individual acting or omitting to act with such intent;

   (ii) if and to the extent that it is so provided by contract (Art.6(f)).

The Government, in enacting subsection 12(3A) clearly regarded requirements (i) and (ii) of the Convention as being alternatives, though the word "or" does not appear there.

Exclusion, extension and reduction of compensation: section 13

3–40

Section 13 contains miscellaneous qualifications to the general principle of compensation provided by section 12. These are considered in the following paragraphs.

Extra-territorial damage: subsection 13(1)

3–41

The ostensibly wide words in section 12 "wherever the injury or damage occurred" are expressly subject to subsection 13(1) which provides that in fact, compensation is not payable in two circumstances:

   (a) where the breach of duty involved matter in the course of carriage within subsection 7(2)(b) or 10(2)(b) and caused injury or damage which is shown to have taken place within the territorial limits of a single relevant territory other than the United Kingdom; or

   (b) the injury or damage was incurred within the territorial limits of a country which is not a relevant territory.

This principle is, however, itself qualified by subsection 13(2) (see para. 3–42 below).

    The first exclusion is presumably on the assumption that rights of action would exist under the legislation of the relevant territory where the damage occurred. The second is consistent with Article 2 of the Paris Convention, which provides that the Convention does not apply to nuclear incidents occurring in the territory of non-Contracting States or to damage suffered in such territory, unless the national legislation of the operator's country provides otherwise.

---

[50] *Hansard*, H.L. Vol. 301, col. 332; and *Hansard*, H.C. Vol. 779, col. 624.

3–42 United Kingdom-registered ships and aircraft: subsection 13(2)

By subsection 13(2), the exclusion of compensation for injury or damage incurred within the territory of a country, which is not a relevant territory, does not apply in the case of breach of duty under sections 7–9 (but not 10) causing injury or damage to persons or property on a ship or aircraft registered in the United Kingdom, or to the ship or aircraft, itself. To take an example, if nuclear matter was being carried on behalf of a United Kingdom nuclear site licence-holder, and an occurrence took place within the territorial waters of a non-party state, the licensee would still be liable for injury to persons (of whatever nationality) on the vessel, for damage to the ship and for damage to property on it. If, however, the carriage was on behalf of a relevant foreign operator and an occurrence falling within section 10 resulted in such damage, subsection (2) would not apply and so liability would be precluded by subsection 13(1).

This provision was made specifically for the protection of crews and of carriers; it is particularly important since, in the absence of any such remedy, enforcement of foreign judgments would be barred by section 17.[51]

3–43 Carriage to foreign operator's site—need for written agreement: subsection 13(3)

Subsection 13(3) adds a gloss to subsection 10(1)(a)(ii) (see para. 3–32 above) in that the agreement of the operator for it to be carried to his site, which is a prerequisite of liability under that subsection, must be in writing. Article 4(b)(iv) of the Paris Convention refers to written consent being necessary in such circumstances.

3–44 Hostile action and natural disaster: subsection 13(4)

Subsection 13(4) provides what is effectively the only general defence to the absolute liability under section 12:

(a) the occurrence constituting breach of duty was attributable to hostile action in the course of any armed conflict, including any armed conflict within the United Kingdom, or the causing of injury or damage by the occurrence was attributable to such conflict.

The subsection goes on to make it clear that natural disaster does *not* constitute a defence and, therefore, liability arises where the occurrence, or the causing of injury or damage by it, was attributable to a natural

---

[51] *Hansard*, H.C. Vol. 702, col. 51.

disaster, even if the disaster was of such an exceptional character that it could not reasonably have been foreseen.

Both the Paris and Vienna Conventions (Arts. 9 and IV(3) respectively) provide that the operator shall not be liable for "damage caused by a nuclear incident directly due to an act of armed conflict, hostilities, civil war or insurrection". This may assist in the interpretation of the term "any armed conflict" in the 1965 Act.

Both Conventions also provide that the operator is not liable for nuclear incidents caused by "a grave national disaster of an exceptional character, unless national law provides to the contrary" which, of course, United Kingdom law does.

Right to recover payments made outside the Act: subsections 13(5) and (5A) 3–45

Subsections 13(5) and (5A) provide a procedure to allow, in limited cases, a third party who has paid compensation for injury or damage caused in breach of sections 7–10 of the 1965 Act to claim compensation from the person subject to that duty, up to the amount of the payment made by the third party. Those cases are:

(a) where the payment was made in pursuance of the international transport conventions referred to at subsection 2(4) (in respect of which the third party's liability is not precluded by section 12 generally) (see para. 3–37 above); and

(b) where the occurrence took place or the injury or damage was incurred within the territory of a non-party, and a person who has his principal place of business in a relevant territory (or is acting on behalf of such a person) was required to make payment by virtue of a law of the country where the occurrence took place. In these cases the amount claimed as reimbursement may not exceed the overall limits applicable under section 16 (see para. 3–49 below) to the person subject to the duty.

The words "or the injury or damage was incurred" at (b) above were inserted by section 3 of the Nuclear Installations Act 1969, so as to bring the wording of the Act fully into accord with Article 6(e) of the Paris Convention (compare the more relaxed wording of Article IX(2)(a) of the Vienna Convention). In introducing the amendment, Lord Stonham referred to the recognition by the Conventions that in some circumstances a person other than the operator might find himself obliged to pay damages under the law of a country which was not a party to the Convention; the example was given of the owner of a ship carrying nuclear material. To cater for such an eventuality, such a person is given the right to make a corresponding claim against the operator.[52]

---

[52] *Hansard*, H.L. Vol. 301, col. 332.

The provisions in (a) above reflect the position under the Paris Convention (Art. 6(d)).

3–46 Contributory malicious or reckless behaviour: subsection 13(6)

Subsection 13(6) provides what is effectively a limited form of contributory negligence, confined to cases where the party suffering the injury or damage acted with the intention of causing harm to any person or property, or with reckless disregard for the consequences. In such circumstances the compensation payable to such person *may* be reduced if, and to the extent that, the causing of the injury or damage was attributable to a malicious or reckless act.

The wording of the subsection provides an interesting example of United Kingdom law which follows the Vienna Convention more closely than the Paris Convention. No such provision appeared in the Nuclear Installations (Licensing and Insurance) Act 1959. Article IV(2) of the Vienna Convention allows the competent court, if its own law so provides, to relieve the operator wholly or partly from payment of compensation to a person suffering nuclear damage caused wholly or partly by their gross negligence or act done with intent to cause damage. The Paris Convention contains no express provision to that effect, though the concept is consistent with the references to persons acting or omitting to act with intent to cause damage found in section 6. The wording in subsection 13(6) is not identical to the Vienna Convention, in that it refers to acts done with reckless disregard for their consequences, rather than gross negligence, as referred to in the Convention.

In relation to unborn children, this subsection must be read in conjunction with the Congenital Disabilities Act 1976, s.3(4) and (5) (see para. 3–24 above).

3–47 Protection for ships and aircraft: section 14

Section 14 provides protection for ships and aircraft against any lien or rights *in rem* where the claim relates to an occurrence mentioned in sections 7(2)(b) or (c) (certain occurrences outside licensed sites), 10 (relevant foreign operators) and 11 (carriers). The provision can be traced back to subsection 4(5) of the Nuclear Installations (Licensing and Insurance) Act 1959.

3–48 Time limitation: section 15

The nature of injury caused by ionising radiations is such that special provision may be needed for limitation periods. Subsection 15(1) (which overrides any other statutory limitation periods) provides that a claim under sections 7–11 shall not be entertained if made after the expiration of 30 years from the "relevant date". This date is either:

(a) the date of the occurrence; or
(b) where the occurrence was a continuing one, or there was a succession of occurrences attributable to a particular happening or operation on a particular site, the date of the last event in the course of the occurrence or succession of occurrences.

As well as the 30 year period, which constitutes an absolute time-bar on claims, it is important to note that by subsection 16(3)(b), a claim which is made after 10 years from the relevant date is made not to the licensee/operator, but to the Government (see below).

The 1959 Nuclear Installations (Licensing and Insurance) Bill was originally drafted with a 10-year limitation period running from the last date on which the relevant ionising radiations were emitted.[53] Amendments were proposed in the Lords to remove that limit on the basis that it was arbitrary and unrealistic. The Government's initial resistance to such an amendment was based upon the difficulties of proof which would arise the longer a claim was delayed, but also because insurers could not, on the current state of claims experience, work with a longer period for the purpose of setting outside reserves:

"Unless some shorter period is fixed licensees would find it difficult and expensive, if not impossible to obtain cover for their liability."[54]

However, after reflection, the Government introduced an amendment extending the period to 30 years, which was explained as follows:

"It would make the time limit for the presentation of claims thirty years instead of ten years. Since, however, claims presented after ten years or in excess of £5 million for each occurrence will be an uninsurable risk, it is necessary to make special arrangements so that the licensee can be reimbursed when he has met them. Therefore, under the Amendment claims presented within ten years or which do not carry the total above £5 million will be dealt with by the licensee or his insurers in the usual way. Claims presented after ten years or carrying the total beyond £5 million are not to be paid unless or until Parliament had decided how the licensee is to be reimbursed for paying them."[55]

Ten years is the period chosen for limitation by the Paris and Vienna

[53] *Hansard*, H.L. Vol. 212, col. 1026.
[54] *Hansard*, H.L. Vol. 212, col. 1034.
[55] *Hansard*, H.L. Vol. 213, col. 336.

Conventions (see Arts. 8(a) and VI(1) respectively). Both conventions allow a longer period to be provided, on condition that measures have been taken to cover liability for actions begun during that period, whether by insurance, other financial security, or state funds; this is of course the case in the United Kingdom.

By subsection 15(2), specific rules apply to nuclear matter which is stolen, lost, jettisoned or abandoned. Here, the claim will not be entertained if the occurrence took place after 20 years beginning with the day when the matter was stolen, lost, jettisoned or abandoned. Any claim made after the 20 year period is to be made to the Government under subsection 16(3)(c) (see para. 3–51 below). This period accords with the Paris and Vienna Conventions (Arts. 8(b) and VI(2)).

3–49 Financial limitation of liability: section 16

As referred to earlier, it was vital for the development of the nuclear industry to have a cap on the amount of potential liability for nuclear incidents, at a level for which insurance could be obtained.

Originally, the legislation provided for a maximum figure of £5 million. To put this figure into perspective, the Windscale accident in 1957 (see para. 1–32 above), the worst at that time in the United Kingdom, generated claims of less than £100,000.[56] The figure of £5 million was itself criticised as being excessive for small reactors.[57]

The figure was increased to £20 million in respect of occurrences after September 1, 1983 by the Energy Act 1983, s.27(1). Following a further increase, subsection 16(1) now provides that in relation to breach of duty under sections 7–9, the person in breach is not required to pay compensation exceeding £140 million in aggregate in respect of any one occurrence, excluding payments in respect of interest and costs. The increase to £140 million was effected by the Nuclear Installations (Increase in Operators' Limits of Liability) Order 1994 (S.I. 1994 No. 909) made under subsection 16(1A), and took effect on April 1, 1994. By subsection 16(1A), the order may not affect liability in respect of any occurrence before (or beginning before) the order came into force.

The Government views the current £140 million figure as being adequate given the high levels of safety within the nuclear industry, but recognises the importance of working with international partners to keep the figure under review.[58]

For licensees of such sites as may be prescribed, the figure is £10 million, increased from £5 million by (S.I. 1994 No. 909) as from April 1, 1994. The Nuclear Installations (Prescribed Sites) Regulations 1983 (S.I. 1983 No. 919) prescribe sites for which a nuclear site licence is in force and for which:

---

[56] *Hansard*, H.L. Vol. 212, col. 505.
[57] *Hansard*, H.C. Vol. 599, col. 901.
[58] *The Prospects for Nuclear Power in the U.K.*, Cm. 2860 (1995), para. 10.35.

(a) the quantity of radionuclides present at any time is such that their total activity does not exceed prescribed limits—these depend on which of a series of groups the radionuclides fall into and whether or not the radionuclides are within a sealed source; or

(b) the only nuclear reactor is a small reactor (defined as a thermal neutron nuclear reactor designed to operate at a thermal power output not exceeding 600 kw) and the quantity of radionuclides outside the reactor, other than associated nuclear fuel, is such that its total activity does not exceed half the limits of (a) above.

In either case, the total mass of fissile materials (other than those comprised in associated nuclear fuel) present at any time must not exceed prescribed mass limits expressed in grammes and referable to plutonium 239, plutonium 241, uranium 233, and uranium 235.

Financial liability of foreign operators   3–50

Special provision is made for the liability of foreign operators under section 10, by subsection 16(2). Liability is limited in two ways:

(a) the foreign operator is not required to make any payment in circumstances where he would not have been required to do so had the occurrence taken place in his home territory and the claim had been made under the relevant foreign law corresponding to sections 7–9. Effectively, therefore, the foreign operator can take advantage of any limitations or exclusions under his home law; and

(b) the foreign operator is not required to make any payment to the extent that the relevant amount is not required to be made available as insurance or other provision, by the relevant foreign law corresponding to section 19 (see para. 3–55 below) and has not in fact been made available by Parliamentary funds under section 18 (see para. 3–61 below) or by a contribution from a foreign government made under a relevant Convention. Thus, the foreign operator may be subject to reduced liability if his home legislation requires lower financial cover than in the United Kingdom.

Claims made to the Government: subsections 16(3) and (4) 3–51

In certain circumstances, a claim for compensation for breach of a duty under sections 7–10 is to be made to the Government, usually to the Secretary of State for Trade and Industry, unless the claim is made under section 8 in relation to a site occupied by a Government department, in which case it is to the Minister for that Department. The instances where claims are to be directed to the Government are:

(a) where the claim exceeds the statutory maximum figure (now £140 million generally or £10 million for prescribed sites), to the extent that it exceeds that figure;

(b) where the claim is made after the expiry of "the relevant period" *i.e.* ten years beginning with the relevant date as defined in section 15 (see para. 3–48 above);

(c) where the claim relates to nuclear matter which has been stolen, lost, jettisoned or abandoned and is made after the 20 year period provided by subsection 15(2) (see para. 3–48 above);

(d) where the claim relates to damage to the means of transport caused by nuclear matter in the course of carriage and full satisfaction is prevented by subsection 21(1) (see para. 3–63 below).

The original rationale for this approach relates to the difficulty of obtaining insurance outside certain parameters and has been referred to above (see para. 3–02 above). Originally, subsection 4(4) of the Nuclear Installations (Licensing and Insurance) Act 1959, provided that all claims were to be made to the operator, but that the licensee was not required to make any payment until Parliament had made provision for his reimbursement. Section 17 of the 1965 Act operates differently in that the claim is made to the Minister, not the operator.

The procedure provided by subsection 16(4) for claims to the Government is that the relevant Minister may refer any question as to the establishment or quantum of the claim to the High Court (or the Court of Session in Scotland, or High Court of Justice in Northern Ireland, as appropriate). If the Government does not refer a question to the court, the claimant may appeal to the court against the Secretary of State's decision on the claim. Whether on a reference or appeal, the court's decision is final.

If the claim is established to the satisfaction of the relevant Secretary of State, then it may be met in whole or in part by money provided by Parliament under section 18, for example, where the claim exceeds the £140 million figure (see para. 3–61 below). There may also be a contribution from foreign Governments pursuant to the relevant Conventions (see para. 3–10 above). If the claim cannot be satisfied from those sources, then by subsection 16(3) it is essentially for Parliament to determine to what extent the relevant Government department shall meet the claim, and by what means the funds for that purpose are to be provided.

Jurisdiction, shared liability and foreign judgments: section 17

3–52

Section 17 contains procedural provisions dealing with two separate issues:

(1) jurisdiction and foreign judgments; and
(2) joint and several liability.

These issues are considered in paras. 3–53 and 3–54 below.

Jurisdiction and foreign judgments: section 17

3–53

Jurisdiction of the United Kingdom courts may be ousted by a certificate from the Secretary of State for Trade and Industry to the effect that the claim or question falls, under any relevant international agreement, to be determined by the court of another party state (subs.17(1)). Such a certificate may also determine conclusively that a question or claim falls to be dealt with in the court of a particular part of the United Kingdom (subs.17(2)). This has the effect of ousting the jurisdiction of the courts of the other parts (subs.17(1)).

Enforcement of the judgments of foreign courts is dealt with by subsection 17(4). The Secretary of State may certify that the judgment is a "relevant foreign judgment" (*i.e.* a judgment which under the Paris and Brussels Conventions is to be enforceable anywhere within the territories of contracting parties), in which case by subsection 17(4), the Foreign Judgments (Reciprocal Enforcement) Act 1933 applies.

Subsection 17(5) provides a defence to proceedings in the United Kingdom to enforce a foreign judgment which is effectively that the injury or damage in question is subject to a relevant international agreement, and that the country in question is not bound by that agreement. By subsection (5A), the defence cannot be used where the judgment is enforceable in the United Kingdom pursuant to an international agreement (not necessarily a *relevant* international agreement relating to nuclear liability). The concern which prompted the defence under subsection 17(5) was that laws of non-contracting states might impose liability on a United Kingdom operator to a higher figure than under the Conventions, leaving a claimant there better off than in the United Kingdom; alternatively, such laws might impose liability on someone other than the operator, thereby defeating the principle of channelling. The barring of such enforcement was also thought likely to help provide the widespread acceptance of the Conventions.[59]

Finally, subsection 17(6) deals with claims against foreign operators under section 10 who turn out to be a government of a state party. The

---

[59] *Hansard*, H.C. Vol. 702, col. 51.

foreign government is deemed to have submitted to the jurisdiction of the relevant United Kingdom court, but the subsection does not authorise execution against the property of that Government.

3–54 Joint and several liability

In the light of the principle of channelling liability, cases where two or more persons are liable under the 1965 Act for the same injury or damage ought to be rare. However, where that is the situation, subsection 17(3) provides that for the purposes of proceedings in the United Kingdom (including enforcing foreign judgments) all of those persons shall be treated as jointly and severally liable. By subsection 17(3)(b), the maximum liability limits for each liable party are stacked, and funds from Parliament under section 18 are not required until the aggregate maximum is exceeded (see para. 3–61 below as to s.18).

3–55 Cover for licensee's liability: section 19

Section 19 places an obligation on the holder of a nuclear site licence to make provision for sufficient funds to be available at all times to ensure that duly established claims made under section 7 or any relevant foreign law corresponding to section 10 (under which, of course, the licensee would be a foreign operator) are satisfied to the extent required by the section. The provision may be made by insurance or by some other means. Insurance is discussed below (see para. 3–56 below). Insurance or other financial security is a requirement under Article 10 of the Paris Convention, and Article VII of the Vienna Convention.

The provision must relate to claims exclusive of interest or costs. The provision required is for the "required amount" in respect of each severally of:

> (a) the current "cover period" if any. "Cover period" is defined by subsection 19(2) to mean the period of the licensee's responsibility. This, in turn, by subsection 5(3) means the period from the grant of the licence until either a new licence is granted or the Health and Safety Executive gives written notice of their opinion that there has ceased to be danger from ionising radiations from the site. The period is also deemed to include any time after the expiration of that period during which it remains possible to incur liability in relation to material in the course of carriage or which was previously on the site (subs. 19(2));
>
> (b) any cover period which ended less than 10 years before the time in question (*i.e.* in respect of which a claim could still be made);

(c) any earlier cover period in respect of which a claim remains to be disposed of, having been made within the relevant limitation period.

The "required amount", by subsection 19(1A), means an aggregate amount equal to the amount applicable under subsection 16(1) in respect of an occurrence within the period, *i.e.* £140 million or £10 million (see para. 3–49 above). It should be noted that the amount is an aggregate one relating to claims within the cover period, not an amount per occurrence, as is the statutory maximum under subsection 16(1). The consequences of this distinction are discussed at para. 3–62.

Subsections 19(2A) and (2B) deal with changes occurring during a cover period. By subsection (2A), if the amount of cover required is changed by order or by the site becoming, or ceasing to be, a prescribed site, then a new cover period begins. However by subsection (2B), the grant of a new licence to the same licensee for the same or an enlarged site will not cause a new cover period to begin.

Failure to make financial provision as required by section 19 is a criminal offence (subs.19(5)).

Provision by insurance   3–56

The development of the insurance market in relation to nuclear risks has been well charted by James C. Dow in his book, *Nuclear Energy and Insurance* (1989). By the end of 1954, it had become clear in the United States and the United Kingdom that the insurance industry was going to have to tackle the issue of providing cover to the fledgling nuclear industry in respect of property damage and third party liability. The U.S. Insurance Association set up an Atomic Energy Committee in 1955, and in the same year the British Insurance Association set up its Atomic Energy Liaison Sub-Committee, which shortly thereafter became known as the Atomic Energy Committee and included representatives of Lloyd's. This Committee was to be highly influential in considering the extent to which the United Kingdom market could cover nuclear risks and in shaping the early United Kingdom legislation; the Committee was, for instance, made aware at an early stage of the importance of limiting liability.[60]

There also evolved a syndicate or pool of insurance companies run by a Committee of Management which was set up in 1956; this became the British Insurance (Atomic Energy) Committee and is still the body which administers nuclear insurance. A formal agreement was signed in December 1957 between the Management Committee, participating companies and Lloyd's Underwriters. Membership of the Pool is renewable annually; as a voluntary association, members may leave by

[60] James C. Dow, *Nuclear Energy and Insurance* (1st ed., London, 1989), p. 176.

giving six months' notice. The Agreement defines the nature of business to be contracted by the Pool and the powers of the Management Committee. The creation of the Pool was a bold step; although risks of a catastrophic accident might be slight, there was no relevant underwriting experience and the nuclear sector could have generated heavy claims before having contributed significant premiums.

The Atomic Energy Committee, referred to above, remained in being as an Advisory Committee to the Pool, and in April 1957 presented a lengthy and detailed report on the issue of nuclear risks and their insurance. This report was remarkably successful in identifying the key risks and issues without the benefit of actual underwriting experience.[61]

The standard form of liability policy developed by the Pool is linked closely to liability under the 1965 Act, or any relevant foreign law as defined in section 26 of that Act. A 10 year limitation is ensured by excluding claims made against the insured more than 10 years after the relevant date (as defined in section 15 of the 1965 Act). This is vital, since it is of the nature of nuclear liability policies that they must be written on an occurrence basis (*i.e.* responding to incidents occurring during the policy) and not on a claims-made basis. Without the 10 year limitation, the insurer would remain indefinitely exposed.

It is also important that exclusions from cover do not reduce the cover provided below the level required by the legislation. Whilst insurers are not, in general, willing to cover liability resulting from intentional releases of radioactivity (*e.g.* permitted emissions), this cannot lawfully be excluded. The approach is, therefore, to cover the legal liability from such emissions, but to negotiate a right to recover such sums from the insured.

Part I of the Liability Policy will cover statutory liability under the 1965 Act; other forms of liability are covered in Part II. The other main aspect of cover by the Pool is material damage, which will include damage to the insured's property not only from conventional risks such as fire, lightning, explosion, storm, tempest, etc., but also "excessive temperature" and irradiation or contamination by radioactivity.

As mentioned below (see para. 3–60 below), the exercise of rights of subrogation against negligent contractors would defeat the principle of channelling liability. Consequently, a standard special condition contains a warranty by the insured not to claim indemnity from any other person, and an agreement by the insurers not to enforce any rights or remedies against other parties to which they would otherwise become entitled or subrogated.

---

[61] *ibid.* at pp. 186–196 for detailed analysis.

Provision by other means 3–57

No indication is given in section 19 as to what "other means" of making provision might involve, although the provision is subject to the overriding requirement of approval by the Minister, with the consent of the Treasury; this form of wording goes back to the Nuclear Installations (Licensing and Insurance) Act 1959, s.5(1). Both the Paris and Vienna Conventions (Arts. 10(a) and VII(1) respectively) refer to the possibility of "other financial security". The most obvious possible form of security is some type of indemnity or guarantee, but the difficulty may be whether this is likely to provide adequate security over the long timescale involved, unless given by an institution such as the Government.

Special provision is made where a licensee is required to make provision for three or more sites and does so otherwise than by insurance (subs. 19(3)). The statutory requirements will be deemed to be satisfied if funds are available for all sites collectively and such funds would, for the time being, be sufficient to meet the requirements for those two sites in respect of which the requirements are highest. For example, if a licensee operated five separate sites, three of which were subject to the £140 million figure, and two of which were prescribed sites subject to the lower £10 million figure, the collective provision for all five sites could be £280 million. However, the subsection is subject to a proviso by which the Secretary of State can either disapply it completely, or can direct provision at some figure between that for two sites and that for all sites.

Claims exceeding the aggregate figure 3–58

The cover required by section 19 relates to claims in the aggregate, whereas the section 16 limit on liability is per occurrence. It could therefore be possible, if a number of occurrences took place within the cover period, for the amount of insurance available to be exceeded before the section 16 limit was reached.

The 1965 Act deals with this by subsection 19(4), which allows the Secretary of State to direct the licensee by notice in writing that a new cover period is to begin, thereby requiring the £140 million cover, which may have been eroded by previous claims, to be reinstated. The notice may be issued when the Secretary of State thinks it proper to do so, either by reason of the gravity of any occurrence which has taken place and may result in claims, or having regard to previous occurrences which have resulted in claims or which may do so. This power is complemented by the requirement under section 20 to furnish information to the Secretary of State on aggregate claims received (see para. 3–59 below).

An important issue for the licensee is its position if a new cover period is not directed, resulting in an excess of claims over cover available. This is discussed below in the context of section 18 (see para. 3–61 below).

3–59 Information relating to cover: section 20

Section 20 imposes various requirements to provide information as to cover and claims. These are:

(1) for each licensed site, notice in writing must be given to the Secretary of State forthwith if it appears to the licensee that the aggregate amount of claims in any cover period (see above) has reached 60 per cent of the cover required (subs.20(1)). This will obviously give the Secretary of State the opportunity to direct a new cover period under subsection 19(4) (see para. 3–58 above). Once such notice has been given, the licensee may not settle further claims in relation to the cover period except after consultation with the Secretary of State and in accordance with any written direction given by the Secretary of State in relation to any particular claim;

(2) in relation to any licensed site where the cover period has ended, the licensee must, not later than January 31 in each year, send a statement in writing to the Secretary of State showing the date when the cover period ended and particulars of aggregate claims reviewed, established and satisfied (subs.20(2));

(3) the Secretary of State in turn is under an obligation to lay notices received under subsection (1) and reports received under subsection (2) before Parliament (subs.20(3));

(4) any person providing the funds required by section 19 (*e.g.* the insurer) must give at least two months' written notice to the Secretary of State before ceasing to keep the funds available (*e.g.* before withdrawing cover or allowing it to lapse) (subs.20(4)). In relation to nuclear matter which is in the course of carriage, cover may not cease while the carriage continues. This reflects the requirements of Article 10(b) of the Paris Convention and Article VII(4) of the Vienna Convention.

3–60 Effect of the 1965 Act in preventing subrogation

On normal principles, an insurer or other financial guarantor of a nuclear operator might claim to be subrogated to any rights and remedies of the operator against a third party in the event of a nuclear occurrence causing injury or damage. However, the effect of the Nuclear Installations (Licensing and Insurance) Act 1959 and subsequent legislation was to extinguish the right of operators to take action in negligence against the suppliers of faulty components or contractors for defective work. Part of the reasoning behind this curtailment of normal legal remedies was to avoid the litigation which would evolve from subrogation and, in this respect, a distinction was originally drawn between the rights of private operators as licensees and the Government:

"It has been necessary to extinguish this right in the case of a licensee in order to avoid litigation which would arise if insurers or other financial guarantors of a licensee tried to exercise any rights they might have against contractors or sub-contractors responsible for building reactors or supplying parts. But the Authority or a Government Department does not normally insure, and, seeing that the taxpayers' money is involved, it is considered proper that in their case the right to sue a negligent contractor is preserved."[62]

That distinction between the Crown and the Authority and private operators no longer exists; sections 7–9 place the Crown and Authority under similar duties, and subsection 12(1)(b) has the same effect for all of them in preventing liability being incurred by any other person (see para. 3–37 above). Standard forms of nuclear insurance contain a waiver of any subrogation rights by the insurers (see para. 3–56 above).

Cover from public funds: section 18 3–61

The requirements of section 19 on insurance or other cover are to be read together with section 18. The two sections provide a scheme by which claims arising from an occurrence are to be met up to an aggregate amount, being the equivalent in sterling of 300 million special drawing rights (SDR) (subs. 18(1A)). Beyond this figure, Parliament decides whether it wishes to vote funds to meet claims. The figure accords with the requirements of the Brussels Supplementary Convention (see para. 3–10 above). "Special drawing rights" means such rights as defined by the International Monetary Fund (IMF) (s.25B). This depends upon the sum in sterling fixed by the IMF as being the equivalent to one SDR on the day in question, *i.e.* the day or first day of the occurrence, or some other day fixed in accordance with a relevant international agreement. A certificate from the Treasury is conclusive as to these matters (subs.25B(2)). To give an indication of the amount, as at May 4, 1994 the SDR was £0.943906, making the sterling equivalent of 300 million SDR approximately £283 million.

Subsection 18(1) requires any necessary sums to be made available from funds provided by Parliament so as to ensure that duly established claims in the case of any occurrence (excluding interest and costs) are satisfied up to the 300 million SDR figure. The sum required from Parliament is that which is necessary to reach the 300 million figure when aggregated with funds available from the following sources:

(a) funds provided pursuant to section 19 or any relevant foreign law made for the equivalent purpose;

---

[62] *Hansard*, H.L. Vol. 212, col. 509.

(b) where the claim is made by virtue of any relevant foreign law (*i.e.* one implementing the Paris and Brussels Conventions) any sum falling to be paid by any relevant foreign government;

(c) where the Atomic Energy Authority has incurred liability, any amounts payable under insurance or other arrangements applying to the Authority.

Compensation up to 175 million SDR is funded exclusively by the United Kingdom Government. Funds between 175 million and 300 million SDR are provided from the Pool contributed to by contracting parties under the Brussels Supplementary Convention (see para. 3–10 above).

Where a claim is satisfied wholly or partly out of money provided by Parliament under section 18, there shall also be provided from such funds, sums necessary to satisfy claims for interest and costs.

Complex provisions apply in relation to liability under foreign law under subsections 18(4A) and 18(4B). The complexity of these provisions reflects the fact that Parliamentary funds may be called upon under section 18 to meet claims not only under the 1965 Act, but also under any relevant foreign law made for equivalent purposes. This may result in adjustment of the 300 million SDR figure; for example, if claims are made under a foreign law which provides a lower aggregate figure than 300 million SDR, or if the foreign law does not provide for any sums to be made available from public funds.

3–62 Gap between occurrence-based and cover period-based provisions

As mentioned at para. 3–55 above, one potential difficulty with sections 18 and 19 is the possible gap between claims made on an occurrence basis and cover available on an aggregate cover period basis.

On considering the relevant Parliamentary debates, it seems reasonably clear that the Government recognised the potential discrepancy between the occurrence-based limit on liability and the cover period-based requirement of insurance, and intended that there should be no gap between the funds available from insurance and the money provided by Parliament.

The issue is complicated by virtue of the drafting changes made to nuclear installations legislation since 1959, but, in relation to the original Act of 1959, it was simply provided that the licensee was not required to meet any claim for the satisfaction of which funds were not required to be made available (the requirement being £5 million per cover period). Subsection 4(4) relieved the licensee of any liability to pay such claims unless and until Parliament had made provisions for the necessary funds to be available. This was stated by the Minister of Power to be in recognition of the fact that "claims presented after ten years or in excess of the £5 million for each occurrence will be an uninsurable risk".[63] This was reiterated in the Commons debates by the Parliamentary Secretary to the Minister of Power (Sir Ian Horobin):

"... beyond £5 million and a period of ten years, the matter becomes uninsurable. Therefore the Government have said that if there were some inconceivable catastrophe—if ... some colossal damage did occur, Parliament and the country would have to meet it. It is not an insurable risk, and therefore it is dealt with in the Bill in a different way."[64]

The matter became more complicated in 1965, when the Nuclear Installations (Amendment) Bill was introduced, since by then the questions of liability and insurance were also covered by the relevant international Conventions. The Joint Parliamentary Under-Secretary of State for the Home Office (Lord Stonham) on the Second Reading drew attention to the fact that "... liability imposed by the Conventions is on a per incident basis, whereas the Act deals with damage caused during the period covered by insurance."[65] He went on to say that the intention was to avoid any gap which might arise between a per incident and per cover-period basis:

"Insurance for nuclear liability is available only on a per installation per cover period basis; so far as we are aware this is so throughout the world. But, in order to give full protection to the public, the Conventions call for cover for the liability to be on a per incident basis. In common with other countries wishing to ratify the Conventions, we must therefore provide backing from public funds for any difference there might be between the per cover period and per incident basis. Such a difference could arise only if two incidents engaging the responsibility of the operator of one installation occurred simultaneously or in very quick succession and together caused damage exceeding £5 million. The position under the Act is that, in this respect, liability and insurance cover correspond; under the Bill, provision is made in Clause 8(1) and (4) to cover any gap there might be. Within our jurisdiction the matter is not significant, because of our system of meeting all established claims in full. But for a United Kingdom operator abroad (outside the coverage of the Supplementary Convention), had the system of the Act been preserved, insurance cover alone would have provided the full cover required by the Conventions."[66]

[63] *Hansard*, H.L. Vol. 213, col. 336.
[64] *Hansard*, H.C. Vol. 599, col. 932.
[65] *Hansard*, H.L. Vol. 263, col. 1276.
[66] *Hansard*, H.L. Vol. 263, col. 1277.

Clause 8(1) of the 1965 Nuclear Installations (Amendment) Bill is sufficiently similar to subsection 18(1) of the Nuclear Installations Act 1965 (as amended) to make this statement of intention relevant. It seems clear, therefore, that if there were successive occurrences within the same cover period which together exceeded £140 million, then the excess would fall to be met from funds provided by Parliament, and not the licensee or other liable party. During the Parliamentary debates, the assumption was always that this would be likely only if the occurrences were so close together in time that there would be no opportunity for the Minister to declare a new cover period.

3–63 Damage to means of transport

Subsection 21(1) makes special provision for cases where there is an occurrence involving nuclear matter in the course of carriage, and a claim is established in respect of damage to the means of transport. The effect is to prevent payment of such claims out of funds under section 18 or 19 or any relevant foreign law, if the effect would be to prevent satisfaction of other types of claim up to an aggregate amount of 5 million special drawing rights.

As has been mentioned above (para. 3–37), the normal rule of excluding other causes of action is disapplied by subsection 21(2), where a claim for damage to the means of transport is established against a foreign operator, but that operator escapes liability under subsection 16(2)(a) because of an exclusion or defence in the law of his home state (see para. 3–50 above). In such cases the foreign operator may be subject to claims outside the Act, *e.g.* in contract or negligence.

3–64 Carriage of nuclear matter—particulars of cover: section 21

By subsection 21(3), where nuclear matter is to be carried, the person who may incur liability under sections 7–10 or any equivalent foreign law must, before the carriage is begun, deliver to the carrier a document issued by "the guarantor" containing prescribed particulars of the responsible party, the nuclear matter, and the funds made available to satisfy claims. These particulars, which are prescribed by the Nuclear Installations (Insurance Certificate) Regulations 1965 (S.I. 1965 No. 1823), are: the name and address of the responsible party, a description of the nuclear matter, the place of departure and destination, the amount of funds available to satisfy claims and the period they cover, the type of security, and a statement by, or on behalf of, the appropriate authority that the responsible party is an operator of a relevant site.

The guarantor is debarred from disputing in court the particulars contained in such a document. The "guarantor" in this sense will be the insurer, relevant Government minister or other person providing such funds (subs.21(4)).

If there is a wilful failure to comply with such requirements, the responsible party who failed to give the particulars may be criminally liable, as may the carrier, if he knew or ought to have known the nature of the matter and the circumstances of its carriage.

Reporting of occurrences: section 22   3–65

Section 22 applies to prescribed types of occurrence on licensed sites or in the course of carriage of nuclear matter. The Nuclear Installations (Dangerous Occurrences) Regulations 1965 (S.I. 1965 No. 1824) prescribe those occurrences. Four classes or descriptions of occurrence are prescribed, as follows:

(a) occurrences on a licensed site involving the emission of ionising radiations or the release of radiations or toxic substances in such circumstances as to cause or be likely to cause death or serious injury to persons on the site at the time of the occurrence, or outside the site;

(b) occurrences in the course of the carriage of nuclear matter (other than on behalf of the Authority or a Government Department) being an occurrence which the person on whose behalf the matter is being carried has reason to believe has caused or may be likely to cause death or serious injury to any person by reason of radioactive properties of such matter; or which involves the breaking open of any outside container in which such matter is being carried;

(c) any explosion or outbreak of fire in a licensed site, affecting or likely to affect the safe working or safe condition of the nuclear installation;

(d) any uncontrolled criticality excursion.

For such occurrences, the licensee or responsible person must cause the occurrence to be reported forthwith in the prescribed manner to the Health and Safety Executive. Failure to do so is an offence (subs. 22(2)).

By subsection 22(4), the Secretary of State may direct an inspector to make a special report on the occurrence and the report may be published, subject to the interests of national security. By subsection 22(5), the Secretary of State may, where he thinks it expedient, direct an inquiry to be held into the occurrence and its causes, circumstances and effects. This shall be held in public except where or to the extent this appears to the Secretary of State inexpedient in the interests of national security. The procedure for such inquiries is set out at Schedule 2, which confers substantial powers of entry, compelling witnesses to give evidence, compelling production of documents and generally of regulating its own procedure.

3–66 Registration of potential claimants: section 23

Given the potentially widespread and delayed effects of nuclear occurrences, it may be very difficult to establish who was affected by an occurrence. Section 23 allows the Secretary of State (or if the occurrence is one where a claim falls to be made against the Crown, the Minister of the Department concerned) to make an order for enabling particulars of persons shown to have been within the area during the relevant period to be registered. Such registration will then be evidence of such presence unless the contrary is proven; however, failure to register will not prejudice the right to make a claim.

3–67 Extension to British territories

Some of the provisions of the 1965 Act have been extended to territories outside the United Kingdom, for which Her Majesty's Government is responsible for foreign relations. Section 28 of the 1965 Act allows directions to this effect to be made by Orders in Council. Such Orders have generally applied modifications to sections 10–17, 20, 21 and 26 together with relevant provisions of the Congenital Disabilities (Civil Liability) Act 1978, effectively creating a scheme of liability and compensation applying only to nuclear matter in the course of carriage. The relevant Orders are as follows:

- The Nuclear Installations (Gibraltar) Order 1970 (S.I. 1970 No. 116), as amended by (S.I. 1985 No. 752).

- The Nuclear Installations (Bahamas) Order 1972 (S.I. 1972 No. 121).

- The Nuclear Installations (British Solomon Islands Protectorate) Order 1972 (S.I. 1972 No. 122).

- The Nuclear Installations (Cayman Islands) Order 1972 (S.I. 1972 No. 123), as amended by (S.I. 1983 No. 1889).

- The Nuclear Installations (Falkland Islands and Dependencies) Order 1972 (S.I. 1972 No. 124).

- The Nuclear Installations (Gilbert and Ellice Islands) Order 1972 (S.I. 1972 No. 125).

- The Nuclear Installations (Hong Kong) Order 1972 (S.I. 1972 No. 126), as amended by (S.I. 1983 No. 1890) and (S.I. 1986 No. 2018).

- The Nuclear Installations (Montserrat) Order 1972 (S.I. 1972 No. 127), amended by (S.I. 1983 No. 1891).

- The Nuclear Installations (St Helena) Order 1972 (S.I. 1972 No. 128), as amended by (S.I. 1983 No. 1892).

- The Nuclear Installations (Virgin Islands) Order 1973 (S.I. 1973 No. 235), as amended by (S.I. 1983 No. 1873).

- The Nuclear Installations (Isle of Man) Order 1977 (S.I. 1977 No. 429), as amended by (S.I. 1983 No. 666) and (S.I. 1987 No. 668).

- The Nuclear Installations (Jersey) Order 1980 (S.I. 1980 No. 1527), as amended by (S.I. 1983 No. 2207) and (S.I. 1987 No. 2207).

- The Nuclear Installations (Guernsey) Order 1978 (S.I. 1978 No. 1528), as amended by (S.I. 1985 No. 1640).

CHAPTER FOUR

RADIOACTIVE SUBSTANCES: CONTROLS OVER THEIR USE AND TRANSPORT

Introduction 4–01

The Radioactive Substances Act 1993 ("the 1993 Act") distinguishes at the outset three types of substances, materials and equipment:

1. *radioactive material* — defined by section 1 (as described at para. 4–05 below);

2. *radioactive waste* — defined by section 2 (as described at para. 5–19 below); and

3. *mobile radioactive apparatus* — defined by section 3 (as described at para. 4–06 below).

Radioactive material and mobile radioactive apparatus are subject to a system of registration under the 1993 Act if they are to be used lawfully. This Chapter deals with the system of registration and also with the separate controls over the transport of radioactive substances, principally under the Radioactive Material (Road Transport) Act 1991 ("the 1991 Act").

Radioactive waste is subject to control over both its disposal and its accumulation; authorisation is required for both activities by sections 13 and 14 respectively of the 1993 Act. The subject of radioactive waste is dealt with in Chapter 5.

It is important to be aware that the systems for registration and authorisation are intimately linked. As described below, the registration system for radioactive materials was drafted with the control of the resulting waste very much in mind. Also, the two systems share common provisions on appeals, enforcement and offences. These are described initially in this Chapter, with appropriate cross references and additional comment in Chapter 5.

The relationship between controls over radioactive materials, radioactive waste and the licensing system applicable to nuclear installations should also be noted; this relationship is described in this Chapter, in Chapter 5 and in Chapter 2 dealing with the site licensing procedures for nuclear installations. Whilst many radioactive substances and wastes will be associated with the nuclear industry in some form, whether in power generation, fuel production and reprocessing, or defence, radioactive materials are used in, and radioactive wastes are produced by, many other types of operations and premises. These include research laboratories, schools, universities, hospitals and clinics, and many types of manufacturing and service industries.

Particularly familiar are medical applications such as the use of X-rays in diagnostic procedures and the use of radiotherapy in the treatment of certain types of cancer. Medical research and diagnosis frequently make use of the technique of introducing radioactive tracers which may subsequently be distinguished from non-radioactive substances in the material tested. More routinely, radiation can be used to sterilise syringes, other surgical materials and even food for patients on special diets; similar applications arise in the dental and veterinary fields.

In industry, the use of radiation provides a highly effective means of monitoring wear and tear inside machinery. Such techniques are likely to save considerable time, avoiding the unnecessary dismantling of parts, and also saving money where a part might otherwise be replaced prematurely. Industrial processes in the paper, steel, mineral and coal industries all make use of such monitoring and measurement procedures. There are additionally potential agricultural applications involving pest control and the irradiation of food to slow down deterioration in the product prior to consumption and thus lengthen "shelf-life". Radiation is also widely used for security purposes, to scan baggage and freight at airports and docks. Radioactive substances may also be used in lightning conductors, fire alarm systems and emergency lighting devices in many types of premises, such as places of entertainment, and industrial or commercial premises.

With such varied applications, it is clear that radioactive substances will be present in significant quantities on sites up and down the country. Given the inherent hazards associated with most, if not all, radioactive materials, it is not surprising that operations involving radioactive materials are subject to stringent regulation.

USE OF RADIOACTIVE SUBSTANCES

4–02 The statutory background

The first statute to provide reasonably comprehensive control over radioactive substances was the Radioactive Substances Act 1948. This Act allowed the Minister of Supply (later the Prime Minister or First Lord of the Treasury, and subsequently the Minister of Science and Secretary of State for Education and Science) to make orders prohibiting or regulating the import and export of radioactive substances. The definition of "radioactive substance" was a straightforward one, referring to any substance consisting of or containing any radioactive chemical element, whether natural or artificial. Further clarification was provided in that "substance" was defined to mean any natural or artificial substance, whether in solid or liquid form, or in the form of a gas or vapour, and also to include any manufactured article, or an article subjected to any treatment or process (s.12).

The sale and supply of substances containing more than the prescribed quantity of a radioactive element were also controlled when they were intended to be taken internally by, injected into, or applied to, any human being. Various exemptions applied to registered pharmacists, medical practitioners, proprietors of hospitals and similar institutions, to wholesale dealings, and so on. The use of irradiating apparatus for therapeutic purposes was also controlled. These controls were ultimately superseded and repealed by the Medicines Act 1968, which provided general controls over medicinal products.

The appropriate minister was empowered by section 5 of the 1948 Act to make safety regulations applying to premises or places where radioactive substances were manufactured, used, stored or treated, or where irradiating apparatus was used, to secure that radioactive waste products were disposed of safely, and to prevent injury to employees and the public from ionising radiations. Regulations could also cover the transport of radioactive substances and the structure and layout of the relevant premises. Before making such regulations, the minister or ministers were required to consult an Advisory Committee appointed under section 6 of the Act. The Atomic Energy Authority, constituted under the Atomic Energy Authority Act 1954 also had an important role in research, advice and the distribution of information (see also para. 1–07 above).

The 1948 Act was seen by the post-war Labour Government, which introduced it, as marking a critical point in the use of radioactivity; as the Minister of Health (Aneurin Bevan) noted when moving the Bill's Second Reading, the need for legislative safeguards was not accepted by everyone. The only controls which existed at that stage were of an advisory nature—for example, the safety code of the X-ray and Radium Protection Committee appointed by the Medical Research Council—and these were not universally accepted.

The Minister said:

> "The primary purpose of the Bill is to secure protection for the health of work-people and of the public generally against the harmful effects of undue exposure to dangerous radiation.... The further and much more ample need for the Bill arises from the fact that we are now on the verge of a vastly extended use of radioactivity. Basic nuclear research has produced its most dramatic achievement in the atomic bomb, but of course, there has been amazing progress in other branches of physics, and in industry and medicine, and the powers of apparatus used and the quantities of radioactivity available have increased enormously."[1]

---

[1] *Hansard*, H.C. Vol. 451, cols. 555–556.

Examples of this "amazing progress" that Bevan referred to included the enormously increased power of apparatus such as particle accelerators and X-ray sets, and "artificial radioactive substances" which were becoming available as by-products of nuclear research for use for therapeutic purposes.

In introducing the Bill, the Minister also drew attention to the fact that it gave ministers novel and unusual powers. These powers were regarded as justified to protect workers and the public; in particular, there were concerns that the public might be harmed by the unrestricted addition of radioactive substances to products such as soap and face cream.

The powerful Advisory Committee—including representatives from the Medical Research Council, the Royal Society, the Physical Society, the Royal Medical Colleges, the Faculty of Radiologists, the Department of Scientific and Industrial Research, the British X-Ray and Radium Protection Committee, the British Employers Federation and the TUC— was intended to provide a safeguard against any abuse of ministerial powers by scrutinising any draft regulations. In fact, the regulation-making powers were used very little; regulations on the carriage of radioactive substances by road, for example, were not made until 1974.

4–03 The Radioactive Substances Act 1960

At the stage of enacting the Radioactive Substances Act 1948, the issue of controlling radioactive wastes was regarded as something of an after-thought, though their environmental significance was certainly recognised:

> "The regulations also deal with effluents where radioactive substances are produced, such as raising the heights of chimneys or dealing with the treatment or the segregation of effluents where they are pouring into rivers or where they are being poured out on land. This, I am informed by the experts, is an extremely important aspect of it, because the by-products from these piles and indeed many of the elements with which they get into contact, can be quite considerably radioactive for a considerable time, and unless proper protection is given, very serious dangers might result."[2]

By the time of the passage of the Radioactive Substances Act 1960, attention had shifted more closely to the management of radioactive wastes, due partly to the work of the expert panel of the Radioactive Substances Advisory Committee, which was appointed in 1956 to consider the issue. The report of that Panel was incorporated into the White Paper entitled *The Control of Radioactive Wastes*[3] which was accepted by the

---

[2] *Hansard*, H.C. Vol. 451, col. 558.
[3] Cmnd. 884 (1959).

Government and embodied in the Radioactive Substances Bill. Temporary provision had been made in the Atomic Energy Act 1954, and the Nuclear Installations (Licensing and Insurance) Act 1959, to deal with some forms of waste, but these needed to be put onto a permanent and more general footing.[4]

The 1960 Act instituted the two-fold system of first registration for users of radioactive material (s.1) and for mobile radioactive apparatus (s.3); and secondly, authorisation for the disposal (s.6) and accumulation (s.7) of radioactive waste. However, it was clear from the outset that the registration system was not an end in itself, but rather a means for the better control of radioactive waste. This was clear from subsection 1(5) of the Act, which stated that in exercising the registration system the registering authority was to regard exclusively the amount and character of the radioactive waste likely to arise from the keeping or use of radioactive material on the premises in question. That the registration system was not intended to provide general control over all aspects of the use of radioactive material is clear from debates.[5] In introducing the Bill at its second reading, the Parliamentary Secretary to the Minister of Housing and Local Government (Sir Keith Joseph) referred to the potential for greater quantities of radioactive waste to be produced by greater numbers of users of radioactive substances, and indicated the main reasons for registration as:

"... first, to prevent users or processes which might throw up excessive or uncontrolled waste, and, secondly, to know in advance of all potential sources of waste and thus help the planning of disposal, which is the main object of the Bill."[6]

The requirement of registration was also seen as a means by which "persistently careless or deliberately cheese-paring users" of radioactive substances could be prevented from causing radioactive contamination by having their registration withdrawn.[7]

Whereas the 1948 Act was an enabling statute and utilised a wide definition of "radioactive substances", a narrower and more complex definition had to be adopted in the 1960 Act to avoid bringing under the Act "... a great variety of familiar articles—wood, bricks, granite, and even, in certain circumstances, Hon. Members of this House".[8]

---

[4] *Hansard*, H.L. Vol. 219, col. 874.
[5] *Hansard*, H.C. Vol. 619, col. 324.
[6] *Hansard*, H.C. Vol. 619, cols. 323–324.
[7] *Hansard*, H.L. Vol. 219, col. 877; quoting the Report of the Expert Panel.
[8] *Hansard*, H.C. Vol. 619, col. 327.

4–04 The enforcing authority for radioactive substances

One of the key principles underlying the 1960 Act, and the White Paper preceding it, was that control should be central rather than local. This decision appears to have been partly a function of the risks involved; long-term genetic hazards as well as the immediate public health risks with which local enforcement authorities were more familiar.[9] There was also the pragmatic reason that adequately qualified people were in short supply, and it was doubtful whether local authorities, water authorities and sewerage authorities would be able to obtain sufficient competent staff to operate on equal terms with the employees of hospitals, factories and research establishments producing waste; nor was any single authority likely to have sufficient problems to justify the employment of a full-time expert.[10]

Accordingly, the 1960 Act placed control in England and Wales in the hands of the Minister of Housing and Local Government (later to become the Secretary of State for the Environment), in association with the Minister of Agriculture, Fisheries and Food for the authorisation of waste disposal from nuclear installations; in Scotland, control was placed with the Secretary of State for Scotland. In the course of debate, the question was raised as to whether the inspectorate to administer the 1960 Act might be merged with the inspectors appointed by the Minister of Power under the Nuclear Installations (Licensing and Insurance) Act 1959. The Government investigated that possibility but concluded that it would not be a wise course on the basis that different specialisms would be required; inspectors under the 1959 Act were principally radio-physicists, whereas those under the 1960 Act dealing with radioactive wastes would mainly be radio-chemists.[11]

In fact, the inspectorate under the 1960 Act became known as the Radiochemical Inspectorate. Upon the subsequent creation of Her Majesty's Inspectorate of Pollution (HMIP), the Radiochemical Inspectorate became part of HMIP, comprising a branch under the Chief Inspector, Radioactive Substances, Dr F. S. Feates. The First Annual Report of HMIP[12] referred to the Secretaries of State for the Environment and for Wales looking to HMIP for technical and policy advice relating to the implementation of the 1960 Act, HMIP being responsible for authorisation and registration in England, whilst in Wales, those functions were undertaken by the Welsh Office with the support of HMIP. It was not until the 1960 Act was amended by the Environmental Protection Act 1990 that reference was made on the face of the legislation to the appointment of a chief inspector with specified functions.

As from April 1, 1996, the functions of HMIP were taken over by the Environment Agency in England and Wales, and by the Scottish

---

[9] *Hansard*, H.C. Vol. 619, col. 324.
[10] *Hansard*, H.L. Vol. 219, col. 875; citing the Report of the Expert Panel.
[11] *Hansard*, H.C. Vol. 322, col. 369.
[12] (1987–1988) p. 23.

Environment Protection Agency (SEPA) in Scotland. In the case of the Environment Agency, those functions include those of the chief inspector for England and Wales appointed under section 4 of the Radioactive Substances Act 1993.[13] Similarly, the functions of the chief inspector for Scotland under section 4 of the 1993 Act are transferred to SEPA.[14] The 1993 Act is consequently amended to replace references to the chief inspector with references to the appropriate Agency.[15]

The general environmental duties of the Environment Agency are set out in sections 4–8, and those of SEPA in sections 31–35 of the 1995 Act. Section 39 places both Agencies, in certain circumstances, under a further duty to take into account the likely costs and benefits of the exercise and non-exercise of their statutory powers.

For the Environment Agency, section 5 distinguishes between pollution control and non-pollution control powers and functions, making general provision in relation to the Agency's pollution control functions. Subsection 5(1) provides that the Agency's pollution control powers are exercisable for the purpose of preventing or minimising pollution of the environment, or remedying or mitigating its effects. This would appear to be a broader duty on the regulator than was previously applicable to HMIP or the Minister under the Radioactive Substances Act 1993, which made no reference to pollution of the environment and under which authorisations were granted simply subject to such conditions or limitations as HMIP or the Minister "think fit". In the absence of a definition of "pollution of the environment" in the 1995 Act, it is at this stage uncertain how the duty is to be interpreted for the purposes of the 1993 Act.

For SEPA, the focus of the Agency's functions and powers is on pollution control. Section 33 of the 1995 Act provides a single statutory purpose for all SEPA's pollution control powers, *i.e.* preventing or minimising pollution of the environment, or remedying or mitigating its effects. As with the Environment Agency, this duty appears to be broader than the duty under which HMIP or the Minister operated in issuing authorisations under the Radioactive Substances Act 1993.

The meaning of "radioactive material"   4–05

"Radioactive material" is defined by subsection 1(1) of the 1993 Act to mean anything which is not waste (see para. 5–19 below) and which is either:

(1) a substance to which the subsection applies; or
(2) an article made wholly or partly from, or incorporating, such a substance.

---

[13] see Environment Act 1995, s.2(1)(e).
[14] see Environment Act 1995, s.21(1)(e).
[15] see Environment Act 1995, s.120, Sched. 22, para. 200.

Subsection 1(2) provides that subsection (1) applies to two descriptions of substance:

(a) substances containing specified elements, listed in tabular form in Schedule 1, such that, the number of becquerels of the element divided by the weight of the substance in grams is greater than a specified number; these numbers vary depending on whether the substance is in solid, liquid or gaseous form; and

(b) substances possessing radioactivity which is wholly or partly attributable to a process of nuclear fission or other process of subjecting a substance to bombardment by neutrons or ionising radiations, not being a process occurring in the course of nature, or in consequence of disposal of radioactive waste or by way of contamination (in which case the material would probably be radioactive waste under section 2).

Schedule 1 lists eight elements, giving in each case the relevant level of activity expressed in becquerels per gram ($Bq\,g^{-1}$) for solid, liquid and gas or vapour. The substances are:

1. actinium;
2. lead;
3. polonium;
4. protoactinium;
5. radium;
6. radon (gas only, since the substance occurs only as a gas);
7. thorium;
8. uranium.

The Secretary of State is given power by subsection 1(5) to vary Schedule 1 by adding, altering or deleting entries; to date, no such amendment has been made.

Originally the intention was to define radioactive material as any material rendered radioactive by artificial means, or, naturally occurring radioactive material which, as a result of any process, contained greater levels of radioactivity than would occur in nature. This definition was, however, seen as defective in that it would be difficult to establish precisely what the levels of radioactivity in the parent substance would have been, hence the decision to specify particular substances and levels.[16] The levels in the table were set at about one tenth of the levels regarded as capable of being ingested for a lifetime without harm.

In relation to limb (b) of the definition, the Secretary of State may prescribe *de minimis* levels, though no substances have been so far prescribed under this provision.

It will be appreciated that the definition is a wide one, in the sense that

---

[16] *Hansard*, H.L. Vol. 220, col. 208.

any article, however large, which contains a relevant substance, or is made wholly or partly from such a substance, will constitute radioactive material for the purposes of the 1993 Act.

Definition of "mobile radioactive apparatus" 4–06

Section 3 defines "mobile radioactive apparatus" to mean any apparatus, equipment, applicance or other thing which is radioactive material (see para. 4–05 above) and:

(a) is constructed or adapted for being transported from place to place, or

(b) is portable and is designed or intended to be used for releasing radioactive material into the environment or introducing it into organisms.

The definition as set out originally in the 1960 Act referred simply to limb (a), but was amended by the Environmental Protection Act 1990 to cover portable apparatus more clearly.

Prohibition on use of radioactive material 4–07

Section 6 imposes a general prohibition on any person keeping or using, or causing or permitting to be kept or used, on any premises used for the purpose of an undertaking carried on by him, radioactive material of any description, knowing or having reasonable grounds for believing it to be radioactive material. The prohibition does not apply where:

(a) the relevant person is registered under section 7 in respect of the premises (see para. 4–08 below); or

(b) he is exempted from registration (see para. 4–10 below); or

(c) the radioactive material consists of mobile apparatus in respect of which he is either registered under section 10 or is exempt from the need of registration.

Contravention of section 6 is an offence under section 32, as is failure to comply with any limitation or condition applying to registration or the relevant exemption. The maximum penalties are a fine of up to £20,000 and six months' imprisonment on summary conviction, and an unlimited fine and up to two years' imprisonment in the Crown Court. Section 36 makes the normal provision for directors and senior officers to be prosecuted personally (see annotations at Appendix B below), and section 37 allows for the prosecution of any other person due to whose act or default the offence was committed. In England and Wales, proceedings relating to any offence under the Act may only be instituted by the

Secretary of State or the Agency, or by or with the consent of the Director of Public Prosecutions (s.38).

There are a number of examples of prosecutions for breach of section 6 of the 1993 Act, or, correspondingly, section 1 of the 1960 Act. These include:

Plessey GEC Semiconductors Ltd who pleaded guilty to holding a radioactive source without a certificate of registration and were fined £1,000 plus £2,261 costs in May 1993;

Imperial Tobacco Ltd who pleaded guilty to holding radioactive strontium-90 sources at their Radford premises in Nottingham without a certificate of registration and were fined £1,500 plus £860 costs in August 1988;

Xidex UK Ltd who pleaded guilty to keeping two unregistered radioactive sources and holding greater than the permitted activity of tritium and were fined £6,000 plus costs of £5,000 in September 1992.

There are aspects to the prohibition which require separate mention:

1. It applies to the keeping and use of radioactive material on "premises"; this term is defined by section 47 to include any land, whether covered by buildings or not, including any place underground and any land covered by water. It may not apply to public places such as the highway.[17] However, in the case of mobile apparatus, section 9 may provide an alternative means of prosecution.

2. The prohibition relates to premises used for the purposes of an undertaking. By section 47, this term includes any trade, business or profession; in relation to a public or local authority it includes any of its powers or duties, and in relation to any other body of persons, any of its activities. It was clearly the Government's intention that research laboratories would be covered.[18] Indeed, part of the problem leading to the 1960 Act arose because research laboratories were not governed by factories legislation.[19] Whilst the prohibition applies to the persons carrying on the undertaking, others who cause an offence to be committed by their act or default may also be subject to prosecution (s.37).

3. The prohibition applies to the keeping and use of material. The natural meaning of those words is modified by subsection 47(3) so that no account is to be taken of radioactive material used or kept in,

---

[17] *Tower Hamlets L.B.C. v. Manzoni and Walder* (1984) 148 J.P. 123.
[18] *Hansard*, H.L. Vol. 219, col. 910.
[19] *Hansard*, H.L. Vol. 219, col. 884.

or on, any railway vehicle, road vehicle, vessel or aircraft if the vehicle, vessel or aircraft is on premises in the course of a journey, or if the material is kept on a vessel as fuel. The 1960 Act was not intended to regulate the transport of radioactive material, which was left to other legislation.[20] For transport by road, the primary legislation is now the Radioactive Material (Road Transport) Act 1991 (see paras. 4–25 to 4–29 below) with secondary legislation in the form of the Radioactive Material (Road Transport) (Great Britain) 1996 (S.I. 1996 No. 1350); the secondary legislation (see paras. 4–31 to 4–33 below) also covers transport by rail. Nor was the 1960 Act intended to cover the means of propulsion of nuclear powered vessels, leaving the possible implications of this for future regulation, if necessary.

4. The prohibition covers situations where the person carrying on the undertaking uses or keeps radioactive materials, or causes or permits them to be used or kept. The distinction here between "permit" and "cause" is relevant, given that two distinct offences are apparently envisaged. To "permit" something is generally considered to be "looser and vaguer" than to "cause".[21] The term "cause" does not imply any intention or negligence and its meaning should be approached in an everyday commonsense way.[22]

5. To "cause" the keeping or use of radioactive material therefore implies some degree of positive participation or control on the part of the defendant.[23] Somebody can only be said to have "caused" another person to have undertaken a particular course of action when he either knew or deliberately chose not to know what was being done.[24]

6. To "permit" an act to occur requires either express or implied permission for that act. Some knowledge of the facts constituting the offence is necessary to establish permitting, although turning a blind eye, or allowing a course of action during which the commission of an offence would be likely, but not caring whether such an offence occurs or not, is sufficient.[25]

7. The section 6 prohibition applies only where the person carrying on the relevant undertaking knew that the material in question was radioactive material, or had reasonable grounds for such belief. To make out an offence, the prosecution would also need to prove that the defendant either had requisite knowledge,[26] or that he had

[20] *Hansard*, H.L. Vol. 220, cols. 1073–1074.
[21] *McCleod v. Buchanan* [1940] 2 All E.R. 179, 187, *per* Lord Wright.
[22] *Alphacell v. Woodward* [1972] A.C. 824.
[23] *McCleod v. Buchanan* [1940] 2 All E.R. 179, 187, *per* Lord Wright.
[24] *James & Son v. Smee; Green v. Burnett* [1955] 1 Q.B. 78.
[25] *ibid.*
[26] *Gaumont British Distributors Ltd v. Henry* [1939] 2 Q.B. 711.

reasonable grounds for believing and actually believed.[27] Knowledge may be imputed to the person who turns a blind eye and there is authority to suggest that where a person deliberately does not make enquiries because he does not wish to know the results, this may constitute actual knowledge of the facts in question.[28] In the case of a company, the relevant knowledge or grounds of belief must be related to an individual sufficiently senior to be equated with the "directing mind" or "will" of the company.[29] This element of the offence may be of importance where, for example, a person receives metal or other materials which, unknown to him, have at some point been exposed to radiation. It has been known for radioactive metal to be imported unwittingly from the former Soviet bloc countries.

4–08 Registration of users of radioactive material

Application for registration is made under section 7 to the appropriate Agency. It must include the information specified at subsection 7(2) and be accompanied by the charge prescribed under the appropriate charging scheme made under section 41 of the Environment Act 1995 (see para. 4–19 below). The information required includes a description of the radioactive material and the maximum quantity likely to be kept or used. Where it is not possible to state precisely the amount of radioactivity involved, the best estimate possible should be given.[30]

The appropriate Agency is required to send a copy of the application to each local authority in whose area the premises are situated (subs.7(3)). Similarly, where a certificate of registration is granted it is to be copied to the relevant local authorities (subs.7(8)). In both cases, however, the Secretary of State may direct otherwise on grounds of national security (s.25). Failure to determine an application within four months, or such longer period as may be agreed, constitutes deemed refusal (subs.7(5)).

Registration may be subject to conditions or limitations, which may include requirements: (a) in respect of the premises, or any associated apparatus or equipment; (b) the furnishing of information with regard to removal of radioactive material; and (c) prohibition on sale or supply of radioactive material unless appropriately labelled and marked (subs.7(6)). Except in respect of conditions of the type mentioned at (b) and (c) above, the Agency in determining whether to impose conditions may have regard only to the amount and character of radioactive waste likely to arise (subs.7(7)).[31]

An appeal procedure to the Secretary of State under sections 26 and 27 is available in various circumstances, including occasions where an

---

[27] *R. v. Banks* [1916] 2 K.B. 621.
[28] *Knox v. Boyd* [1941] J.C. 82; and *Mallon v. Allon* [1964] 1 Q.B. 385, 394.
[29] *Tesco Supermarkets v. Nattrass* [1972] A.C. 153; and *R. v. Boal (Francis)* [1992] 1 Q.B. 591.
[30] Guide to the Administration of the Radioactive Substances Act 1960, para. 25.
[31] *ibid.*, para. 27.

application for registration is refused or the conditions or limitations attached to the registration are disputed.[32]

Relationship with Environmental Protection Act 1990, Part I   4–09

The potential for overlap between registration under the Radioactive Substances Act 1993, and Part I of the Environmental Protection Act 1990 should be mentioned here. The existence of prescribed processes involving uranium, for example, as set out in Chapter 6, paragraph 6.4 of Schedule 1 to the Environmental Protection (Prescribed Processes and Substances) Regulations 1991 (S.I. 1991 No. 472), as amended, suggests the possibility of duplication of control. The matter is resolved by section 28(2) of the 1990 Act which provides that where activities comprising a prescribed process are regulated both by an authorisation under Part I of the 1990 Act, and by a registration under the Radioactive Substances Act 1993, and the two consents include different obligations in respect of the same matter, those imposed by the 1990 Act are not binding on the person carrying on the prescribed process. The primacy of the provisions in the 1993 Act relating to such activities is therefore clearly established in relation to conditions that would otherwise be inconsistent.

Exemptions from registration   4–10

Section 8 sets out the basic exemptions from the need for a registration under section 7 and also empowers the Secretary of State to grant by order further exemptions for particular types of premises and specific descriptions of radioactive material.

Subject to certain qualifications, a licensee holding a nuclear site licence is exempted from the need for registration in relation to any premises situated on the site covered by that licence and in respect of the keeping and use of radioactive material of every description (subs.8(1)). The exemption applies while the nuclear site licence is in force and during any subsequent period of responsibility (see para. 2–42 above). By subsection 8(2), the exemption may however be made subject to certain conditions by means of a direction. An exemption also exists in respect of the keeping and use on premises of clocks and watches which are radioactive material, although this exemption does not extend to premises on which clocks or watches are manufactured or repaired by processes involving the use of luminous material.

The list of exemptions is increased considerably by a number of statutory instruments made by the Secretary of State under subsection 8(6) or its predecessor, subsection 2(6) of the Radioactive Substances Act

---

[32] *ibid.*, para. 28.

1960. Exemptions are thereby applied to such things as "substances of low activity", smoke detectors, gaseous tritium light devices, electronic valves and prepared uranium and thorium compounds. The relevant statutory instruments are listed at the annotations to section 8 in Appendix B and also in the Annex to this Chapter.

4–11 Prohibition of use of mobile apparatus

Section 9 prohibits any person from:

(a) keeping, using, lending or letting on hire mobile radioactive apparatus of any description; or

(b) causing or permitting mobile radioactive apparatus of any description to be kept, used, lent, or let on hire;

unless he is either registered under section 10 or is exempt from registration.

The prohibition applies only to the activities referred to at subsection 9(2), *i.e.* testing, measuring or investigating the characteristics of substances or articles, or releasing quantities of radioactive material into the environment, or introducing such material into organisms. As with section 6, breach of the prohibition is an offence under section 32.

Like section 6, the prohibition includes causing and permitting as separate elements, as to which, see the discussion at para. 4–07 above. However, unlike section 6, the prohibition of section 9 does not require knowledge or grounds for belief as to the character of the apparatus as radioactive; presumably because this will generally be obvious, whereas the radioactive nature of material may not be at all apparent.

4–12 Registration of mobile apparatus

Application for registration is made to the appropriate Agency under section 10. It must specify the relevant apparatus and the purpose for which it is to be used, and must be accompanied by the appropriate fee under the relevant charges scheme. As with registration of keeping and use under section 7, a copy of the application and the certificate of registration must be sent to the relevant local authorities—in this case each authority in whose area it appears to the appropriate Agency the apparatus will be kept, or will be used for releasing radioactive material (subss.10(3) and (5)). This obligation, as with a section 7 registration, is also subject to the national security restriction under section 25.

Failure to determine the application within the prescribed period of 4 months, or such longer period as may be agreed, constitutes a deemed refusal under subsection 10(4).

The registration can be unconditional, refused, or, subject to such conditions and limitations as the appropriate Agency thinks fit. As with a refusal of an application or disputes over conditions or limitations in relation to a registration under section 7, an appeal procedure is available under sections 26 and 27.

Exemptions from registration of mobile apparatus 4–13

Exemptions from the requirement for registration can be made under subsection 11(1). Orders making exemptions apply to certain types of electronic valves and testing instruments. These are listed at the annotations to section 11 in Appendix B.

Relationship of registration under sections 7 and 10 4–14

Given that mobile radioactive apparatus is, by definition, a sub-set of radioactive material, there is potential for overlap and confusion between the requirements of sections 6 and 9. However, it is clear that registration of mobile apparatus under section 10 does not involve identifying all premises or areas where it may be used, and that the presence of registered or exempt mobile apparatus on premises does not require registration under section 7 (subs.6(c)). The Government regarded any restriction on where the mobile apparatus could be used as involving a contradiction in terms and potential for administrative overload; the idea was that in registering mobile apparatus the Minister would satisfy himself that the apparatus was of a nature and construction "as to produce no predictable waste hazard wherever it was used".[33] It would also plainly be absurd if premises containing mobile apparatus which was exempt were themselves subject to the requirement of registration by virtue of the presence of the radioactive material within the mobile apparatus, and a Government amendment was passed to ensure this would not be the case.[34]

Cancellation and variation of registration 4–15

Registration under sections 7 and 10 may at any time be cancelled or varied, by adding, revoking, or varying limitations or conditions (s.12). Cancellation or variation is effected by notice given to the person to whom the registration relates; where a copy of the certificate of registration has been sent to a local authority, a copy of the notice must be sent to the local authority also. Appeal against cancellation or variation lies to the Secretary of State under sections 26 and 27.

[33] *Hansard*, H.L. Vol. 220, col. 1074.
[34] *Hansard*, H.L. Vol. 220, col. 790.

During October 1995, a process of cancellation of registrations associated with the use of radioactive attachments to lightning conductors was initiated by Her Majesty's Inspectorate of Pollution in England and Wales, and by Her Majesty's Industrial Pollution Inspectorate in Scotland. In what amounted to an action against a class of radioactive material, the use of the regulator's power to cancel was based on the availability of non-radioactive alternatives so that the continued use of radioactive sources was no longer justified. A few thousand of these devices were originally installed on industrial and commercial buildings in England, Scotland and Wales at the top of lightning conductor rods; their use, however, does not appear to have extended to Northern Ireland.

From the operator's point of view, cancellation of the registration probably carries with it the added obligation of disposing of the source safely to an authorised recipient at its own expense.

4–16 Display of certificate of registration and retention of records

For a section 7, but not a section 10 registration, section 19 requires that a person to whom the registration relates exhibits copies of the certificate in suitable places on the premises. Failure to do so is an offence under section 33, carrying a penalty of up to £5,000 in the Magistrates Court and an unlimited fine in the Crown Court.

For both section 7 and section 10 registrations, the appropriate Agency may use powers under section 20 to require the retention of records and the continuing availability of these documents for inspection by the regulator. Such requirements may be imposed by notice. The records in question are those required to be kept by conditions attached to the registration and may include "site records" relating to the condition of the premises on which the relevant activities are carried on, or where the mobile apparatus is kept. Retention may be required for a specified period after the relevant activities cease to be carried on. It is an offence under section 33 to fail to comply with this requirement; the perceived seriousness of the offence is indicated by penalties which include the possibility of imprisonment (up to three months in the magistrates' court and up to two years in the Crown Court), a sentence which is not available as a penalty for contravening section 19.

4–17 Enforcement notices

Section 21 provides enforcement powers to the appropriate Agency to regulate the activities of an operator registered under either section 7 or section 10. The necessary trigger for action by the Agency is its opinion that a person holding a registration under section 7 or section 10 is

174

failing, or is likely to fail, to comply with any condition in its registration. The Agency is then entitled to issue an enforcement notice setting out the matters constituting the breach and specifying the steps to be taken to remedy this together with the period within which they are to be taken. A copy of the notice must also be served on the relevant local authority if a certificate of registration has previously been sent to that authority.

A key aspect of the enforcement notice procedure concerns the accuracy with which matters are specified on the face of the notice. The most widely accepted test on the precision of the content of an enforcement notice relates to planning enforcement notices and is whether the notice tells the recipient "fairly what he has done wrong and what he must do to remedy".[35] It is very likely that the same test should be applied to notices served under section 21. An appeal procedure to the Secretary of State against enforcement notices is contained in sections 26 and 27. The Secretary of State may, on appeal, cancel or affirm the notice, either in its original form or with such modifications as he thinks fit (subs.27(5)). This power can, no doubt, be used to cure some defects in notices, but it is questionable whether it would be possible to cure fundamental defects that make the notice a nullity.

If a person fails to comply with any requirements of a notice served on him under section 21 he is guilty of an offence under subsection 32(1). Penalties in the magistrates' court amount to a fine not exceeding £20,000, or to imprisonment for a term not exceeding 6 months or both; in the Crown Court, to an unlimited fine, or imprisonment for up to 5 years or both. There is no daily incremental fine under the 1993 Act for continued non-compliance with an enforcement notice following conviction.

Prohibition notices 4–18

A prohibition notice under section 22 may be served where the appropriate Agency believes that continuation of the relevant activity involves an imminent risk of pollution of the environment or of harm to human health. The fact that the operator may be complying in full with any limitations or conditions of his registration is irrelevant. Section 22 sets out the details that must appear on the face of the notice; this includes a direction that the registration, either in whole or in part as specified on the face of the notice, ceases to have effect until the notice is withdrawn.

The provisions relating to events following the service of a prohibition notice are identical to those for the service of an enforcement notice (see para. 4–17 above). An appeal procedure to the Secretary of State is provided under sections 26 and 27, while provisions for sending a copy of the notice to the relevant local authority apply under subsection 22(6).

---

[35] *Miller-Mead v. Minister of Housing and Local Government* [1963] 2 Q.B. 196, 226 *per* Upjohn L.J.

Additionally, the criminal offence provision under section 32 for failure to comply with the terms of a prohibition notice carries the same penalties as for the breach of an enforcement notice.

4–19 Fees

Fees for registration are now fixed by reference to a charges scheme made under the Environment Act 1995 (subsections 7(1)(c) and 19(1)(c) of the 1993 Act, as amended). Section 41 of the 1995 Act allows the Agency to prescribe the charges to be paid in relation to "environmental licences" by charging scheme. Section 56 of the 1995 Act defines "environmental licences" to include registrations under the Radioactive Substances Act 1993.

Charge schemes made under subsection 41(2) of the 1995 Act may relate to the grant, variation and subsistence of authorisations, and a wide discretion is provided as to how the scheme is to be framed. Schemes must be submitted to the Secretary of State for approval under section 42. The Scheme in force prior to the introduction of section 41 was made under section 43 of the Radioactive Substances Act 1993, and came into effect on April 1, 1995.[36] On February 1, 1995, HMIP issued consultative proposals and the scheme came into operation on April 1, 1996; the scheme being substantially the same as that previously existing. The scheme includes a charge "per inspector day" and divides application, variation and subsistence fees into bands depending on the type of premises, *e.g.* nuclear sites, research laboratories and hospitals. Copies of current charging schemes may be obtained from the Agency's regional offices.

4–20 Appeals

Section 26 provides a right of appeal to the Secretary of State against:

    (a) refusal of application for registration under section 7 or section 10 (deemed or actual);

    (b) attachment of limitations or conditions to registration;

    (c) variation of registration, otherwise than by revoking a limitation or condition;

    (d) cancellation of registration; and

    (e) service of an enforcement or prohibition notice under section 21 or section 22.

The right of appeal is conferred on "the person directly concerned" (*i.e.* the applicant or holder of the registration) in cases (a)–(d) and the person on whom the notice is served in case (e).

---

[36] The Radioactive Substances Authorisations and Registrations Fees and Charges Scheme (England and Wales) revised 1995.

The appeals procedure is dealt with by section 27, which allows the Secretary of State to refer the appeal to an inspector appointed for that purpose. Either party may require that the appeal be in the form of a hearing, which may be held in private if, and to the extent that, the person holding it so determines (subs.27(2)). The effect of bringing an appeal is dealt with by subsection 27(6); in the case of cancellation of registration, the cancellation is suspended until the determination of the appeal unless the Secretary of State directs to the contrary; but otherwise, the validity of the decision or notice in question is not affected, again unless the Secretary of State directs otherwise.

The Radioactive Substances (Appeals) Regulations 1990 (S.I. 1990 No. 2504) provide further detail as to appeals procedure and in particular:

(a) provide a general time limit of two months for bringing an appeal, or such longer period as may be allowed by the Secretary of State; for appeals against cancellation of registration, the period is 28 days (reg. 3);

(b) prescribe the matters to accompany the written notice of appeal (reg. 2);

(c) deal with the action to be taken by the Secretary of State on receipt of an appeal (reg. 4);

(d) provide a framework and timetable for appeals to be determined by written representatives (reg. 5);

(e) provide a procedure for the fixing of hearings, including publicity by newspaper advertisement (reg. 6); and

(f) require the Secretary of State's determination to be notified in writing and for the reasons to be given, together with a copy of the inspector's report if a hearing took place (reg. 7).

These procedures are explained more fully in a Guidance Note on Appeals produced by HMIP; this includes elements of a code of practice to ensure smooth running, similar to that advocated in planning appeals.

Powers of the Secretary of State   4–21

The Secretary of State occupies a highly influential position in the administration of the 1993 Act. It has already been pointed out that appeal procedures to him are available under sections 26 and 27 where applications for registration under section 7 or section 10 are refused, limitations or conditions to such registrations are disputed, and cancellations or certain variations in the terms of a registration are made.

Under section 23, the Secretary of State is empowered to give directions to the appropriate Agency on a variety of matters including applications for registration under section 7 or section 10, the service of notices under sections 21 or 22, and the provision of specified written particulars to relevant local authorities of activities carried on pursuant to registrations.

Under section 24, the Secretary of State also has the power to require certain applications for registration to be determined by him; he may then cause a local inquiry to be held in relation to that application. Under section 25, the Secretary of State may direct the appropriate Agency to restrict knowledge of applications for registration under section 7 or section 10 on grounds of national security. The effect of such a direction is to prevent the Agency from sending a copy of specified information contained in, or relating to, registrations or to applications for registration to any local authority under any provision of either section 7 or section 10.

In principle, the powers of the Secretary of State are extended under section 40 of the Environment Act 1995. This section empowers the Secretary of State to exercise considerable control over any activity of the Agencies, including those that apply under the Radioactive Substances Act 1993. This is to be achieved by giving them directions as to the carrying out of any of their functions (subs.40(1)), with which the relevant Agency must comply (subs.40(8)).

4–22 Powers of entry, inspection, etc.

Powers of entry and inspection which were formerly contained in section 31 of the 1993 Act are now to be found in the Environment Act 1995 sections 108–110, and are the general powers applying to the Environment Agency and SEPA.[37]

4–23 Access to information

A number of separate issues arise in relation to public access to information on activities regulated by the 1993 Act. Given the concern and suspicion with which the public tends to view issues involving radioactive substances, this matter is of considerable practical importance.

1. *Public Registers*

A recurring feature of environmental statutes is the requirement of a relevant regulator to keep a register of operators, their addresses and the type of operations conducted for which regulation is required. These registers are likely to contain extensive information about the operator, such as a copy of an application for any authorisation to operate, details of the conditions and limitations on the authorisation, any variations in the terms of conditions and limitations, together with details of any prosecutions that the regulator has conducted against the operator.

---

[37] For detailed commentary, see: S. Tromans, *The Environment Acts 1990–1995—Text and Commentary* (Sweet & Maxwell, 1996).

Typically, these registers are available for public scrutiny within office hours and copies of any inclusions on the register may be taken by interested parties. Such attempts at transparency in the regulation of certain activities and operations reflect an increasing concern by Government, the regulators and the general public as to the potentially damaging effects to the environment and to human health of industrial processes. The inclusion of material on any particular register is usually subject to national security and justified commercial confidentiality considerations. In practice, operators have found it difficult to have material excluded from a register purely on grounds of commercial confidentiality.

Section 39 of the Radioactive Substances Act 1993 (see point 3) requires the regulator to keep a register and to ensure the availability of public access to the contents of that register.

2. *The Environmental Information Regulations 1992*

These Regulations (S.I. 1992 No. 3240) came into force on December 31, 1992, implementing Directive 90/313/EEC, on the Freedom of Access to Information on the Environment.[38] The Regulations set out a definition of environmental information (reg. 2(2)) by reference to information that is not otherwise the subject of statutory disclosure (reg. 2(1)). In other words, the Regulations extend the range of environmental information to which an interested party may gain access beyond whatever is already carried in public registers under specific environmental statutes.

Where such environmental information is held by "relevant persons" (including Ministers of the Crown, Government departments, local authorities and any other body with public responsibilities for the environment), there is a basic obligation to make such environmental information available to any person requesting it (reg. 3). A relevant person is also under a duty to make adequate arrangements for the provision of the requested environmental information; these arrangements include a speedy response to a reasonable request and the imposition of a reasonable charge to meet the costs of that response. However, a relevant person is entitled to refuse a request for information where a request is clearly unreasonable or "formulated in too general a manner" but in making a refusal, the relevant person must give a response in writing and specify the reasons for the refusal.

The right to information is circumscribed under the Regulations by reference to confidential information. Regulation 4(2) lists the circumstances in which information is capable of being treated as confidential; the 1992 Regulations would not require the disclosure of these types of information. The relevant categories of information are restricted and include information relating to national defence or public security, legal proceedings or commercial/industrial confidentiality. Regulation 4(3)

---

[38] [1990] O.J. L158/56.

provides further exceptions to the right to information by setting out the circumstances in which information must be treated as confidential. Again, the circumstances when regulation 4(3) would apply are restricted and would apply only, following a request for information under regulation 3, where disclosure of the information would contravene any statutory provision, or the information is personal containing records of an individual who has not given consent to its disclosure, or for certain information provided voluntarily by third parties. The list under regulation 4(3) also includes disclosure of information which would increase the likelihood of damage to the environment affecting anything to which the information relates.

Regulation 5 provides for comparable arrangements to those required by the 1992 Regulations to be created for the provision of environmental information requested under any other statutory provision, and therefore otherwise outside the ambit of the 1992 Regulations.

3. *Public Access to Documents and Records under the Radioactive Substances Act 1993*

Section 39 of the 1993 Act sets out the relevant provisions relating to public access to documents and records. These provisions were added originally to the 1960 Act by Part V of the Environmental Protection Act 1990. DoE Circular 21/90 (Local Authority Responsibilities for Public Access to Information under the Radioactive Substances Act 1960, as amended by the Environmental Protection Act 1990), taken with DoE Circular 22/92, explain the provisions on public access to information and give guidance to local authorities on their obligations in this area. Section 13A of the 1960 Act is now consolidated into section 39 of the 1993 Act which therefore needs to be read in association with the two DoE Circulars.

The keeping and availability of documents is required at two levels. The appropriate Agency is required to keep copies of:

(a) all applications made to it (or originally to the Chief Inspector) under any provision of the Act;
(b) all documents issued by the appropriate Agency (or the Chief Inspector) under any provision of the Act;
(c) all other documents sent by the appropriate Agency (or the Chief Inspector) to any local authority in pursuance of directions of the Secretary of State; and
(d) specified records of convictions under the Act or under Regulations made under the Act.

The Agencies are under a duty to make copies of these documents available to the public subject to certain trade secret or national security restrictions.

In addition, local authorities are required to make available to the general public copies of all the documents sent to them by the appropriate Agency (or previously the Chief Inspector) unless they have been directed that a particular document or part of a document must not be made available for inspection. Such a direction would normally involve preventing the disclosure of information relating to any relevant process or trade secret. Otherwise, the public has the right to inspect copies of documents required to be made available under section 39 at all reasonable times and, on payment of a reasonable fee, to be provided with a copy of any such documents.

Except for special rules relating to the rehabilitation of offenders under the Rehabilitation of Offenders Act 1974, local authorities should keep documents sent to them by the appropriate Agency (or previously the Chief Inspector) for a minimum period of four years after the documents cease to have effect. In relation to records of convictions, the Radioactive Substances (Records of Convictions) Regulations 1992 (S.I. 1992 No. 1685) sets out the prescribed information to be made available to the public. In relation to each conviction this is: the details of the offence, the name of the offender, the date of conviction, the penalty imposed and the name of the court. Spent convictions under the Rehabilitation of Offenders Act 1974 must be removed from the register.

Application to Crown 4–24

The starting point in relation to Crown Immunity for the provisions of the 1993 Act is set out in subsection 42(1) to the effect that subject to certain exemptions, the Act binds the Crown. In other words, subject to any section 42 exemptions, an operator operating mobile radioactive apparatus or keeping or using radioactive material on land owned by the Crown, will need either a section 7 or section 10 registration, or will require an exemption from registration under either section 8 or section 11. Additionally, subject to a section 42 exemption, the same requirement for registration would apply where the Crown is the relevant operator.

The main exemption giving immunity to the Crown is provided in subsection 42(2). The Act does not bind the Crown in relation to any premises (whether or not owned by the Crown) if they are occupied:

(a) on behalf of the Crown for naval, military or air force purposes or for the purposes of the Secretary of State at the Ministry of Defence; or

(b) by or for the purposes of a visiting force.

Of these possibilities for immunity, (a) is the more likely to be contentious. While it seems reasonably clear that premises occupied and operated by personnel in the armed services or employees of the Ministry of Defence will come within (a), the legal position of premises occupied by a private contractor operating under a commercial contract with the

Ministry of Defence (*e.g.* at atomic weapons establishments or naval dockyards) is not so clear.

To obtain immunity from the provisions of the 1993 Act under (a) above, such a contractor would need to show that the site at which his activities involving radioactive material or mobile radioactive apparatus were conducted, was occupied:

    (a) on behalf of the Crown for armed service purposes; or

    (b) for MoD purposes.

In the absence of any definitions in the 1993 Act of "on behalf of the Crown" or "for the purposes of . . ." and any case law on the interpretation of the subsection, a contractor should probably work on the assumption that the Act does apply to its activities. Registration under section 7 or section 10 would therefore be necessary, subject to any available exemptions under section 8 or section 11.

Subsection 42(3) also provides that where the Crown contravenes any provision in the 1993 Act, it will not be criminally liable. Some limited sanction is available to the appropriate Agency in that it can apply to the High Court (in England and Wales) or the Court of Session (in Scotland) for a declaration from the Court that the act or omission constituting a contravention by the Crown is unlawful. Any benefit enjoyed by the Crown under subsection 42(3) does not extend to persons "in the public service of the Crown" who will be subject to the provisions of the 1993 Act in the same way as any other persons with no association with the Crown (subs.42(4)).

TRANSPORT

4–25 Transport by road

Provisions relating to the transport of radioactive material by road are to be found in the Radioactive Material (Road Transport) Act 1991. It applies to England, Scotland and Wales but, with the exception of section 8, does not extend to Northern Ireland.

The definition of "radioactive material" in subsection 1(1) of the 1991 Act is expressed in purely quantitative terms (material with a specific activity in excess of 70 kilobecquerels per kilogram), or by reference to orders made by the Secretary of State for Transport; no distinction is drawn between waste and other types of material in this regard. As yet, no orders have been made under the power given to the Secretary of State.

For the purposes of the operation of the Act, a number of other key definitions are provided in section 1. Packaging is defined in subsection 1(2) in very broad terms by reference to a wide range of packaging components including not only receptacles but absorbent materials, shielding and insulation; the designation "radioactive package" is defined both in terms of the radioactive material contained within and the packaging enclosing it; a further essential element of a "radioactive package" is that it has been consigned for transport, defined as transport by road only.

The Secretary of State for Transport has powers under section 2 to make regulations which he believes to be necessary or expedient to avoid injury to human health, damage to property or the environment arising out of the transport of radioactive material, and to give effect to relevant international regulations published by the International Atomic Energy Agency in Vienna. The powers provided under section 2 are very extensive and may relate to various aspects of any packaging, the preparation, transport and delivery of radioactive packages, the keeping of records and provision of information. The Radioactive Material (Road Transport) (Great Britain) Regulations 1996 (S.I. 1996 No. 1350) ("the 1996 Regulations") were made under section 2 and came into force on June 20, 1996, replacing the Radioactive Substances (Carriage by Road) (Great Britain) 1974, as amended. For the purposes of the 1996 Regulations, "radioactive material" does not include radioactive material which forms an integral part of the means of transport of that material (reg. 2).

With certain exceptions, the 1996 Regulations apply to the transport in Great Britain of any radioactive material (reg. 3(1)). These exceptions include the transport of radioactive material in a vehicle engaged in international transport operations as defined in the 1995 edition of the ADR Agreement ("European Agreement Concerning the International Carriage of Dangerous Goods by Road"), the transport of radioactive material wholly or partly constituting an instrument of war and certain radioactive materials of low-activity (reg. 3(2)). There is no obligation imposed on a person in relation to radioactive material if that person does not know or does not have reasonable grounds to believe that the material is radioactive (reg. 4).

The 1996 Regulations set criteria for persons transporting radioactive material or causing radioactive material to be transported before attempting to transport the material by road (regs. 5–9). There is a general obligation on the consignor and the carrier of radioactive material and on the driver of a relevant vehicle to exercise reasonable care to ensure that the material does not cause any injury to health or damage to property or to the environment during transport (reg. 10). Further duties are imposed on the driver during transport (reg. 36), while regulation 38 sets out the duties of the driver and carrier in the event of a specified incident during transport.

The various obligations are designed to minimise the potential harm from carrying radioactive material in vehicles by road. Key elements in these obligations include the standards which must be used for the design and testing of packagings used in the transport of radioactive material and the maximum amount of radioactive material that may be carried in each package type, or, in certain circumstances, when it may be transported unpacked (regs. 8 and 13). Some types of packaging design need to be approved by the Secretary of State (the competent authority) and may only be used where a current approval certificate is held; certain shipments of radioactive material also require approval by certificate issued by the Secretary of State (regs. 15, 17–20 and 23).

The 1996 Regulations extend the use of quality assurance programmes from the design and manufacture of a package or packaging to include the roles of consignors and carriers, including the maintenance and use of such packagings (reg. 24). Regulation 26 imposes a duty on the designer, manufacturer, owner and user of any package or packaging to retain information in its possession relating to the design, manufacture, testing, use and maintenance of that package or packaging for as long as it is used for the transport of radioactive material; an inspector (see para. 4–27) may require the relevant person to produce such information in his possession (reg. 26(3)). In addition, the consignor of radioactive material must retain for two years, any information in its possession regarding contamination measurements of that consignment (reg. 26(1).

Another key obligation in the 1996 Regulations concerns the documents that must accompany the transport of radioactive material by road (regs. 27 and 28). Under Regulation 27, unless the package contains radioactive material of low-activity or no longer contains radioactive material ("excepted packages"), a consignor is under a duty to give the carrier (or if the consignor and the carrier are the same person, to the driver of the vehicle) a document meeting the requirements of Schedule 20 and a statement in accordance with Schedule 21. Schedule 20 to the 1996 Regulations lists the various components to be incorporated in the transport document for each package included in that consignment. Schedule 21 sets out the required contents of any statement to the carrier, including actions that need to be taken by the carrier, supplementary operational requirements for loading, stowage, etc., restrictions on the mode of transport and emergency arrangements appropriate to the consignment.

For excepted packages, the relevant transport document is much reduced and should contain details of the consignor, the name and address of the consignee and a few further items from the list in Schedule 20, including the signed declaration set out at para. 20 of the Schedule. Regulation 27 also contains provisions relating to any certificates of approval from the Secretary of State required for the transport of the consignment; the consignor must have in his possession a copy of each necessary certificate of approval before preparing a package for shipment and these must be made available to the carrier for inspection on request. There are also certain record-keeping requirements imposed on the consignor for two years from the date on which the transport of a consignment begins (reg. 27(5)).

Where the same packaging with the same radioactive contents is consigned as a package on a regular basis by the same consignor and the consignor is also the carrier of that package, the Secretary of State may issue a regular consignment certificate (reg. 28(1)). The existence of such a certificate imposes an obligation on the consignor/carrier to carry a document in the vehicle containing a statement that the consignment is covered by such a certificate. The document should include the date and contents of the certificate and a record (including destinations and dates) of all consignments made under that certificate. There is a further obligation on the consignor to keep a record of these documents for two years (reg. 28(3)).

Amongst other provisions in the 1996 Regulations, particularly important is Regulation 30 which specifies requirements for the labelling of packagings and the placarding of vehicles, including the provision for the use of smaller placards in certain circumstances.

In many ways, the 1996 Regulations may be seen as a substantial tightening up in the regulation of transport by road of radioactive material in Great Britain. For provisions relating to enforcement of both the 1991 Act and these Regulations made under section 2, see para. 4–27 below.

Relationship with controls over carriage of waste   4–26

As stated in para. 4–25 above, legislation relating to the carriage of radioactive material does not distinguish between whether or not that material is waste. The perception is that radioactive substances are potentially hazardous as a source of radioactivity whether or not they may be considered as raw materials or as substances ready for disposal. The provisions in the 1996 Regulations and in the Radioactive Material (Road Transport) Act 1991 will be equally applicable (for a more detailed account of the carriage of radioactive waste see paras. 4–27 to 4–37 below).

One interesting anomaly in the regulation of the transport of radioactive waste compared with the transport of non-radioactive waste relates to a significant difference in the regulation of the carriers. With some specific exceptions, carriers of "controlled waste" are subject to a registration procedure with the relevant local authority (the relevant Agency since April 1, 1996) under the Control of Pollution (Amendment) Act 1989. The details and matters arising from such a registration are set out in the Controlled Waste (Registration of Carriers and Seizure of Vehicles) Regulations 1991 (S.I. 1991 No. 1624), but only apply to carriers of controlled waste. The definition of this term is extracted from its meaning in Part II of the Environmental Protection Act 1990 and specifically does not include radioactive waste. It would therefore appear that carriers of radioactive waste do not need to register and additionally, if the level of activity in the radioactive waste is below the level at which this waste would be designated radioactive material (*i.e.* 70 kilobecquerels per kilogram) these carriers would not be caught by the provisions of either the 1991 Act or the 1996 Regulations.

4–27 Transport by road: enforcement

Policing the Radioactive Material (Road Transport) Act 1991 and the 1996 Regulations is largely conducted by "examiners" and "inspectors". An examiner is an appointee under subsection 68(1) of the Road Traffic Act 1988, while inspectors are appointees of the Secretary of State for Transport under subsection 1(3) of the 1991 Act. Inspectors and examiners have the power to prohibit the driving of a vehicle used to transport radioactive packages under circumstances set out in subsection 3(1); these include occasions where the vehicle or radioactive package transported by the vehicle fails to comply with any regulations made under section 2. An inspector also has the power to prohibit the transport of any radioactive package, or the use of a packaging component for the packaging of radioactive materials if the package or packaging component fails to comply with any regulations made under section 2.

Where a prohibition is imposed under section 3, an inspector or examiner must "forthwith" give notice of that prohibition to the person in charge of either the vehicle, the package, or the packaging component, whichever is relevant to the prohibition (subs.3(5)). The notice must specify the reason for the prohibition, whether it applies absolutely, or for a specified purpose, and whether it is temporary or permanent in its effect. Use of the term "forthwith" implies that the notice must be given as soon as possible in the circumstances.[39] There may, however, be a degree of latitude in that, failure to act forthwith may not necessarily invalidate the action if the person subject to the action did not suffer any detriment from the failure.[40] Any prohibition comes into force as soon as the notice has been given appropriately (subs.3(6)). In imposing a prohibition, an inspector or examiner also has the power to direct in writing (subs.3(4)) that the person in charge of the relevant vehicle take it to a specified place under specified conditions; the prohibition clearly does not apply while movement of the vehicle complies with the direction.

Prohibitions under section 3 may be removed under subsection 3(7) by an inspector or examiner, as appropriate; notice of the removal of the prohibition must be given to the person in charge of the vehicle, or package or packaging component, whichever has been the cause of the prohibition. Any person who contravenes or fails to comply with the 1996 Regulations or future regulations made under section 2, or who contravenes a prohibition under section 3 or fails to comply with the direction under subsection 3(4) is guilty of an offence.

Inspectors are also empowered to serve enforcement notices under section 4 where a person has failed or is likely to fail to comply with the 1996 Regulations or future regulations made under section 2 relating to

[39] see: *Re Soutram, ex p. Lamb* [1881] 19 Ch.D. 169 at 173.
[40] see: *Hillingdon L.B.C. v. Cutler* [1968] 1 Q.B. 124.

the manufacture, or requiring the maintenance, of packaging components. The notice must contain a statement of the inspector's opinion concerning the failure to comply, a list of the failures or likely failures, the steps that need to be taken to remedy matters and the time within which the remedy must be applied. Anyone who fails to comply with an enforcement notice is guilty of an offence.

Transport by road: powers of entry and inspection   4–28

The powers of inspectors and examiners under section 4 are augmented by further powers under section 5 which permit entry into any vehicle used to transport radioactive packages so that they may determine whether any vehicle or any radioactive package carried by that vehicle fails to comply with the 1996 Regulations or future regulations made under section 2. These powers also extend to a determination of whether any radioactive package or its contents have been lost or stolen and whether the vehicle or any radioactive package contained in it has been involved in an accident. For an inspector only, there is the additional power under subsection 5(1)(b) to enter premises to ascertain whether there is on the premises any radioactive package, packaging component or vehicle used to carry radioactive packages which fail to comply with the 1996 Regulations or future regulations made under section 2. If necessary, entry of a vehicle or premises may be secured by a warrant signed by a justice of the peace. Powers of entry include the power to seize anything which an inspector or examiner has reasonable grounds for believing is evidence in relation to an offence (subs.5(4)) and any person who intentionally obstructs an inspector or examiner is himself guilty of an offence (subs.5(5)).

A key aspect in the application of these powers is that entry by inspectors or examiners is confined to "all reasonable hours". This would certainly include a time during normal business hours,[41] but the question of what is a reasonable hour will necessarily depend upon the circumstances and is, therefore, a question of fact.

In relation to the power given to inspectors to enter premises, it should be pointed out that "premises" is not defined in the Act. In previous cases, it has been construed in such a way as to indicate a whole property which may be subject to a single occupation or a single ownership, whichever is applicable in the circumstances.[42]

---

[41] *Davies v. Winstanley* [1930] 144 L.T. 433.
[42] *Cadbury Bros. Ltd v. Sinclair* [1934] 2 K.B. 389.

4–29 Transport by road: offences

The seriousness of the crimes that may be committed under the 1991 Act is reflected in the level of penalties that are applicable to most offences. An exception is to be found in relation to a person who intentionally obstructs an inspector or examiner exercising powers under section 5. Obstructive behaviour does not always require the threat of physical harm but case law suggests that behaviour which makes it more difficult for a person to carry out his or her duty may be construed as obstruction.[43] Under the 1991 Act, the obstruction must be intentional and therefore deliberate and conscious rather than accidental. Such an offence carries a maximum fine of £1,000 and is only triable summarily. All other offences under the Act are triable either way. In the Crown Court, the maximum penalty is an unlimited fine and imprisonment for up to two years, or both; on summary conviction, the maximum penalty is £5,000 and up to two months imprisonment, or both.

Section 6 contains the familiar provision relating to potential director/officer liability where an offence has been committed by "a body corporate". The prosecution would need to prove the consent, connivance or neglect of "any director, manager, secretary or other similar officer", or any person who purports to act in any such capacity (see para. 4–07 above).

In relation to offences by persons who fail to comply with any regulations under section 2, contravene a prohibition, or fail to comply with a direction under section 3, the courts may order the destruction or disposal of any relevant radioactive material and require the guilty party to meet the reasonable cost of carrying out destruction or disposal (subs.6(4)).

4–30 Revised legislation on the transport of dangerous goods

A package of new secondary legislation came in to force on September 1, 1996, the purpose of which was to implement Directive 94/55/EC on the approximation of the laws of Member States with regard to the transport of dangerous goods by road (the ADR Framework Directive) and Directive 96/49/EC on the approximation of laws with regard to the transport of dangerous goods by rail (the RID Framework Directive), These Directives are effectively transpositions of the relevant European Agreements on the subjects, but are applicable to domestic as well as international carriage. The implications of the new regulations for transport of radioactive substances and waste are as follows:

1. New Regulations, the Packaging, Labelling and Carriage of Radioactive Material by Rail Regulations 1996 (S.I. 1996 No. 2090), dealing with carriage by rail (see paras. 4–31 to 4–33 below).

[43] *Hinchcliff v. Sheldon* [1955] 3 All E.R. 406.

2. The existing legislation on transport of radioactive materials by road is generally not affected. The Carriage of Dangerous Goods (Classification, Packaging and Labelling) and Use of Transportable Pressure Receptacles 1996 (S.I. 1996 No. 2092) are not generally applicable to radioactive material as defined in the Radioactive Material (Road Transport) Act 1991 (reg. 3(1)(q)). They apply only to radioactive material which is being carried in accordance with the conditions specified in Schedules 1–4 of marginal 2704 to the ADR Agreement (Current edition 1995), *i.e.* limited quantities in excepted packaging, and articles or empty packaging which offer very low radiation risk. The same applies to the Carriage of Dangerous Goods by Road Regulations 1996 (S.I. 1996 No. 2095) which similarly have only very limited application to material falling within the 1991 Act (Sched. 2, para. 2(f)).

3. The Carriage of Dangerous Goods by Road (Driver Training) Regulations 1996 (S.I. 1996 No. 2094) apply to radioactive material (other than that falling within Schedules 1–4 of marginal 2704 of the ADR Agreement (see para. 2 above) which is carried:

    (a) in a road tanker with a capacity exceeding 1,000 litres;
    (b) in a tank container with a capacity exceeding 3,000 litres; or
    (c) in or on any other vehicle (reg. 22(1)(c)).

Certain exceptions apply under Schedule 1 (for example, vehicles not being used in connection with work, broken-down vehicles, vehicles being tested). The Regulations require instruction and training for drivers and the holding by drivers of vocational training certificates, which are to be available during carriage. The Regulations are enforced by the HSE.

Carriage of radioactive material by rail: background   4–31

Directive 96/49/EC on the approximation of laws with regard to the transport of dangerous goods by rail was adopted in July 1996 and applies generally to prohibit the transport of certain dangerous goods by rail and to require the transport of other dangerous goods by rail to be authorised (Art. 3). Member States are free to regulate or prohibit the transport of dangerous goods in their territory for reasons other than safety (*e.g.* national security) and to continue to apply more stringent provisions for transport within their territory than are provided by the Directive (Art. 5). The Directive draws directly on the IAEA's Regulations for safe transport of radioactive material.[44]

In the process of consulting on the proposed implementing regulations for this Directive, the HSC referred to the background of existing

[44] (1985 ed.) *Safety Series—No. 6.*

controls within the United Kingdom applying to the carriage of radioactive materials by rail. Control was originally provided by requirements in the British Railways Board's *List of Dangerous Goods and Conditions of Acceptance by Rail on Freight Services and Parcels Services.* These requirements provided that the materials had to comply with the IAEA Regulations for the Safe Transport of Radioactive Material.[45] On privatisation of the rail industry, new statutory controls over the carriage by rail of all dangerous goods (including radioactive materials) were provided by the Carriage of Dangerous Goods by Rail Regulations 1994 (S.I. 1994 No. 670). These Regulations did not include details corresponding to the RID Agreement and proposed Directive, but the Regulations were structured so as to allow later amendment to implement those requirements. Insofar as radioactive materials were concerned, interim arrangements were included to provide short-term control, coupled with a commitment to introduce stand-alone Regulations to deal with the carriage of radioactive materials by rail as time permitted. The Packaging, Labelling and Carriage of Radioactive Material by Rail Regulations 1996 (S.I. 1996 No. 2090; "RAMRail") have two underlying objectives: to implement the RID Framework Directive, and to institute changes based on experience of the existing 1994 Regulations.

4–32 The RAMRail Regulations: scope

The Packaging, Labelling, and Carriage of Radioactive Material by Rail Regulations 1996 (RAMRail) came into force on September 1, 1996. They apply to the carriage by rail of radioactive material as defined in the 1991 Act (reg. 2). The parameters of carriage by rail are defined more precisely by regulation 1(7), with the effect that for a wagon or container already loaded with radioactive material, carriage begins at the point at which it is brought onto the railway, and ends at the point where it is removed from the railway. Where the wagon or container is loaded or unloaded while on the railway, carriage begins when loading commences and ends when unloading is complete and any necessary purging or decontamination operations have been carried out. In this connection a "railway" means any system of transport employing parallel rails to support and guide vehicles on flanged wheels, but excludes any tramway within the meaning of section 67 of the Transport and Works Act 1992 or any railway operated wholly within a factory, harbour area, military establishment, mine or quarry. The Regulations therefore cover the former British Rail network, London Underground and the metropolitan railway systems. Where the railway transport is wholly within a factory, harbour, mine, etc., then other regulations will be relevant: the Factories Act 1961, the Dangerous Substances in Harbour Areas Regulations 1985 and the Mines and Quarries Act 1954. If the railway is not wholly operated within the factory,

---

[45] *ibid.*

etc., *i.e.* the material is to be carried beyond the confines of the factory, etc., the Regulations will apply to the railway within the factory.

The RAMRail Regulations: requirements   4–33

The Regulations are technically extremely complex, but essentially comprise the following requirements:

APPROVALS — Various types of approval are required for the design of materials and packaging (regs. 5–11). Requirements are set out in the Approved Document published by the HSC under regulation 2, and the Secretary of State issues an approval certificate stating that the packaging meets the relevant requirements.

PACKAGING REQUIREMENTS — A general duty to maintain adequate quality assurance programmes is placed on various parties including designers, manufacturers and consignors of packaging or packages, the operators of wagons and containers, train operators and rail infrastructure controllers (reg. 12). Packaging must be assigned a unique serial number (reg. 14) and inspection requirements are to be met before shipment (reg. 15 and Scheds. 8 and 9).

GENERAL CARRIAGE PROHIBITIONS AND OPERATING REQUIREMENTS — Regulations 17–27 contain a number of requirements prohibiting the carriage of radioactive material which does not conform to the packaging requirements, or material exceeding certain limits of activity. Wagons and containers must be suitable for their purpose and adequately maintained, must not be overfilled, and must have all openings, valves and caps securely closed. Restrictions apply to the carriage of radioactive material in bulk.

LOADING AND UNLOADING — Regulations 28–29 require segregation of radioactive material from certain incompatible materials, and impose a general requirement on the operator of the wagon or container to take reasonable steps in loading, stowing and unloading the material to avoid signficant risks or significantly increased risks. Further technical requirements are imposed by Schedule 12.

SECURITY AND EMERGENCIES — Regulations 30–33 require all concerned in the carriage of radioactive material to take all reasonable steps to ensure that nothing is done during the carriage to create a significant risk or significantly increase any existing risk, and to prevent unauthorised access to the material. Similar duties apply to the marshalling of trains and to prevention of fire, explosion or leakage. Train operators, facility owners and infrastructure controllers must draw up and give effect to safety systems and procedures for emergencies, and co-operate with each other so as to ensure effective co-ordination of their respective safety systems.

INFORMATION — Regulations 34–39 deal with the display of information during carriage and Schedule 13 provides details of the relevant signs and symbols to be used. Consignors of radioactive material are required to compile and provide to the train operator a package of Carriage Information, as are the operators of wagons or containers. Such information is to be kept for a period of two years from the time carriage commenced. Train operators and infrastructure controllers are to provide information, instruction and training for relevant train crews and staff.

DEFENCES — Regulation 42 contains a defence against prosecution for contravention of the Regulations that the commission of the offence was due to the act or default of another person (not an employee of the accused company) and that the defendant took all reasonable precautions and exercised all due diligence to avoid commission of the offence. To rely on the defence, notice must be given to the prosecutor giving such information to identify the "other person" as the defendant has within his possession.

RAMRAIL: APPROVED DOCUMENT — By regulation 3 of the RAMRail Regulations, the HSC is to approve and publish an Approved Document containing details as to requirements for packaging, test procedures, information, tank requirements and explanatory notes. The relevant parties under the Regulations are to ensure that the requirements of the Approved Document are complied with. The relevant document published by HSC is *Approved Requirements for the Packaging, Labelling and Carriage of Radioactive Material by Rail* (Ref. L94).

4–34 E.C. Law — shipments of radioactive substances

Regulation 93/1493/EURATOM[46] deals with shipments of radioactive substances between Member States. The Regulation was originally seen as an urgent stop-gap measure to ensure adequate controls with the removal of internal frontier controls as part of the Single Market; however, the current intention appears to be to retain the regulation as a separate measure, rather than incorporating controls into revisions to the Basic Safety Standards Directive 80/836/EURATOM (the BSS directive).[47] The Regulation also provided a means of controlling transfrontier movements of radioactive waste pending implementation of Directive 92/3/EURATOM on that subject (see para. 4–35 below).

Since the Regulation is directly applicable, no changes to United

[46] [1993] O.J. L148.
[47] see: N. Haigh, *Manual of Environmental Policy—The E.C. and Britain*, p. 8.3–3.

Kingdom law have been made to implement it. It applies, by Article 1, to shipments between Member States of sealed sources (defined by the BSS directive as a source incorporated in other materials, or in a sealed container, so as to prevent dispersion of radioactivity under normal use) and other relevant source (*i.e.* radioactive substances intended for direct or indirect use of the ionising radiations they emit for medical, veterinary, industrial, commercial, research or agricultural applications).

Article 3 applies controls for the purpose of radiation protection in a non-discriminatory manner. Article 4 requires a holder of sealed sources who intends to carry out a shipment to obtain a prior written declaration by the consignee to the effect that the consignee has complied with applicable national provisions implementing the BSS Directive and any other relevant national requirements for the safe storage, use and disposal of that class of source. Standard documents for the declaration are set out in Annex I to the Regulation, and must be used. By Article 6, shipments are to be notified on a quarterly basis to competent authorities in the Member State of destination.

E.C. Law — shipments of radioactive waste   4–35

Article 3 of the Basic Safety Standards Directive 80/836/EURATOM requires that Member States should make compulsory the reporting of the transport of radioactive substances, but leaves it to Member States themselves to decide whether such activities should be subject to prior authorisation. When the original Waste Shipments Directive 84/631/ EEC was introduced it did not include radioactive waste. However, serious concern arose in 1988 following the "Mol/Transnuklear" Affair involving alleged movements between Belgium and Germany.[48] This led to a resolution of the European Parliament calling for comprehensive Community Rules.[49] At the same time, the International Atomic Energy Agency published its Code of Practice on International Transboundary Movements of Radioactive Waste (see para. 4–37 below) which all Member States accepted.

The outcome was agreement of Directive 92/3/EURATOM.[50] The Directive, which is implemented in the United Kingdom by the Trans-frontier Shipment of Radioactive Waste Regulations 1993 (S.I. 1993 No. 3031), applies to shipments of radioactive waste between Member States and into and out of the Community (but not, as was originally proposed, internal movements within Member States) where the relevant quantities and concentrations exceed the levels laid down in Article 4(a) and (b) of the BSS Directive. "Radioactive waste" is defined simply to mean any material which contains or is contaminated by radionuclides and for which no use is foreseen.

[48] *ibid.*
[49] [1988] O.J. C235.
[50] [1992] O.J. L35.

By Article 4, a holder of radioactive waste who intends to carry out a shipment within the Community must submit an application for authorisation to the competent authority of the country of origin; this competent authority must in turn send an application for approval to the competent authorities of the countries of destination and transit, using standard documentation. Those authorities may then, under Article 6, notify their acceptance (which may be conditional) or their refusal; approval is generally deemed to be given if no reply is received within the relevant time. The competent authority of origin is then entitled to authorise despatch under Article 7 if all necessary approvals have been obtained.

Shipments into and out of the Community are dealt with in Title III of the Directive. Where waste is to enter the Community, an application for authorisation must be submitted to the Member State of destination, which must act as if it were the country of origin in relation to any state of transit. By Article 11, shipments from Member States may not be authorised to a destination south of latitude 60° south or to a state party of the Fourth ACP-EEC (Lomé) Convention which is not a member of the E.C., or to a country which, in the opinion of the competent authority of the country of origin, does not have the technical, legal or administrative resources to manage the waste safely.

The Directive does not apply to the return of sealed sources by their users to their suppliers, so long as the sealed source does not contain fissile material (Art. 13). Nor does it affect the right of a Member State to which waste is exported for processing, to return the waste after treatment to the country of origin, or to return waste or other products of irradiated nuclear fuel reprocessing (Art. 14).

4–36 International shipment of radioactive materials and waste: United Kingdom Law

The shipment of non-waste radioactive materials is governed by Regulation 93/1493/EURATOM (see para. 4–34 above) which is directly applicable and therefore needs no United Kingdom implementing legislation.

By contrast, the shipment of radioactive waste is governed by the Transfrontier Shipment of Radioactive Waste Regulations 1993 (S.I. 1993 No. 3031) which implement Directive 92/3/EURATOM (see para. 4–35 above). Radioactive waste is defined to mean any material which contains, or is contaminated by, radionuclides and for which no use is foreseen.

Part II of those Regulations deals with the following shipments of radioactive waste:

(a) from a place of origin in the United Kingdom to another country (whether a Member State or not);
(b) imports into the Community from a third country where the United Kingdom is the country of destination; and

(c) shipments into the Community from a third country where the country of destination is not a Member State and the United Kingdom is the point of entry.

By regulation 6, such shipments are prohibited unless carried out under, and in accordance with, the conditions and requirements contained in an authorisation. In relation to the grant of such authorisations, regulation 7 requires notification of competent authorities in the other relevant countries and consultation with the HSE in the case of shipments from or to nuclear sites in the United Kingdom. The authorisation must be granted using standard documentation (with any additional requirements attached) and is valid for the period specified in the authorisation, which cannot exceed three years.

Part III prohibits the carrying out of shipments falling outside Part II, except where carried out under, or in accordance with, an authorisation granted by the competent authority of another Member State, and an approval given under regulation 11. Since the Regulations apply only to shipments between Member States and shipments into and out of the Community (reg. 3(1)), effectively the prohibition relates to "shipments" within the United Kingdom where the United Kingdom is a country of transit for material which has originated in another Member State or which has entered the Community from a third country by way of another Member State. The term "shipment" is defined by regulation 1(1) to cover all transport operations between the places of origin and destination, including loading and unloading of the waste.

Regulation 18 creates various offences; these include the carrying out of shipments in the United Kingdom in breach of regulations 6 and 10. It is also an offence to fail to comply with the requirements of the Regulations relating to notifications and documentation, *i.e.*:

1. in the case of exports from the United Kingdom to third countries, the person who made the application for shipment must notify the Agency of its arrival at its destination within two weeks of such arrival (reg. 9(2));

2. no shipment may be made within the United Kingdom unless accompanied by the application made under the directive, the necessary approvals and the authorisation (in the case of shipment by rail, these documents need not accompany the shipment but must be made available to the Agency (reg. 15(1)));

3. for shipments where the United Kingdom is the country of destination, the consignee must, within 15 days of receipt of the waste, send the Agency an acknowledgement of receipt using the appropriate document (reg. 15(2)).

The penalties for offences are a fine of up to £20,000 on summary conviction and on conviction on indictment, an unlimited fine and up to 2 years' imprisonment.

Part IV of the Regulations deals with re-shipment. Essentially, these provisions ensure that where radioactive waste is exported from the United Kingdom for processing (or reprocessing in the case of irradiated fuel) the Regulations do not affect any rights to return such materials, or the products of reprocessing, to their country of origin after treatment.

4–37 The Basel Convention and radioactive waste

The Basel Convention on the Control of Transboundary Movements of Hazardous Waste (Cm. 3108) excludes "wastes which, as a result of being radioactive, are subject to other international control systems, including international instruments, applying specifically to radioactive materials" (Art. 1(3)). It has been pointed out that this wording is open to interpretation, particularly the words "international control systems".[51] At the time of negotiating the Convention, the IAEA was working on its Code of Practice on transboundary movements of nuclear wastes, which was adopted in September 1990 and which embodied most principles of the Basel Convention, albeit in weaker form[52] and a number of negotiating parties argued that the Convention should apply to radioactive waste. The term "control system" is obviously broader than simply legal instruments and thus could, on that basis, include the IAEA systems. In any event, some radioactive wastes will not be subject to IAEA or CIMO control systems because of their low levels and will accordingly be covered by the Basel Convention; the position of radioactive wastes arising from military activity is also the subject of ongoing discussion.[53]

[51] K. Kummer, *International Management of Hazardous Wastes* (O.U.P., 1995), p. 51.
[52] *ibid.*, p. 85.
[53] *ibid.*, p. 51.

ANNEX TO CHAPTER 4

Exemption Orders under section 8 of the Radioactive Substances Act 1993 for England and Wales

- The Radioactive Substances (Exhibitions) Exemption Order 1962 (S.I. 1962 No. 2654).

- The Radioactive Substances (Storage in Transit) Exemption Order 1962 (S.I. 1962 No. 2646).

- The Radioactive Substances (Phosphatic Substances, Rare Earths, etc.) Exemption Order 1962 (S.I. 1962 No. 2648).

- The Radioactive Substances (Lead) Exemption Order 1962 (S.I. 1962 No. 2649).

- The Radioactive Substances (Uranium and Thorium) Exemption Order 1962 (S.I. 1962 No. 2710).

- The Radioactive Substances (Prepared Uranium and Thorium Compounds) Exemption Order 1962 (S.I. 1962 No. 2711).

- The Radioactive Substances (Geological Specimens) Exemption Order 1962 (S.I. 1962 No. 2712).

- The Radioactive Substances (Schools, etc.) Exemption Order 1963 (S.I. 1963 No. 1832).

- The Radioactive Substances (Precipitated Phosphate) Exemption Order 1963 (S.I. 1963 No. 1836).

- The Radioactive Substances (Electronic Valves) Exemption Order 1967 (S.I. 1967 No. 1797).

- The Radioactive Substances (Smoke Detectors) Exemption Order 1980 (S.I. 1980 No. 953), as amended by (S.I. 1991 No. 477).

- The Radioactive Substances (Gaseous Tritium Light Devices) Exemption Order 1985 (S.I. 1985 No. 1047).

- The Radioactive Substances (Luminous Articles) Exemption Order 1985 (S.I. 1985 No. 1048).

- The Radioactive Substances (Testing Instruments) Exemption Order 1985 (S.I. 1985 No. 1049).

- The Radioactive Substances (Substances of Low Activity) Exemption Order 1986 (S.I. 1986 No. 1002), as amended by (S.I. 1992 No. 647).

- The Radioactive Substances (Hospitals) Exemption Order 1990 (S.I. 1990 No. 2512), as amended by the Radioactive Substances (Hospitals) Exemption (Amendment) Order 1995 (S.I. 1995 No. 2395).

Similar orders have been issued in relation to exemptions from Scotland and Northern Ireland.

Exemption Orders under section 11 of the Radioactive Substances Act 1993 for England and Wales

- The Radioactive Substances (Electronic Valves) Exemption Order 1967 (S.I. 1967 No. 1797).

- The Radioactive Substances (Testing Instruments) Exemption Order 1985 (S.I. 1985 No. 1049).

Similar orders have been issued in relation to exemptions for Scotland and Northern Ireland.

CHAPTER FIVE

RADIOACTIVE WASTE

"How to dispose of radioactive waste safely in perpetuity is one
of the most intractable problems currently facing industrial
countries."

British Government Panel on Sustainable Development, Second Report (1996),
p. 18.

Introduction   5–01

The various types of radioactive waste which arise from industrial and
other activities, together with the main problems which they present,
have been outlined in Chapter 1. This Chapter considers the controls
applied by the Radioactive Substances Act 1993 to the accumulation and
disposal of radioactive waste, but begins with a short overview of
underlying domestic and international policy in this area. Whilst it is not
possible to deal with a complex area in more than the broadest outline,
some appreciation of this policy and its evolution is necessary for a full
understanding of how the legislation itself operates.

The history of the various attempts to grapple with the problem of
nuclear waste disposal in the United Kingdom does not always make for
happy reading. It may be some small comfort, however, to realise that the
problems are no less intractable in other democracies, such as the United
States and Germany. As in the United Kingdom, no final repository for
intermediate-level waste is likely to be ready in the USA (at Yucca
Mountain in Nevada) until at least 2010, and in the meantime the
problems of interim storage are increasing and becoming more politically fraught. A useful explanation of the various technical, social and
political complexities of radioactive waste management from an international perspective can be found in the 1996 Report of the Nuclear
Energy Agency of the O.E.C.D. *Radioactive Waste Management in Perspective.*

Classifying radioactive waste   5–02

For some years attempts have been made to classify radioactive waste into
various levels according to its characteristics. In the United Kingdom, the
classification currently designates four types, as outlined in the Government's *Review of Radioactive Waste Management Policy*[1]:

[1] *Review of Radioactive Waste Management Policy—Final Conclusions,* Cm. 2919 (1995), para. 53.

1. *High-level or heat-generating wastes* — the temperature of such wastes may rise significantly as a result of their radioactivity.

2. *Intermediate-level waste* — radioactivity levels exceed the upper bounds for low-level wastes, but do not generate heat in sufficient quantity that it requires to be taken into account in the design of storage or disposal facilities.

3. *Low-level wastes* — these contain radioactive materials other than those acceptable for disposal with ordinary refuse, but not exceeding 4 GBq/te (gigabecquerels per tonne) of alpha activity or 12 GBq/te of beta activity. Under existing authorisations, they can be disposed of at the BNFL Drigg facility or at the UKAEA's facility at Dounreay.

4. *Very low-level wastes* — these can be safely disposed of with ordinary refuse. The criteria are that each 0.1 cubic metre of material should contain less than 400 kBq (kilobecquerels) of beta/gamma activity or comprise single items containing less than 40 kBq of beta/gamma activity.

It may be that in due course, these criteria will be further revised and refined in the light of Government research and relevant studies of the IAEA or European Commission.[2]

EVOLUTION OF POLICY ON RADIOACTIVE WASTE

5–03 The Flowers Report and the Government's response

It was the Government's 1959 White Paper, *Control of Radioactive Waste*,[3] which prompted the enactment of the Radioactive Substances Act 1960, with its requirement that the disposal and accumulation of radioactive waste be authorised. However, public attention to radioactive waste disposal practices remained very limited until the mid-1970s.

The Sixth Report of the Royal Commission on Environmental Pollution, *Nuclear Power and the Environment*,[4] the so-called "Flowers Report", considered in detail the various disposal methods applied at the time. In relation to gaseous discharges of radioactive substances to the atmosphere from, for example, nuclear reactors and fuel reprocessing plants, it appeared to the Royal Commission that the standards being applied were such that there was no significant problem (para. 343). The Royal

---

[2] *ibid.*, para. 54.
[3] *The Control of Radioactive Waste*, Cmnd. 884 (1959).
[4] Royal Commission on Environmental Pollution, Sixth Report, *Nuclear Power and the Environment*, Cmnd. 6618 (1976).

Commission referred to the principles set out in the White Paper,[5] namely, that strict upper limits should be given to the doses that may be received by an individual or population, below which doses should be reduced to as low a level as was reasonably practicable "having regard to cost, convenience and the national importance of the subject".

Whilst agreeing with this approach, the Royal Commission suggested that in the case of gaseous effluents the agreed maximum levels should be regarded as a presumptive standard, forming part of the overall authorisation to discharge (at that time, standards were only generally fixed numerically for aqueous effluents). As to liquid waste, the Royal Commission noted that by far the largest aqueous discharge of radioactivity to British waters originated from BNFL's Windscale works. No particular comment was made upon this practice, other than to note that discharges of plutonium by this route could be a serious issue for the future. With respect to low-level solid waste deposited on municipal and "special precautions" refuse tips, the Royal Commission concluded that the present practices of the Radiochemical Inspectorate were "perfectly satisfactory" (para. 356) and that moves to have such slightly contaminated material moved across the country at great expense for ocean-dumping were misguided. With respect to the low-level solid wastes disposed of by burial at Drigg in Cumbria, the Royal Commission was satisfied that the practice presented no danger to the public, but were doubtful that it should be repeated around the country to deal with waste from an expanded nuclear programme; they considered that more thought needed to be given to the arrangements which might be needed in the future.

Difficulties really began to emerge when the Royal Commission considered intermediate-level solid waste, as they were unable to discover any clearly formulated policy for its future disposal (para. 364). Such waste was at that stage kept at nuclear sites, and the Commission gained the impression that there was a lack of clarity about where responsibility lay for determining the best strategy for dealing with such wastes. The Commission referred also to the policy of BNFL, expressed in their evidence to the Commission, that solid waste should be disposed of by deep ocean dumping. The 1972 London Dumping Convention, which had entered into force in September 1975, restricted the sovereignty of the United Kingdom with regard to ocean dumping and prohibited the dumping of high-level waste (see further, paras. 5–05 and 5–14 below). At that time the United Kingdom took part in an annual international operation to dump packaged, low-level solid waste in the deep North Atlantic some 900 kilometres off Land's End; such operations had been conducted and controlled by the Nuclear Energy Agency of the OECD since 1967. The Commission did not feel this scale of dumping, amounting to about 7,000 tonnes per year, gave rise to any worry, but felt it unlikely that an expanded programme would be acceptable to the

---

[5] *The Control of Radioactive Waste*, Cmnd. 884 (1959).

United Kingdom authorities and internationally, and that it was therefore necessary to begin work on developing a national disposal facility. In relation to the interim management strategies for high-level waste, the Commission considered the various possible options which were open and stressed the need for a clear strategy, which was lacking under contemporary arrangements.

Overall, the picture which emerged from the Commission's review of radioactive waste management was "in many ways a disquieting one, indicating insufficient appreciation of long term requirements either by Government Departments or by other organisations concerned" (para. 427). In view of the long lead times required for the development of appropriate disposal facilities, the Royal Commission was convinced that a much more urgent approach was needed and that the responsibilities for both devising and executing policy needed to be more clearly defined.

5–04 Response to the Flowers Report

The Government responded to the Flowers Report in Cmnd. 6820 (1977). This response was generally endorsed by a further White Paper in 1982.[6] These policy documents set out six main objectives of Government policy:

(1) to minimise the creation of waste from nuclear activity;
(2) to deal with waste management problems in principle before any large scale programme of nuclear power was undertaken;
(3) to carry out the handling and treatment of waste with due regard to environmental considerations;
(4) to dispose of wastes at nuclear sites in accordance with a programme;
(5) to provide adequate research and development on methods of disposal; and
(6) to dispose of wastes in appropriate ways, at appropriate times and at appropriate places.

These objectives were sufficiently vague to mean very little in practice, though the Government did take positive action by creating the Radioactive Waste Management Advisory Committee (RWMAC) in 1978, and the Nuclear Industry Radioactive Waste Executive (Nirex) in 1982; additionally, primary Government responsibility in this area was passed to the Department of the Environment. Other concerns expressed in the Flowers Report prompted the Department to commence a programme of test drilling in potential rock structures for the disposal of highly active wastes. These plans were abandoned in December 1981, following opposition from the rural communities involved in Ayrshire, mid-Wales

[6] Cmnd. 8607 (1982).

and Northumberland (see the planning appeal decision at [1981] E.G. Vol. 261, p. 144—where the UKAEA unsuccessfully appealed against refusal by Northumberland County Council to sink test boreholes in the Northumberland National Park). By the time of the decision, plans for further investigation had been shelved by a Parliamentary announcement two days earlier and, in view of this, the Secretary of State decided that a loss of amenity in the National Park, even in the short-term, was not justified. By this time, it had been decided from available scientific evidence that high-level waste could be vitrified and stored for up to 50 years prior to its disposal; later work advocated such storage of 100 years to allow for radioactive decay and reduction in heat output. In its 1982 White Paper, the Government felt able to assert that problems of waste management did not present a barrier to the foreseeable development of nuclear power.

Intermediate-level waste disposal problems   5–05

Once the RWMAC had pointed to the urgent need for suitable disposal options for intermediate-level wastes in its first Annual Report in 1980, attention switched to securing such options. The Government's 1982 White Paper identified the lack of suitable disposal facilities for intermediate-level wastes as a major gap in waste management policy and Nirex was given the role of identifying such facilities. The Government was shortly to find out how difficult this task would be. At the same time, the practice of sea dumping assumed a high profile internationally when in February 1983, the London Dumping Convention resolved to prohibit the practice until scientific evidence could demonstrate conclusively that no harm to the marine environment would result. The possibility that Britain would ignore this decision led to a boycott of radioactive waste dumping at sea in the summer of 1983 by the National Union of Seamen and other transport unions. This action resulted in the creation of an independent panel of scientists chaired by Professor F. G. T. Holliday to investigate the safety of the North Atlantic Sea dump. The report, published in December 1984,[7] recommended a continued moratorium until further evidence could substantiate a case for removal of the ban; it was also recommended that sea dumping should be compared with land-based alternatives on the basis of identifying the best practicable environmental option.

Against this background, in 1983, Nirex announced its selection of sites at Billingham in Cleveland and Elstow in Bedfordshire as potential repositories for intermediate and low-level wastes. This process of selection was seen as seriously flawed in the absence of consideration of alternative sites for additional disposal options. Major campaigns were mobilised against the choice of both sites. In the midst of the controversy,

---

[7] *The Report of the Independent Review of the Disposal of Radioactive Waste in the North-East Atlantic,* HMSO (1984).

the Government published its *National Strategy for Radioactive Waste Management* (DoE, 1984) which contained only a general statement suggesting that, for low-level and intermediate-level waste, there was no requirement for any lengthy research and development into disposal methods. Both sites were ultimately withdrawn in the face of considerable opposition.

The withdrawal of Billingham as a potential site was announced in January 1985, and was mainly prompted by the refusal of the owner of the mine (ICI) to sell it. In the statement announcing this decision, the Secretary of State indicated that at least two further sites additional to Elstow would be investigated to provide a comparative assessment, and that to facilitate the comparative evaluation the Government was to introduce a special development order to allow experimental drilling to take place. This latter aspect was important, given the strategy followed by Bedford County Council in seeking injunctions to prevent site sampling work at Elstow.[8]

In February 1986, the Secretary of State announced that three further sites would be considered, namely South Killingholme in South Humberside, Fulbeck in Lincolnshire and Bradwell in Essex. Planning permission for investigation was given by the Town and Country Planning (Nirex) Special Development Order 1986 (S.I. 1986 No. 812). Local opposition at all sites led to protests with Nirex having to resort to court action to gain access by obtaining injunctions against protestors and all those "associated or affiliated with" the local protest groups, a description said to have been enough to cover: "The Bishop of Lincoln, Mr Austin Mitchell, MP for Great Grimsby, and the Women's Institute".[9] An additional political embarrassment was that all three sites were in Conservative constituencies, including that of the then Government Chief Whip, Mr John Wakeham.

In March 1986, *Assessment of Best Practicable Environmental Options for Management of Low and Intermediate-Level Solid Radioactive Wastes* (HMSO) was published. This significant study by the DoE Radioactive Waste Management (Professional) Division concluded that the BPEO for most LLW and some short-lived ILW was near-surface disposal, as soon as practicable, in appropriately-designed trenches. Where levels of alpha emissions made this method unacceptable for ILW, deep underground disposal would be required, with no technical preference between deep cavity disposal or offshore borehole disposal. Economic, radiological and social impacts could be used to distinguish between these opinions only on a site-specific basis.

[8] see: *Bedfordshire County Council v. CEGB, SSEB, UKAEA, BNFL Plc* [1985] J.P.L. 43 (Nirex not having been formally incorporated at that stage).
[9] *The Times*, September 30, 1986.

The House of Commons Environment Committee Report   5–06

Meanwhile, the House of Commons Environment Select Committee had published its first report on *Radioactive Waste.*[10] The Report was deeply critical of many aspects of policy:

> "It has become apparent to us that far from there being a well-defined, publicly debated policy on the creation, management and disposal of radioactive waste, there was confusion, and obfuscation among the various organisations entrusted with its care . . . In short, the U.K. Government and nuclear industry are confused. On the one hand, bold announcements about prospective new disposal sites are issued. They are then withdrawn, left hanging in the air, or modified ad hoc. On the other hand, a very large proportion of radioactive waste goes on being produced unquestioned and a sequence of different studies shows that the U.K. is still only feeling its way towards a coherent policy. For an issue which is of such great public concern, this is regrettably inadequate."

Whereas the Flowers Report had been satisfied with much of the detail of then current waste management practice, the Commons Select Committee expressed dissatisfaction with many aspects. In relation to disposal of low-level waste at Drigg, the conclusion was that much valuable space was wasted by the unnecessary disposal of material which had been subject to even the remotest contact with radioactive matter; conversely, the Committee was not assured that harmful higher activity wastes could not leak out of the site into groundwater. The Committee also critically referred to the fact that public perception was at the heart of the problem in this respect, and concluded that the "industry must radically change its present attitudes and its relationship with the public; that a 'Rolls-Royce' approach must be embraced and is not significantly more expensive than less cautious solutions".

The importance of public perception in this area was also strongly emphasised by the 1985 DoE Research Report, *Social Impacts of Radioactive Waste Disposal,*[11] which considered the reactions of various interest groups to different methods of disposal.

The Government's *volte-face*   5–07

Then in May 1987, in an amazing turnaround in policy, the Secretary of State announced the abandonment of the search for a near surface disposal repository at the four sites then under consideration, in favour of a multi-purpose facility for both low-level and intermediate-level waste.

[10] Session 1985–1986, 191–I.
[11] Final Report (November 1985).

This decision was attributed to cost considerations, in particular that the increase in shallow repository costs had narrowed any cost differential between that and a deep repository. It had also become clear to the Government that the public were no less concerned about the creation of a site for the disposal of low-level waste as for intermediate-level waste, notwithstanding the very different tangible risks involved. The increase in costs relating to near surface facilities was referable to the House of Commons Committee's suggestion that a "Rolls-Royce" solution was required to meet the public's perception of the problems. Therefore, it became more economic to consider disposing of low-level and intermediate-level waste at a single facility, rather than creating two separate facilities.

These costing arguments were met with some scepticism; suggestions were made that the real reason was related to the atmosphere of heightened concern about nuclear issues following the Chernobyl disaster.[12]

Attention then became focused on three possible options: a deep mine on land, a repository under the seabed accessed by tunnels from the shore, and a repository under the seabed accessed by a vertical shaft from a rig or similar structure. Nirex published in 1988 its discussion document, *The Way Forward*, aimed at assisting general understanding of the issues and stimulating comment. The document invited comments in particular on the issue of site selection factors, selection process and local liaison—including the issue of whether ". . . an adequate site which enjoys local support should be preferred to a superior site which does not?"

5–08 The Government's 1995 review of policy

In May 1994, the Secretary of State for the Environment announced a fresh review of radioactive waste management policy, which was carried out in parallel with the Government's review of the future prospects for the nuclear power industry, also announced in May 1994. Preliminary conclusions were published by the Government in August 1994. The final outcome of the review was published in July 1995: *Review of Radioactive Waste Management Policy, Final Conclusions*.[13] The purpose of the review was to examine current policy in the light of changes which had taken place since the 1984 National Strategy was published (see para. 5–05 above). In legal terms it forms part of the guidance given by Government to the Environment Agency and SEPA, providing a policy framework within which those bodies will carry out their regulatory activities in relation to radioactive waste disposal. It is also stated as being intended to inform the United Kingdom's approach to negotiations in bodies such as the IAEA

[12] *The Times*, May 3, 1987.
[13] Cm. 2919 (1995).

and EURATOM on radioactive waste management and to help determine the Government's research programme in future years (para. 5).

Policy aims   5–09

The consultation paper and consultative responses which had preceded
the 1995 Review document indicated widespread support for updating
the six principles or responsibilities for radioactive waste management set
out in the 1977 White Paper (Cmnd. 6820) (see para. 5–04 above).
Suggestions included placing policy on radioactive waste more firmly
within the context of general environmental policy, making reference to
the precautionary principle, greater emphasis placed on sustainable
development, and a firmer line taken on minimising the creation of
waste.

Paragraphs 50–52 of the Review Paper set out new principles for
radioactive waste management policy. These took the starting point that
such a policy should be based on the same central principles as apply
more generally to environmental policy, particularly the principle of
sustainable development. Reference is made to the formulation of that
policy in *Sustainable Development—the U.K. Strategy*,[14] and also to the various
supporting principles contained in that document. These include the
basing of decisions on the best possible scientific information and risk
analysis; the application of the precautionary principle where there is
uncertainty and where potentially serious risks exist; the consideration of
ecological impact, particularly where resources are non-renewable or
effects may be irreversible; and that the cost implications should be
brought home directly to those responsible for impacts (the "polluter
pays" principle).

More specifically, the policy goes on to provide at paragraph 51 that
radioactive waste should be managed and disposed of in ways which
protect the public, the workforce and the environment. In applying the
principle of risk minimisation, the Government recognises that there is a
point where additional costs of further risk reduction will exceed the
benefits arising from improvements in safety. The level of safety, and the
resources required to achieve it, should not be inconsistent with those
accepted in other spheres of human activity. Within this approach, the
Government intends to maintain and develop a policy framework to
ensure that:

(a) radioactive wastes are not unnecessarily created;
(b) such wastes as are created are safely and appropriately managed
    and treated; and
(c) they are then safely disposed of at an appropriate time and in
    appropriate ways,

[14] Cm. 2426 (1994).

so as to safeguard the interests of existing and future generations and wider environments, and in a manner which commands public confidence and takes due account of costs.

The task of ensuring that such a framework is properly implemented falls to the Environment Agencies, and within that framework, the producers and owners of radioactive waste are responsible for developing their own waste management strategies in consultation, as appropriate with the Government, regulatory bodies and disposal organisations. Producers and owners are responsible for bearing the costs of managing and disposing of radioactive waste, including costs of regulation and related research. The Government's policy is that producers and owners should cost radioactive waste management and disposal liabilities before these are incurred and should make appropriate financial provision for meeting them. They should regularly review the adequacy of these provisions. More specifically, the Government advises that they should ensure that:

(a) they do not create waste management problems which cannot be resolved using current techniques or techniques which could be derived from current lines of development;

(b) where it is practical and cost effective to do so, they should characterise and segregate wastes on the basis of their physical and chemical properties and store them in accordance with the principles of passive safety (*i.e.* immobilisation of the waste and minimisation of the need for maintenance, monitoring or other human intervention); and

(c) they should undertake strategic planning, including the development of programmes for the disposal of waste accumulated at nuclear sites, within an appropriate timescale, and for decommissioning of redundant plant and facilities. Such programmes should be discussed with the regulators and must be acceptable to them.

5–10 Waste categories

Reference has been made in Chapter 1 and at paragraph 5–02 above to the various types of radioactive waste arisings, and to the practice of attempting to categorise such wastes. Reactions were mixed to the suggestion made in the Government's consultation document that the current categorisation should be revised to take account of, for example, plant life and activity. The Government had indicated that consideration will be given to the possibility of refining the categories and in particular to the suggestion that shortlife intermediate-level waste might be disposed of at Drigg (which currently only accepts low-level waste), provided that the overall safety case of the site was not compromised.

Radiological protection principles 5–11

The issue of radiological protection is considered in greater detail in Chapter 1. The radiological protection principles underpinning the Government's policy take account of ICRP Publication 60 (1990) and the NRPB's advice to the Government in its 1993 *Board Statement*. The system is based on the following principles:

1. *Justification of a practice* — no practice involving exposure to radiation should be adopted unless it produces sufficient benefit to the exposed individuals or to society to offset the radiation detriment it causes.

2. *Optimisation of protection* — in relation to any particular source within a practice, the magnitude of individual doses, number of people exposed, and the likelihood of incurring exposures should all be kept as low as reasonably achievable, taking account of economic and social practice. In order to limit unfairness or inconsistency which might result from applying inherent economic and social judgments on a case by case basis, the procedure is constrained by restrictions on doses to individuals or on risks to individuals in the case of potential exposures.

3. *Individual dose and risk limits* — exposure of individuals resulting from a combination of all relevant processes should be subject to dose limits, ensuring that no individual is exposed to radiation risks which are judged to be unacceptable in any normal circumstances. Since not all sources are susceptible to control by action at source, it is necessary to specify those sources that are relevant before setting a dose limit.

The Government's Review Paper points out that all these principles, with the exception of the new concepts of dose and risk constraints referred to at (2) above, already form the basis of radiological protection in the United Kingdom. The new methodology recommended by ICRP 60 for calculating doses has been adopted for authorisation of discharges in the United Kingdom, and formal implementation is now possible following the adoption of the revised EURATOM Basic Safety Standards Directive 96/29 in May 1996. The ICRP principles recognise the need to adopt different approaches. This stems from the fact that where radioactive waste is routinely released in liquid or gaseous form into the environment, radiation exposure of the public is certain to result, even if at low levels; an estimate of radiation dose will therefore be used in the process of setting the limits. In the case of storage and disposal of solid waste however, exposure of the public is not certain. It is therefore important to consider the measure of risk involved, based upon radiation doses, the likelihood of the event giving rise to it, and to have risk criteria in the licensing and authorisation of such facilities (para. 57). The criteria adopted are considered separately in greater detail below in relation to the authorisation process.

5–12 Specific policies

Chapter 5 of the Government's 1995 Review[15] contains specific policies relating to particular aspects of radioactive waste management from which the following points emerge:

1. *Spent fuel management*

   The Government's policy is that the question of whether to re-process spent fuel, and if so when, is a matter for the commercial judgment of the owner of the fuel. The specific issue of the siting of dry stores for spent fuel was raised at a public inquiry into an application made by Scottish Nuclear Limited under section 36 of the Electricity Act 1989, to construct a dry store for spent fuel at its AGR Power Station at Torness, East Lothian. The Reporter (Inspector) at the public inquiry concluded that the proposal represented a sound engineering solution, but recommended that before any consent was issued for a dry store, the Government should consider the need for a national strategy, and in particular whether a single site or multi-site approach should be taken. The Secretary of State accepted this recommendation. The Secretary of State announced on February 21, 1995 that the results of an appraisal of the alternative structures did not point to conclusive benefits deriving from a single central store, as compared with a number of stores sited beside nuclear generating stations. The siting of dry stores for spent nuclear fuel therefore remains a matter for the commercial judgment of individual operators (paras. 88–90).

2. *High-level waste*

   Whilst reaffirming its policy that high-level waste should be stored for a minimum of 50 years to allow for cooling and the decay of short-lived radionuclides, the Government believes that positive steps should not be taken to consider the ultimate destination of such waste. The Government's view is that disposal of high-level waste to geological formations on land remains the favoured operation for the long term, once the waste has been allowed to cool. In this respect, the direct disposal of spent fuel does not present fundamentally different technical problems from the disposal of other high level radioactive waste. The Government has indicated its intention to put in hand steps to develop and implement the necessary research

---

[15] Cm. 2919 (1995).

strategy for the United Kingdom, its programme of geological studies relating to deep underground disposal having been discontinued in 1981 (see para. 5–04 above). The Government's objective is to produce a United Kingdom national statement of future intent, setting out the decisions to be taken and the milestones to be achieved in developing a high level waste repository. Whilst the statement of policy is a matter for the Government, its implementation will fall to the owners of the waste and to the regulators (paras. 91–93).

3. *Partitioning and transmutation*

This is an alternative means which has been proposed for dealing with some types of waste whereby certain long-lived and toxic radionuclides will be partitioned or separated and transmuted, using portable accelerators or reactors, into radionuclides with a shorter half life. By this means, the long term hazards presented by the waste may be reduced. Whilst the United Kingdom carried out various studies into the subject in the late 1970s and early 1980s, the results were not encouraging. However, research is currently being carried out elsewhere, and the United Kingdom Government will continue to watch with interest the results of this work. The Government has no plans to initiate its own research into this subject (para. 94).

4. *Storage and intermediate-level waste*

The Government continues to favour a policy of deep disposal rather than indefinite storage for intermediate-level waste and considers it appropriate that Nirex should continue with its programme to identify suitable sites. The Government says that there is no advantage to be gained from delaying the development of a repository, and that once a suitable site has been found, it should be constructed as soon as is reasonably practicable. The Government will follow its earlier decision to hold a public inquiry into any disposal (paras. 95–110); but the issue of site selection is considered more fully elsewhere (see para. 5–59 below). The Government's decision to press forward towards the construction of a repository is underpinned by the view of the RWMAC, that safe final disposal of radioactive waste can be achieved on the United Kingdom mainland and that responsibilities towards future generations are best discharged by final rather than deferred disposal. However, given the length of time needed to develop this facility and the period during which it would remain operational, the option to retrieve material would not be lost until final closure of the repository, probably 50 years or so after it came into operation. In considering the safety case for any such repository, it would need to be demonstrated that the continued safety of future generations would not depend upon monitoring, surveillance, preventive or remedial actions after closure of the facility.

5. *Interim storage of intermediate and low-level waste*

Estimates by Nirex suggest that the earliest time by which a repository for intermediate-level waste could be available is the year 2010. Such waste, therefore, will have to remain in interim storage for some time to come, as will such types of low-level waste as are inappropriate for disposal at Drigg. The 1984 National Strategy contained a presumption against the conditioning of stored waste, on the grounds that treatment could prove incompatible with the possible options of long-term disposal. The Government has reviewed this presumption, and believes that where the demands of safety are overriding, waste should be treated as necessary to improve storage conditions. Thus, the general presumption against treatment may be relaxed where early treatment will secure either worthwhile safety benefits, or worthwhile economic benefits without prejudicing safety (paras. 111–113).

6. *Controlled burial*

The review refers to the practice of "controlled burial" whereby some types of low-level waste are buried at suitable landfill sites, used mainly for other kinds of waste or, (more rarely) at the site where the waste is produced. This form of disposal is used by non-nuclear industries processing raw materials containing natural radioactivity, by hospitals and universities for their relatively more active waste streams, and by BNFL for waste from its uranium enrichment plants and for lightly contaminated excavation spoil. For hospitals the position is complicated by the fact that the waste may also be clinical waste, which may restrict available disposal options. The practice is a politically controversial one, especially at a local level, as demonstrated by initial press coverage in October 1994 of the possibility of waste from nuclear power stations being disposed of at the 39 sites already licensed to take radioactive waste from hospitals and research establishments.[16]

The Government's consultation documents suggest that, in order to relieve pressure on the disposal capacity of the low-level waste facility at Drigg, waste producers could be encouraged to make greater use of controlled burial. This proposal sparked a difference of opinion between, on the one hand, the RWMAC and the NRPB, both of whom felt that controlled burial could be expanded subject to adequate safeguards and, on the other hand, local authorities and environmental groups who were opposed to any extension of this route. The Government takes the view that there are sound economic and radiological grounds for encouraging greater use of controlled burial, but also recognises the genuine anxieties that such proposals arouse amongst local residents. For that reason, the Government has

[16] *The Observer*, October 9, 1994.

decided not to encourage greater use of controlled burial by the nuclear industry—a significant example of public perception taking precedence over technical and economic considerations. The Government believes that controlled burial should continue to be available as a disposal route, particularly for users of materials such as hospitals, universities, research laboratories and for non-nuclear industry. It also believes that no change is necessary for a system where primary control over such disposal is exercised through authorisation under the Radioactive Substances Act 1993 and concludes it is unnecessary to duplicate this authorisation procedure by imposing additional controls on the relevant landfill site operators (paras. 114–119). This issue is considered further below at para. 5–21. However, the Government accepted that it could be difficult to identify from public registers which sites were taking radioactive waste and invited proposals for making such information more transparent.

7. *Decommissioning strategies*

The Government's general policy is that in general, the process of decommissioning nuclear plant should be carried out as soon as is reasonably practicable, taking account of all relevant factors. The Government intends to ask all nuclear operators to draw up strategies for decommissioning their redundant plant, including justification of the timetable proposed and demonstration of the adequacy of the financial provision being made to implement the strategies. The decommissioning process itself will be subject to conditions attached to the nuclear site licence (see para. 2–43 above), whereas the wastes arising through decommissioning are subject to regulation under the Radioactive Substances Act 1993 (paras. 121–126). The Government also recognises that in addition to nuclear power stations, a variety of other nuclear facilities are in the process of being decommissioned, or will be decommissioned in the future. As with power stations, decisions relating to these facilities will be taken on a case by case basis (para. 127). In relation to decommissioned nuclear-powered submarines, the current policy of the Ministry of Defence is that they be stored afloat in safe and secure facilities at the naval bases of Devonport and Rosyth; indeed a number of submarines are already stored in this way. Ministry of Defence plans for long-term disposal of the radioactive waste arising from reactor compartments are predicted on the availability of the Nirex intermediate-level repository in about 2010. This policy is being kept under review.

8. *Decommissioning—financial provision*

The Government's consultation document which preceded the review noted the considerable concern that existed about the likely cost of decommissioning, but concluded that it should be for the industry itself to continue to make its own arrangements for financial provision in this respect. Various responses to the consultation exercise argued that external, segregated funds should be introduced, managed by independent trustees and invested in Government bonds or similar low-risk securities. Having considered these responses, the Government's view is that for those parts of the nuclear industry being privatised, segregated funds for decommissioning should be established. Accordingly, the Government is to examine which improvements can be made to the way in which the unprivatised sections of the industry report on their progress towards decommissioning and on their financial provision policies (paras. 129–131).

9. *Waste substitution*

Since 1976, all BNFL contracts for reprocessing spent fuel from foreign companies have included options for the return of wastes arising during reprocessing. It is Government policy that such options be exercised and that the waste be returned. In relation to high-level waste, the proposal is to return all such waste arising as soon as practicable after the reprocessing. However, since 1986 the Government has been considering the possibility of waste substitution, whereby lower level waste might be substituted for an equivalent quantity, in radiological terms, of higher level wastes. In 1992, BNFL proposed to offer its overseas customers the option of substituting an equivalent amount of high-level waste, and retaining the intermediate and low-level waste in the United Kingdom. This would have the effect of reducing significantly the volume of waste to be returned and would greatly reduce transport costs.

The RWMAC was asked to consider whether substitution could result in environmental detriment to the United Kingdom. Consideration was given to the proposed "integrated toxic potential" (ITP) system proposed by BNFL to establish radiological equivalence between different waste categories. The RWMAC, whilst recognising the many doubts that existed in the area, agreed that ITP was a supportable means of quantifying substitution and that the substitution policy based on ITP would be broadly neutral in terms of radiological impact. The idea, however, received considerable opposition in the review process on the basis that the United Kingdom should not agree to take extra quantities of radioactive waste, and that to do so would run counter to the principle of self-sufficiency in waste management. The Government's ultimate conclusion was that a fully rigorous appraisal of comparative environmental effects could not take place in the case of intermediate-level waste until a specific site and finalised design concept for a proposed deep underground disposal facility had been agreed. Until then, it will be imprudent for the United Kingdom to become irrevocably committed to retaining wastes in respect of which there is the contractual option to return.

The practical effect is that BNFL is engaged in waste substitution for low-level waste; any arrangements entered into with overseas customers in relation to intermediate-level waste must be conditional upon proper calculation of equivalence, following grant of planning permission for the Nirex repository, and on the ability to return intermediate-level waste to the country of origin should the Nirex repository not be established by the time the contractual obligation to take back the waste applies, *i.e.* 25 years after the waste is initially generated (paras. 134–141).

10. *Imports and exports of waste*

The Government's preliminary review suggested that policy towards the import and export of radioactive waste should be broadly similar to that of other waste, namely a presumption of self-sufficiency, but with some flexibility in view of the highly specialised nature of the waste itself. The Government regards such a policy as consistent with the IAEA's *Code of Practice on the International Transboundary Movement of Radioactive Waste* which provides for the sovereign right of every state to prohibit the movement of radioactive waste into, from, or through its territory. As might be anticipated, the issue provoked a clash of views between the nuclear industry (in favour of the receipt of shipments of waste from countries which did not have the technology to deal with radioactive waste adequately) and environmental groups and local authorities (opposed to further imports and viewing the exceptions proposed by the Government as unduly relaxed).

The Government's general policy following the review is that radioactive waste should not be imported to, or exported from, the United Kingdom except for the recovery of useful materials (provided this is the genuine prime purpose), or for treatment which will make the subsequent storage or disposal more manageable, in cases where the processes are at a redevelopment stage, or which involve quantities too small for the processes to be practicable in the country of origin. Where such processes would add materially to the waste needing to be disposed of in the United Kingdom, the presumption should be that they will be returned to the country of origin. Additionally, Government policy is that waste should be imported for treatment and disposal in the United Kingdom where it is in the form of spent sources which were manufactured in the United Kingdom, or it is waste from small users (*e.g.* hospitals) situated in E.C. Member States which produce such small quantities of waste as to make provision of their specialised installations impracticable, or from developing countries which cannot reasonably be expected to acquire disposal facilities. The Government recognises that the difficulty here will lie in refining the detail of the policy, and as a first step proposes to invite the RWMAC to consider what detailed items might be prepared (paras. 142–147).

11.   *Small Users*

The Government's consultation document indicated that small users of radioactive material (*e.g.* hospitals, universities, research laboratories and non-nuclear industries) should be responsible for the safe management of their waste. At the same time, it recognised that such small users sometimes experience difficulties in finding suitable disposal routes. One issue to be considered was whether the Government should take a more prescriptive approach in directing the use of specific disposal routes and, as a result a consultation exercise was carried out in conjunction with small users on this issue.

This exercise revealed concerns over the continuous availability of certain routes used by many small users in the past, but now increasingly in short supply, for example, incinerators and landfill. In the light of these responses, the Government has reaffirmed its preliminary conclusion that small users should not be directed to particular routes, but should remain free to make their own arrangements. At the same time, the Government recognises that careful control is needed and encourages suppliers of radioactive substances to consider what further action they might take to assist small users in finding disposal routes. The Government is to ask the Environment Agencies to maximise the consistency of their regulatory approach to small users and to provide them with advice on the management and appropriate disposal of radioactive wastes. The Government's view is that the creation of any special body to offer advice and promulgate good practice should be a matter for small users themselves rather than being imposed by the Government or by regulatory bodies (paras. 148–154).

12.   *Research*

The Government has confirmed its view that each of the component parts of the industry, regulatory bodies and Government should continue to be responsible for the commissioning and funding of the research and development necessary to support their respective functions in relation to radioactive waste management, and that they should do so on the basis of clearly stated aims and objectives. It was generally agreed that there is a need for suitable liaison to ensure that there are no unnecessary gaps or overlap in the respective research programmes; the existing Radioactivity Research and Environmental Monitoring Committee (RADREM) provides a suitable forum for such liaison. The Government also recognises that in addition to the research support for day to day work, there is a need for basic research of a more strategic and long-term nature, in which the Research Councils should be involved. The Government advises that it is for individual sponsors of research to decide how best to take up any opportunities for obtaining support for their research under the *1994–98 Framework Programme for the European Atomic Energy Committee* (UAEC). At the same time, the lack of any clear research strategy has been an issue subject to criticism over the years for the various bodies, most recently by the British Government Panel on Sustainable Development. In its Second Report (January 1996), the Panel welcomed the existence of a research strategy for high-level waste, but suggested the need for a strategy including intermediate-level waste and exploring all options for disposal, drawing on research carried out elsewhere. The Panel suggested that the Government must assume responsibility for ensuring that any such research strategy is effectively implemented (p. 18).

INTERNATIONAL ASPECTS

The Role of the International Atomic Energy Agency (IAEA)   5–13

The IAEA has taken an interest in developing safety objectives for the management of radioactive waste, through its Safety Series documents. *Safety Series No. 69* provides a *Code of Practice for the Management of Radioactive Waste from Nuclear Power Plants.* The Code of Practice defines the minimum requirements for design and operation of structures, systems and components for the management of radioactive waste from nuclear power plants. Its approach is to emphasise which safety requirements shall be met rather than specifying how such requirements can be accomplished (para. 1.1.1). The Code proceeds on the following established basic principles:

1. *Justification*

   In order to prevent unnecessary exposure, no practice involving exposure to ionising radiations can be authorised unless the introduction of the practice produces a positive benefit. This requirement must be applied to the practice of generating waste, and not to the subject of waste management and isolation.

2. *Optimisation*

   The design, planning and subsequent use of the operation of those forces and practices shall be performed in a manner to ensure that exposure is as low as reasonably achievable (ALARA), economic and social practices being taken into account.

3. *Dose Limitation*

Doses to individuals shall not exceed the limits recommended for the appropriate circumstances by the relevant authorities.

Section 4 of the Code of Practice describes in general terms principles for the design of waste management systems; section 5 deals with their operation; and section 6 with their surveillance and monitoring. The general guidance is underpinned by a number of Technical Reports from the IAEA, for example, *Technical Reports Series No. 198: A Guide to the Safe Handling of Radioactive Waste from Nuclear Power Plants.*

Another important general document is the IAEA *Code of Practice on the International Transboundary Movement of Radioactive Waste* (1989), which has been referred to in the context of United Kingdom policy above (see para. 5–12 above). This Code of Practice takes the form of a resolution of the General Conference of the Agency (GCXXXII) RES/509; its underlying concern is the awareness of the danger of unauthorised transboundary movements of radioactive waste, particularly to the territory of developing countries. The basic principles of the Code are that states should take appropriate steps to ensure safe management and disposal of radioactive waste within their territory; to minimise the amount of radioactive waste, taking into account social, environmental, technological and economic considerations; that it is the sovereign right of every state to prohibit the movement of radioactive waste into, from, or through its territory; and that all states involved in international transboundary movements of radioactive waste should take the appropriate steps necessary to ensure that such movement is undertaken in a manner consistent with international safety standards. Principle 5 relates to prior informed consent, *i.e.* every state should take the appropriate measures necessary to ensure that, subject to the relevant norms of international law, the transboundary movement of radioactive waste takes place only with the prior notification and consent of the sending, receiving and transit states. Principle 7 provides that no receiving state should permit radioactive waste to be received unless that state has the administrative and technical capacity and regulatory structures to manage and dispose of such waste in a manner consistent with international safety standards: (on the control of radioactive waste movements, see further Chapter 4).

A final IAEA document which should be specifically mentioned is *Safety Series No. 99* dealing with *Safety Principles and Technical Criteria for the Underground Disposal of High Level Radioactive Waste.* This document reflects the need for internationally harmonised criteria for the safe underground disposal of high-level waste and sets out a basic safety philosophy for use in planning such disposals. The two overlying objectives for underground disposal of high-level waste set out in the paper are firstly, the isolation of such waste from the human environment over long timescales without relying on future generations to maintain the integrity of the disposal system, or imposing upon them significant

220

constraints by the existence of the facility; and, secondly, to ensure the long term radiological protection of human beings in accordance with current internationally agreed radiation protection principles. The first principle is known as "responsibility to future generations" and proceeds on the premise that since present generations benefit directly from the exploitation of nuclear energy, it is reasonable that they should bear the financial burden of waste disposal. The publication contains technical criteria to assist in ensuring compliance with the safety principles, including the primary criterion of an overall systems approach, assessing the basis of the performance of the disposal system as a whole and relying on the concept of multi-barriers.

An important recent development is the formulation by the IAEA, at the request of member states, of a set of documents containing standards and criteria for the management and disposal of radioactive waste (the RADWASS programme). This entails the preparation and publication of 55 documents covering planning, pre-disposal, near surface disposal, geological disposal, uranium and thorium mining waste, and decommissioning. The documents comprise a general *Safety Fundamentals* document dealing with overall principles of radioactive waste management, and specific *Safety Standards* documents on various subject areas, supplemented by the more specific and practical *Safety Guides* and *Safety Practice Documents*.

International rules on sea disposal   5–14

Whilst the United Kingdom Government believes that sea disposal for low-level, solid and bulky radioactive waste can be the BPEO for wastes arising from the decommissioning of power stations and other nuclear plants,[17] the international community has increasingly set itself against this option. The difficulties encountered by the United Kingdom Government in relation to resolutions under the London Dumping Convention 1972 have been referred to previously. At their 26th meeting in November 1993, the parties to the London Convention decided to ban the dumping of all forms of radioactive waste in the open sea, the ban being effected by resolution LC.51 (16) amending Annexes I and II of the Convention.[18]

The background to this development is that the UN Convention on Environment and Development in Agenda 21, Ch. 22.5(b), had urged the contracting parties to the London Convention to expedite the completion of studies on replacing the previous voluntary moratorium on the disposal of low-level radioactive waste by a legal ban. The parties decided to effect an immediate ban in relation to all levels of radioactive waste at their 26th meeting, rather than wait for wholesale revision of the

---

[17] *Review of Radioactive Waste Management Policy—Final Conclusions*, Cm. 2919 (1995), para. 14.
[18] See further: *Year Book of International Environmental Law* (O.U.P. 1993), Vol. IV, p. 188.

Convention. The changes are, therefore, made by replacing the reference to high-level radioactive waste at paragraph 6 of Annex I to the Convention with a reference to "radioactive waste and other matter". The wastes and other matter listed in Annex I are prohibited from dumping at sea (Art. IV.1(a) of the Convention). Annex II is correspondingly amended by deleting the reference to lower level radioactive wastes and other radioactive matter, the dumping of which had previously required a prior special permit under Article IV.

The United Kingdom was one amongst the five of 42 state parties who abstained in relation to the ban (the others being France, Belgium, Russia and China). The Government's position is as announced in February 1994, *i.e.* it accepts an indefinite ban on the disposal of low and intermediate-level radioactive wastes at sea, but is ready to re-open discussion in the Convention at any time should the weight of opinion change. Meanwhile, the Government will continue its own programme of monitoring and research on the issue.[19]

In September 1992, the United Kingdom signed the Convention for Protection of the Marine Environment for the North East Atlantic (OSPAR), Cm. 2265, which, when ratified, will replace the 1972 Oslo and 1974 Paris Conventions (Art. 31). The Convention bans the disposal of low and intermediate-level radioactive wastes, but includes an option for France and the United Kingdom to resume the practice, subject to certain conditions, after a period of fifteen years from January 1, 1993 (Annex II, para. 3). France and the United Kingdom are, under this provision, required to report to the meeting of the Commission in 1997 on the steps taken to explore alternative land-based options. Unless within that period of 15 years the Commission decides unanimously not to continue the exception, it must take a decision on the prolongation of the prohibition for a further period of 10 years after January 1, 2008. Furthermore, should France and the United Kingdom still wish to retain the option, they are required to submit further reports on the progress of land-based options and studies to show that sea dumping would not result in hazards to human health, harm to living resources or marine ecosystems, damage to amenities, or interference with other legitimate uses of the sea.

EUROPEAN COMMUNITY REQUIREMENTS

5–15 General

Article 37 of the EURATOM Treaty creates obligations concerning radioactive waste, requiring each Member State to provide the Commission with such general data relating to any plan for the disposal of radioactive

---

[19] *Review of Radioactive Waste Management Policy—Final Conclusions*, Cm. 2919 (1995), para. 16; and *MAFF News Release 55/94* (February 1994).

waste in whatever form, as will make it possible to determine whether the implementation of such a plan is liable to result in the radioactive contamination of the water, soil or air space of another Member State. For such an important issue, the wording of Article 37 is excessively vague; this is not altogether surprising, since it was drafted in the 24 hours preceding the signature of the EURATOM Treaty.[20] Opinions under Article 37, as described below, tend to follow a standard format. Whilst the procedure is one of the few means available for the Commission to obtain an up-to-date picture of radiological protection issues nationally, it also has serious weaknesses. The opinion has no legal status and is issued only a short time before the plant is authorised; it is very rare for opinions to conclude that discharge limits for liquid effluent are liable to result in radioactive contamination in other Member States. Ultimately, the Member State to whom the opinion is issued is not legally bound by it, and may decide that it disagrees with the conclusions, or that adequate safeguards are in place.[21]

The application of this requirement was considered in relation to the THORP nuclear reprocessing plant of Sellafield. The Commission issued an Opinion on April 30, 1992 (92/269/EURATOM) that implementation of the proposals for disposal of radioactive waste from THORP was not liable, either under normal operations or, in the case of an accident of the type considered in the general data provided by the United Kingdom, to result in radioactive contamination which was significant under Article 37 (the nearest other Member State being the Republic of Ireland). The Opinion referred to the fact that the distance from the nearest Member State was 180km; that exports of locally produced foodstuffs were of minor importance; that solid wastes produced would be stored on-site pending disposal; and that normal and unplanned discharges of radioactive effluent would not be liable to cause significant exposure in other Member States, taking into account all possible pathways, particularly the consumption of seafoods.[22] Similar considerations were relevant in the decision relating to the planned disposal of waste from the Sizewell B nuclear power station.[23] In Opinions relating to the nuclear power stations at Heysham[24] and Torness,[25] the Commission took note of expert opinion that maximum proposed discharge limits for liquid effluent appeared to be unnecessarily high, and recommended that they be fixed at levels taking into account the ALARA principle.

Council Recommendation 91/4/EURATOM of December 7, 1990 made certain changes to the procedures under Article 37, requiring

[20] *Application of Article 37 of the EURATOM Treaty—Experiences Gained: 1959–1981*, Com. (82) 455 Final, p. 4.
[21] See evidence of: Mr Fraser (European Commission), *Nineteenth Report of the House of Lords Select Committee on the European Communities: Radioactive Waste Management* (Session 1987–1988), H.L. Papers 99–I, p. 510.
[22] [1992] O.J. L138.
[23] Dir 92/537/EURATOM: [1992] O.J. L344.
[24] Dir 87/170/EURATOM: [1987] O.J. L68/33.
[25] Dir 87/530/EURATOM: [1987] O.J. L189/42.

communication of general data not less than six months before the planned date of commencement of disposal of radioactive waste. This period was intended to allow the Commission to issue its Opinion and for the content of the Opinion to be taken into account prior to disposal beginning. In Case C–187/87, *Saarland v. Minister of Industry*,[26] the European Court of Justice ruled that Article 37 must be interpreted as meaning that the Commission of the European Community is to be provided with general data relating to plans for disposing of radioactive waste before such disposal is authorised by the competent authority of the Member State concerned. The case arose from a long-running dispute concerning the French nuclear installation at Cattenom, close to the borders with Luxembourg and the FRG.[27]

5–16 European action plans on radioactive waste disposal

Council Decision 75/406/EURATOM of July 9, 1975 adopted a programme on the management and storage of radioactive waste. This was followed by three important Council Resolutions of February 18, 1980, the first of which relates to the implementation of a Community plan of action in the field of radioactive waste.[28] This resolution referred to the fact that the Council had already approved Community environment programmes and research and development programmes in the field of management and storage of radioactive waste, but pointed out that these programmes needed to be supplemented in relation to matters of a legal, administrative, financial and social nature. This first plan ran from 1980 to 1992 and was based on five points:

(1) continuous analysis of the situation in terms of techniques available, technological research and management practices;
(2) examination at Community level of measures to ensure the long-term or permanent storage of radioactive waste under optimum conditions;
(3) consultation on practices concerning the management of waste;
(4) continuity of Community research and development work during the plan;
(5) provision of regular information to the public.

The second resolution of February 18, 1980 related to the reprocessing of irradiated nuclear fuels[29] and recorded the agreement of the Council that it was in the interests of the Community and its Member States to keep open the option of recovering and re-using spent fuel discharged from nuclear reactors. The third resolution of the same date[30] presented the

---

[26] [1989] 4 C.M.L.R. 529.
[27] Lenaerts, *Nuclear Border Installations: A Case Study* [1988] E.L.Rev. 158.
[28] [1980] O.J. C51/1.
[29] [1980] O.J. C51/4.
[30] *ibid.*

Advisory Committee for the Management and Storage of Radioactive Wastes with the additional task of advising the Commission in connection with implementation of the Community Plan of Action.

The Community's Plan of Action for Radioactive Waste was renewed by a Council Resolution of June 15, 1992.[31] The renewed plan runs from 1993 to 1999 and is reviewable every three years. The initial five points of the plan adopted in 1980 have now become seven points, namely:

1. Continuous analysis of the situation and prospects in the field of radioactive waste management in Member States, including the status of research and technological development, a list of storage installations intended for construction, and the list of management practices and strategies defined by Member States.

2. Development of technical co-operation in the Community in relation to long-term or final disposal of radioactive waste.

3. Concerted action on the safe management and storage of radioactive waste, which should make it possible to develop a common approach and achieve harmonisation on strategies and practices.

4. Consultation on management practices and strategies in the context of abolition of frontier controls within the Community.

5. Continuity of interaction between the research programmes and administrative, legal and regulatory issues. Regular consultations are to be held within the Advisory Committee on the plan to provide a single framework for considering possible improvements to techniques for final storage and on the legal, administrative and social problems to be solved.

6. Information for the public, whereby Member States are required to continue and intensify their efforts to provide the public with regular information on their activities in the field of radioactive waste management and storage by drawing up, as far as possible, a common information strategy.

7. Development of international consensus, thereby promoting concerted action by Member States within international organisations such as the IAEA, NEA and ISO.

The need for continued co-operation is emphasised by the Council Resolution of December 19, 1994, on Radioactive Waste Management.[32] This Resolution takes as its starting point the view that for the protection of Community citizens and the environment, comprehensive policies

---

[31] [1992] O.J. C158/3.
[32] [1994] O.J. C379/1.

covering all stages of the nuclear fuel cycle are needed for the management of radioactive waste, and that such policies should dovetail smoothly with the Community's general waste policy. The Resolution endorses the view that each Member State is responsible for ensuring that radioactive waste produced on its territories is properly managed, but also considers that optimum use should be made of facilities at national level, which may involve co-operation between Member States. The Resolution also considers that recycling and re-use of materials with low levels of radioactive contamination are options to be explored further and calls on the Commission to continue its work in order to help determine the conditions of such recycling and re-use.

5–17 Transfrontier movement of radioactive waste

The 1994 Council Resolution on Radioactive Waste Management reaffirmed that shipments of radioactive waste between Member States and into and out of the Community must continue to be subject to appropriate controls.

Directive 92/3/EURATOM deals with the supervision and control of shipments of radioactive waste between Member States and into and out of the Community. This Directive is considered in Chapter 4 within the context of transport of radioactive materials and waste (see paras. 4–34 and 4–35 above).

CONTROL UNDER THE RADIOACTIVE SUBTANCES ACT 1993

5–18 Introduction

The legislative origins of the Radioactive Substances Act 1993 (the "1993 Act") have been discussed in Chapter 4. Sections 13 and 14 respectively provide control over the disposal and accumulation of radioactive waste by the requirement of authorisation.

As with the system for registration of the use of radioactive material, from April 1, 1996, the relevant enforcing authorities became the Environment Agency (for England and Wales) and SEPA (for Scotland), and the 1993 Act is amended to reflect this. As described below, the 1993 Act has also been amended to change the procedure in relation to authorisations for the accumulation or disposal of radioactive waste on or from nuclear sites.

Both the Environment Agency and SEPA will exercise their functions within the detailed statutory framework of powers and duties under Part I of the Environment Act 1995, including the principal aim under section 4 of so protecting or enhancing the environment, taken as a whole, as to make the contribution that Ministers consider appropriate towards achieving sustainable development. In the context of radioactive waste management this will involve following the guidance contained in the *Review of Radioactive Waste Management Policy.*[33]

Radioactive waste   5–19

Section 2 defines radioactive waste as waste which consists wholly or partly of:

(a) a substance or article which, if it were not waste, would be radioactive material; or

(b) a substance or article which has been contaminated in the course of the production, keeping or use of radioactive material, or by contact with or proximity to other waste falling within paragraph (a) above.

Under subsection 47(1), "waste" includes any substance which constitutes scrap material or an effluent or other unwanted surplus substance arising from the application of any process, including any substance or article which requires to be disposed of as being broken, worn out, contaminated or otherwise spoilt.

There are, therefore, two main components to the definition: first, the material must be waste and second, it must consist wholly or partly of material satisfying the criteria for "radioactive material" in section 1 (see para. 4–05), or of a substance or article contaminated in the course of producing, keeping or using radioactive material, or by contact with radioactive waste. The definition is, in this second aspect, a broad one since it can include materials which are not of radioactive origin (*e.g.* wood, concrete, metal, textiles, rubber, soil, etc.) but which have been exposed to radioactivity. The contamination may take the form of the absorption, admixture or adhesion of radioactive matter, or the effect of the emission of neutrons or ionising radiations; the material must as a result have become radioactive or must possess increased radioactivity (subs.47(5)). The potential problems of adhesion and adsorption of radioactive materials to other substances was clearly recognised when the Radioactive Substances Act 1960 was being framed (*e.g.* the adsorption of radioactive material to sand or sediment, or to sewage sludge).[34] Since waste materials cannot, by definition, be "radioactive material", the definitions in sections 1 and 2 are mutually exclusive.

The more difficult component in the definition of radioactive waste relates to the requirement that the substance or article be waste. Radioactive waste is excluded from the E.C. Framework Directive on waste 75/442/EEC, as amended by 91/156/EEC (Art. 2(1)(b)(I)

---

[33] Cm. 2919 (1995).
[34] *Hansard*, H.L. Vol. 219, cols. 886 & 911.

excludes radioactive waste where such waste is already covered by other legislation). Nor is it within the waste management regulatory scheme of Part II of the Environmental Protection Act 1990.[35] The definition of "waste" provided by section 47 of the 1993 Act (and, before it, by the Radioactive Substances Act 1960) was consistent with the definition contained in section 75 of the 1990 Act and, before it, section 30 of the Control of Pollution Act 1974. However, whereas the definition in the 1990 Act has been replaced with a definition explicitly more consistent with European waste legislation, the definition in the 1993 Act has remained unchanged, and consequently may be interpreted in the light of decisions on the definitions previously applying in the 1974 and 1990 Acts.

In this respect, a number of relevant points are discussed below.

1. Subsection 47(4) of the 1993 Act provides that any substance or article which, in the course of carrying on any undertaking, is discharged, discarded or otherwise dealt with as if it were waste, shall be presumed to be waste unless the contrary is proved.

2. On the corresponding definition in section 30 of the Control of Pollution Act 1974, the courts have held that the correct approach is to regard the material from the point of view of the person who produces it, *i.e.* is it something made as a product or by-product or is it useless? If the latter, it will be waste even if it has economic value or use to some other party. The relevant judicial decisions to that effect were discussed at paras. 14–17 of DoE Circular 14/92 on *The Controlled Waste Regulations 1992,* namely: *Berridge Incinerators v. Nottinghamshire County Council*[36] and *Kent County Council v. Queensborough Rolling Mill Co. Ltd.*[37] To the same effect, see also *Cheshire County Council v. Armstrong's Transport (Wigan) Ltd*[38]; and *Meston Technical Services Ltd v. Warwickshire County Council.*[39]

3. Although Circular 14/92 has been superseded in relation to its original purpose of helping to define non-radioactive waste, the questions suggested in that guidance still seem apposite to the definition in section 47 of the 1993 Act, *i.e.*:

    (a) is the material what would ordinarily be described as waste?
    (b) is it a scrap material?
    (c) is it an effluent or other unwanted substance arising from the application of any process?
    (d) does it require to be disposed of as being broken, worn out, contaminated or otherwise spoilt?

[35] Environmental Protection Act 1990, s.78.
[36] 1987; unreported.
[37] 89 L.G.R. 306; (1990) 154 J.P. 530.
[38] [1994] Env.L.R. D21.
[39] [1995] Env. L.R. D36.

(e) is it being discarded or dealt with as if it were waste?

In line with the wording of section 47, if any of these questions is answered "yes", the material is *waste*.

4. There is a further important presumption contained in subsection 14(4) and relating to the accumulation of waste, which is considered at para. 5–20 below.

5. Subsection 47(1) defines "substance" to mean any natural or artificial substance, whether in solid or liquid form or in the form of a gas or vapour. Radioactive waste can (as has been described at para. 5–02 above) take all of these forms.

6. In practice, there will generally be little doubt as to whether a material is radioactive waste. From the way in which the material is being treated and handled, this will usually be obvious; nor will there generally be much scope for beneficially re-using materials contaminated by radiation without specialist decontamination. Nonetheless, both the E.C. and the United Kingdom Government recognise the possibility of recovering re-usable materials from radioactive waste: see, for example, the Government's policy on imports of waste at paragraph 145 of its *Review of Radioactive Waste Management Policy*.[40]

Disposal and accumulation   5–20

"Disposal" in relation to waste is defined to include its removal, deposit, destruction, discharge or burial (subs.47(1)). It is stated to cover discharges into water or into the air, or into a sewer or drain or otherwise. "Burial" may be underground or otherwise.

"Accumulation" is not defined in the 1993 Act, and the question arises as to how the concept relates to those aspects of disposal which involve accumulating waste as opposed to dispersing it, *i.e.* deposit and possibly burial. The *Guide to the Administration of the Radioactive Substances Act 1960* drew a distinction for practical purposes between "storage" and "disposal". "Storage" was seen as emplacement in a facility with the intention of taking further action at some future time, and in such a way that such action is feasible. "Disposal" on the other hand, involves dispersal into an environmental medium or emplacement in a facility, with the intention of taking no further action apart from any monitoring which may be thought desirable on technical grounds or to provide reassurance.

The most obvious case of accumulation will be where waste is kept at the premises where it is produced, pending its disposal elsewhere. In the context of radioactive waste management, this could be for very

---

[40] Cm. 2919 (1995).

protracted periods. Subsection 14(4) contains an important presumption: where radioactive material (as defined by s.1) is produced, kept or used on any premises, and any substance arising from the production, keeping or use of that material is accumulated in a part of the premises appropriated for the purpose for three months or longer, the substance is presumed unless the contrary is proven:

(a) to be radioactive waste; and
(b) to be accumulated on the premises with a view to the subsequent disposal of the substance.

Waste can also be accumulated on premises other than those where it is produced, *e.g.* waste being subjected to or awaiting treatment or processing, or simply being allowed to cool and decay, could fall into this category. One potentially difficult area arises as to the point at which accumulation becomes disposal. If waste is buried and covered (as at Drigg, for example), there is no intention to retrieve it, though obviously it could be dug up and retrieved if circumstances dictated. The burial is clearly disposal rather than accumulation.

With some types of facility, however, the position may for some time remain ambivalent. An example could be the proposed repository for intermediate-level waste to be constructed by Nirex. As referred to above (see para. 5–12), during the lifetime of the repository, which could be as long as 50 years, there could be the option of retrieving the waste until final closure of the facility. Indeed, retrieval would still be available after closure.[41] Essentially, in such a case there would appear to be two options. First, the facility could be authorised for disposal at the outset, dealing with any interim accumulation within the authorisation; in that case no separate authorisation to accumulate will be needed (subs.14(2)). The second option is to treat the facility initially as a site for the accumulation of waste, until such time as it can be demonstrated that the facility is ready to be closed and that the wastes can be left there with no further action needed other than monitoring; at this point, the accumulation would become disposal and would require authorisation as such.

It should be noted that to avoid duplication of control an authorisation for disposal may also require or permit waste to be accumulated with a view to its subsequent disposal, and that in such a case no separate authorisation to accumulate is needed (subs.14(2)).

5–21 The requirement of authorisation for disposal: section 13

Section 13 contains three separate prohibitions on the disposal of radioactive waste, as follows:

(1) By subsection (1) no person shall, except in accordance with an

---

[41] *ibid.*, para. 100.

authorisation, dispose of any radioactive waste on or from any premises which are used for the purposes of an undertaking carried on by him, or cause or permit it to be disposed of, if he knows or has reasonable grounds for believing it to be radioactive waste. This prohibition is expressly subject to section 15 which creates or allows for the creation of various exemptions; these are discussed at (para. 5–24 below). The wording of the prohibition is analogous in various respects to that relating to the keeping and use of radioactive materials, *i.e.* the reference to causing or permitting, and the requirement of knowledge or reasonable grounds for belief that it is radioactive waste. Reference should therefore be made to the case-law referred to at para. 4–07 above in this respect.

(2)   Subsection (2) provides that any person who keeps mobile radioactive apparatus (as to which see para. 4–06 above) shall not dispose of any radioactive waste arising from such apparatus, or cause or permit it to be disposed of, except in accordance with an authorisation granted for that purpose. Again reference may be made to the case-law at para. 4–07 above.

(3)   By subsection (3) where any person, in the course of carrying on an undertaking, receives radioactive waste for the purpose of his disposing of it, he must not dispose of it, or cause or permit it to be disposed of, knowing or having reasonable grounds for believing it to be radioactive waste, other than in accordance with an authorisation. As with subsection (1), the prohibition is subject to the exception provisions of section 15 (see para. 5–24 below). It is also subject to subsection 13(4), the effect of which is that such disposal does not require an authorisation if the waste falls within an authorisation granted under either subsection (1) or (2) and it is disposed of in accordance with that authorisation.

This last provision raises an important issue as to the relationship between authorisation of the producer of the waste, the disposal of the waste from the producer's premises and ultimate disposal. These associated activities may be dealt with by a single authorisation setting the conditions for disposal. The ultimate disposer can rely on the authorisation granted under subsection (1) or subsection (2) to legitimise ultimate disposal. In debates on the 1960 Act it was already clear that the intention was to issue an authorisation to the person from whose premises the waste was disposed of rather than to the person operating the disposal site to which the waste was taken.[42] Current practice is still to issue a single authorisation to the person disposing of the waste from its premises, attaching conditions as to the site to which it must be taken and the way in which it must be buried or otherwise disposed of.[43] The Government concluded as

[42] *Hansard*, H.L. Vol. 220, col. 1077.
[43] Cm. 2919 (1995), para. 118.

part of its review of radioactive waste management that there was no need to duplicate control by means of a separate authorisation granted to the site operator.

In considering an application for controlled burial (see para. 5–12 above) the regulatory bodies consider the characteristics of the waste and also the containment characteristics of the site to which it is proposed to be sent; conditions imposed may involve, for example, the monitoring of leachate from the site for radioactivity.

5–22 Nuclear sites—disposal of waste

The disposal of waste on, or from, a nuclear site requires authorisation. Subsection 13(5) ensures the requirement of authorisation applies also to nuclear sites which have ceased to be used for an undertaking carried on by the licensee (*e.g.* after closure). Any radioactive waste disposed of by being left *in situ* would therefore require authorisation as if the premises were still used for an undertaking carried on by the licensee. The power of the HSE to attach conditions to the nuclear site licence is without prejudice to this subsection.

5–23 The requirement of authorisation for accumulation: section 14

Subsection 14(1) prohibits the accumulation of radioactive waste by any person (with a view to its subsequent disposal) on any premises which are used for the purpose of an undertaking carried on by him, or the causing or permitting of such accumulation, if the person accumulating or causing or permitting the accumulation knows, or has reasonable grounds for believing, it to be radioactive waste. Reference may be made to the case law noted at para. 4–07 above in relation to the concepts of causing and permitting. It should be noted that the prohibition applies to any premises used for the purposes of an undertaking carried on by the person accumulating the waste; it is immaterial whether those premises are the place where the waste arose. The relationship with authorisation to dispose of radioactive waste, and the fact that no separate authorisation for prior accumulation may be necessary, has already been noted (see para. 5–21 above).

Two further points are relevant. First, the requirement of authorisation is subject to any applicable exemptions under section 15; second, that no authorisation is necessary for the accumulation of radioactive waste on any premises situated on a nuclear site (subs.14(3)). This exclusion was provided in the 1960 Act on the basis that equivalent controls were already applicable under the Atomic Energy Authority Act 1954 and the Nuclear Installations (Licensing and Insurance) Act 1959.[44] Control in

---

[44] *Hansard*, H.C. Vol. 619, col. 325.

this respect must therefore be provided by the nuclear site licence. The implications of this are discussed at para. 2–35 above.

The lack of direct control through the authorisation process raises the need for liaison arrangements between the NII as licensing authority and the Environment Agencies. An interesting example of co-operation is provided by the joint audit carried out by the NII and HMIP between September and December 1994 of the management of solid radioactive waste accumulated at the Sellafield and Drigg sites of BNFL. The audit resulted in a two volume report, *The Management of Solid Radioactive Waste at Sellafield and Drigg: An Audit of BNFL Facilities by HM Nuclear Installations Inspectorate and H.M. Inspectorate of Pollution* (February 1996). The audit demonstrated a lack of uniformity in standards and resulted in various recommendations for improvement which were accepted by BNFL.[45]

Exemptions: section 15   5–24

Subsection 15(1) creates a general exemption from three requirements: authorisation for disposal of waste on or from premises (subss.13(1) and (3)); and authorisation for accumulation (subs.14(1)). The exemption applies to radioactive waste arising from clocks or watches, but does not affect the operation of subsections 13(1) and 14(1) in relation to the disposal or accumulation of radioactive waste on, or from, premises on which clocks or watches are manufactured or repaired by processes involving the use of luminous material.

Subsection 15(2) allows further exemptions from the requirements of sections 13 and 14 to be made by order, either absolutely or subject to conditions. A list of the orders made under this provision is given at the annotations to the subsection in Appendix B. The Orders providing exemption for educational establishments and hospitals are of particular practical relevance, and are outlined below.

Educational establishments   5–25

Although made under earlier legislation, the Radioactive Substances (Schools, etc.) Exemption Order 1963 (S.I. 1963 No. 1832) confers exemptions under section 15 of the 1993 Act by virtue of the Interpretation Act 1978, s.17(2)(b). The Order applies by Article 4 to maintained schools, colleges for teacher training and institutions for further education. By Article 8, radioactive waste arising on such premises and consisting of waste which, immediately before it becomes waste, was an exempted closed source (defined by reference to Art. 5) is excluded from the requirements for authorisation for disposal, subject to the condition that it is disposed of by sending to or collection by:

[45] see *HSE/HMIP News Release—E26:96* (February 1996).

(a) a person authorised under s.13(3) to dispose of radioactive waste of that description; or

(b) a person who in the course of an undertaking, produces radioactive material of that kind.

Effectively therefore, the school or college may send such closed sources to, or have them collected by, an authorised disposer, or a producer of such materials.

By Article 9, further exemptions are given to certain types of liquid and solid waste forming part of exempted open sources (again defined in Art. 5) or contaminated by such sources or by waste therefrom. For solid waste, it must be disposed of either by being removed by a refuse disposal authority or its contractors, or by deposit at a landfill used for the deposit in substantial quantities of non-radioactive waste; the waste must be placed in a container with at least three cubic feet of other refuse; quantitative limits on the level of radioactivity also apply. Such solid waste received by a refuse disposal authority or its contractors is exempt from the requirement of authorisation for disposal under subsection 13(3); as is such waste received by other persons under certain conditions, *i.e.* disposal at a site taking substantial quantities of non-radioactive waste, provided it is not deposited at a part of the site used only for radioactive waste. Liquid waste may be disposed of only into the foul drainage or trade effluent system and again is subject to quantitative limits on levels of radioactivity. Low-level gaseous emissions are also subject to an exemption.

Educational establishments also benefit from an exclusion in relation to authorisation to accumulate waste (Art. 10). This exemption applies to sealed sources falling within Article 8 provided they are accumulated with a view to disposal by the means benefiting from the exemption (that means being available or about to become available) and are disposed of by that means as soon as practicable. Solid and liquid waste falling within Article 9 is also excluded, provided that it is contained in a closed container and is disposed of as soon as practicable, and in any event within two weeks of when its accumulation began. Those receiving solid waste falling within Article 9 also benefit from an exemption in relation to its accumulation, provided that it is disposed of as soon as practicable.

5–26 Hospitals

The Radioactive Substances (Hospitals) Exemption Order 1990 (S.I. 1990 No. 2512) (as amended by the Radioactive Substances (Hospitals) Exemption (Amendment) Order 1995 (S.I. 1995 No. 2395)) replaces previous orders made in 1963 and 1974. It takes effect to create exemptions under section 15 of the 1993 Act by virtue of the Interpretation Act 1978, s.17(2)(b). The Order applies to "hospital premises" which by Article 2(1) include: institutions for the reception and treatment of persons suffering from illness (including mental disorder and any injury or disability); maternity homes; convalescent and rehabilitation institutions; nursing homes or mental nursing homes under the Registered Homes Act 1984, or the equivalent Scottish legislation, as well as clinics, dispensaries and out-patient departments maintained in connection with such institutions.

By Article 4(1) certain disposals of radioactive waste arising on hospital premises are excluded from the requirement of authorisation for disposal. The waste must be solid or liquid, contain no alpha-emitters and must be one of the following:

(a) material which immediately prior to becoming waste fell within Article 3(2) (*i.e.* solid or liquid, containing no alpha emitters, and not a closed source in the case of solid material); or

(b) a substance or article which is radioactive solely because of contamination by radioactive material kept for medical diagnosis, treatment of patients, or supply to another hospital; or is radioactive by contact with or proximity to such waste; or

(c) human excreta; or

(d) residual ash produced by the burning of other waste falling within these categories.

The exemption does not apply where an existing authorisation for disposal from the premises is in force (Art. 4(2)). The exemption is subject to the conditions specified in Schedule 2 to the Order. These require clear and up-to-date record keeping, showing dates of disposal and the sum total of radioactive waste disposed of in any month. The permissible disposal routes for solid waste are collection by a waste collection authority, collection by some other person for disposal at a licensed site with substantial quantities of non-radioactive waste, or burning on the hospital premises or the premises of another hospital. For flammable liquid waste, the permitted disposal route is incineration on the hospital premises or at another hospital. Quantitative limits on levels of activity are applied, and for burning there are special conditions as to the use of a clinical waste incinerator to prevent the entry of emitted gas into any building, and as to the disposal of residual ash. For aqueous liquid waste, the route is discharge into a foul drainage or trade effluent system, subject again to limits on activity levels.

Article 4(3) excludes certain wastes removed from hospital premises from the requirement of authorisation for disposal under subsection 13(3). These are:

(a) solid waste removed by a person other than a waste collection authority for disposal at the same place as substantial quantities of non-radioactive waste pursuant to a waste site licence; or

(b) solid waste removed by a waste collection authority; or

(c) solid waste or flammable liquid waste removed by a person other than a waste collection authority for disposal by burning on other hospital premises, subject to the detailed requirements in Schedule 2.

An exemption is also provided for the accumulation of waste falling within Article 4 prior to disposal (Art. 5). It must be kept in a closed container on the hospital premises and be disposed of within 2 weeks of the date on which the waste arises. Waste which has been removed from hospital premises in accordance with Article 4(3) benefits from an exemption from the requirement of authorisation under subsection 14(1), subject to its being disposed of as soon as practicable after removal.

5–27 Authorisation procedure: non-nuclear sites

Section 16 deals with the granting of authorisations under sections 13 and 14. As from April 1, 1996, the power to grant authorisations is exercisable by the Environmental Agency or SEPA. Applications must be accompanied by the charge prescribed under the relevant charging scheme (see further para. 4–19 above).

By subsection 16(5), the Agency must consult with such local authorities, relevant water bodies or other public or local authorities as appear proper to be consulted. The absence of any clear requirement to consult local authorities and sewerage undertakers was a cause of serious concern and protracted discussion in debates on the 1960 Act, particularly since under other provisions of the Act (see para. 5–49 below) radioactivity was to be disregarded for other forms of control. The Government was unwilling to impose any more specific requirement on the basis that it would be very difficult to find an acceptable criterion for consultation. Viscount Hailsham on behalf of the Government suggested that:

> "... the cases in fact vary so much that it is impossible to define them in the statute without clogging up the administrative machinery by requiring a number of consultations, which if we really went into it might well prove to be a pure waste of time and paper ..."[46]

On that basis, he suggested that it was appropriate to:

> "... follow the practice of accepting from Ministers their assurance that it is their intention and practice invariably to consult those local authorities and other bodies whose interests are directly affected."[47]

Typically, consultation will take place with appropriate local authorities,

[46] *Hansard*, H.L. Vol. 220, col. 1079.
[47] *ibid.*, *Hansard*, H.L. Vol. 219, col. 907; *Hansard*, H.L. Vol. 220, cols. 1082–1083; and *Hansard*, H.C. Vol. 619, col. 372.

parish councils and water plcs, as well as various national and local environmental, farming, fishing and food production, consumer and interest groups.

On an application being made, the Agency must (subject to any direction under section 25, as to which see para. 5–43 below) send a copy of the application to each local authority in whose area the radioactive waste is to be disposed of or accumulated in accordance with the authorisation (subs.16(6)).

By subsection 16(7), an application may be treated by the applicants as refused if it is not determined within the prescribed period of four months.

The authorisation may, by subsection 16(8)(a), be granted in respect of radioactive waste generally or in respect of specified descriptions of such waste; in practice, the waste authorised for disposal will always be specified in some way. Wide discretion is given by subsection 16(8)(b) as to the imposition of limitations or conditions. On granting an authorisation, the Agency must provide the holder with a certificate containing material particulars, and must also (again subject to any direction under section 25) send a copy to each local authority in whose area the waste is to be disposed of or accumulated. Subsection (10) deals with the date from which the authorisation takes effect; this date is to be specified in the authorisation and in fixing it the authorising Agency must allow a period of at least 28 days after the date on which copies are expected to be sent to the relevant local authorities, unless it is necessary to bring the authorisation into effect on an immediate or expedited basis. Debates on the 1960 Act indicate that the Government felt that some flexibility might be required to avoid the 28-day delay in emergency cases, *e.g.* medical applications or "detecting the seat of a failure in an important public utility".[48]

Authorisation procedure: nuclear sites 5–28

"Nuclear sites" are defined by subsection 47(1) to means sites where a nuclear site licence is currently in force or where, after surrender or revocation of such a licence, the statutory "period of responsibility" is still current (see para. 2–42 above as to this period).

Under the 1960 and 1993 Acts until April 1, 1996 in England, Wales and Northern Ireland, the authorising body for nuclear sites was not simply HMIP, but rather HMIP acting jointly with "the appropriate Minister"— defined as the Minister of Agriculture, Fisheries and Food in England; the Secretary of State in Wales; and the Department of Agriculture in Northern Ireland.

This requirement was repealed by amendments made to the 1993 Act by the Environment Act 1995, to coincide with the establishment of the

---

[48] *Hansard*, H.L. Vol. 220, col. 1092.

Environment Agencies. However, there are still some differences between the procedure in relation to waste from nuclear sites:

1. On an application for authorisation under subsection 13(1) in respect of the disposal of waste on or from any premises situated on a nuclear site in any part of Great Britain, the appropriate Agency must—

    (a) consult the "relevant Minister" and the HSE before deciding whether to grant the authorisation and, if so, subject to what limitations and conditions; and
    (b) consult the "relevant Minister" concerning the terms of the authorisation, in particular by sending a copy of any authorisation which it proposes to grant (see subs.16(4A)). The "relevant Minister" is defined by subsection 16(11) to mean the Minister of Agriculture, Fisheries and Food in England; and in Scotland and Wales, the respective Secretary of State.

2. Before granting an authorisation in respect of the disposal of radioactive waste on or from a nuclear site, the appropriate Agency must also consult such local authorities, "relevant waste bodies" (defined in subs.47(1)) or other public or local authorities as appear to the Agency to be proper to be consulted. This will obviously depend on the nature of the disposal under consideration; it would clearly be proper to consult a sewerage undertaker in the case of a proposed discharge of aqueous effluent to sewer, but not a release of gaseous effluent to air.

3. Unlike other types of application, an application for an authorisation for disposal of radioactive waste on or from a nuclear site is not deemed to be refused if it is not determined within the prescribed period (subs.16(7)).

5–29 Public consultation on applications

The 1993 Act makes no provision for general public consultation on applications for authorisation. Nonetheless, HMIP recognised that the accumulation, and more particularly disposal, of radioactive waste are matters of public concern and followed the practice of allowing a period of public consultations on applications. Typically, this would involve a period of six or eight weeks for comments, with copies of the application, draft authorisation and an explanatory memorandum being placed on deposit in public libraries, local authority offices and the appropriate HMIP regional office. The same applied to applications made to HMIPI for authorisations in Scotland; see, for example, the consultation in 1992 on the continued disposal of liquid and gaseous wastes from Hunterston

"A" nuclear power station. There seems no reason to suppose that the practice of the Environment Agencies will differ in this respect.

It was held in *R. v. Inspectorate of Pollution, ex p. Greenpeace Ltd (No. 2)*[49] (see para. 5–30 below) that HMIP was not under a mandatory obligation to undertake consultation either for new authorisations or for variations of existing authorisations. Wide discretion is given, and the legislation does not, in fact, refer to general public consultation at all. In that 1994 case, it was held that it was not illogical for HMIP and MAFF to deal with an application for variation of existing authorisations to allow testing of the THORP plant without consultation, whereas the main application for authorisation was subject to a major consultation exercise.

The consultations carried out in respect of the THORP authorisations give an indication of the scale of the process which may be involved in major and high profile applications; the first round of consultation on draft authorisations lasted for 10 weeks and generated 84,000 responses; the second round dealing with wider policy issues lasted two months and elicited 52,500 responses.

Notwithstanding that public consultation is not mandatory, it may well be that the practice of consulting could give rise to legal effects under the principle of legitimate expectation as "a regular practice which the claimant can reasonably expect to continue".[50]

The issue of consultation also arose in *R. v. Secretary of State for the Environment, ex p. Greenpeace Limited.*[51] As described above, the application by BNFL for full authorisation of discharges from THORP was subject to two rounds of consultation—the first on the terms of the draft authorisation and the second on wider policy issues. Greenpeace argued that on the second round of consultation, the public should have been provided with greater information on economic issues; allegations that the whole exercise was a "sham" were not, however, pursued. Potts J. held that given the way in which the Ministers had approached economic issues, nothing turned on the precise details of contracts and profitability figures. In any event, exhaustive information did not have to be given to ensure effective consultation.[52] The consultation procedure followed was therefore held to be adequate.

Justification: *R. v. Pollution Inspectorate, ex p. Greenpeace Ltd (No. 2)*   5–30

The principle of justification in the context of radioactive waste management policy has already been referred to at para. 5–11 above. The issue has been considered judicially in two separate cases, the first of

---

[49] [1994] 4 All E.R. 329 at 344.
[50] *Council of Civil Service Unions v. Minister for the Civil Service* [1985] A.C. 374, 401.
[51] [1994] 4 All E.R. 352.
[52] *R. v. Rochdale Health Authority, ex p. Rochdale Metropolitan Borough Council* [1992] C.O.D. 320.

which was *R. v. Inspectorate of Pollution, ex p. Greenpeace Ltd (No. 2)*.[53] The case concerned the thermal oxide reprocessing plant (THORP) operated by BNFL at Sellafield. As referred to previously, BNFL obtained outline planning permission in 1978 to construct the plant by way of special development order following the lengthy inquiry chaired by Parker J. (see para. 1–13 above). Full planning permission was granted in 1983.

Following completion of the plant in 1992, BNFL applied for new authorisations for the discharge of radioactive waste to air and to sea, in order to operate the plant. Whilst these applications were being processed, BNFL sought approval from HMIP to commence the uranium commissioning or testing phase for THORP. BNFL requested that such approval be given simply by adding new discharge points to the "Implementation Documents" under the existing authorisations, on the basis that emissions overall would stay within the limits of such authorisations. HMIP and MAFF took the view that formal variation of the authorisation to refer to the new plant was required. BNFL accordingly made formal applications for the variations necessary to carry out the testing. The conclusion of HMIP and MAFF was that the radiological impact of emissions from testing would be very small. Greenpeace, who were opposing the variations, were asked whether they wished to request a hearing, but declined to respond. The variations were granted in August 1993, following which Greenpeace sought judicial review to quash the decision.

Leave for judicial review was given by Brooke J. who refused, however, to impose a stay on implementation of the variations so as to prevent testing. This aspect of the decision was appealed unsuccessfully by Greenpeace.[54] The Court of Appeal held that the judge had directed himself correctly in taking into account when refusing to give a stay, the financial position of BNFL and the absence of any cross undertaking in damages from Greenpeace.

The substantive application to quash the variations was heard by Otton J. The attack on the decision by Greenpeace rested on two main grounds. The first related to the lawfulness of the way in which the variation procedure was used and is discussed at para. 5–36 below. The second related to the alleged absence of justification of the process from which the waste discharges would arise. In this respect, Greenpeace relied upon paragraph 46(a) of the DoE's 1982 publication, *The Radioactive Substances Act 1960: A Guide to the Administration of the Act.* This paragraph stated that "all practices giving rise to radioactive waste must be justified, *i.e.* the need for the practice must be established in terms of its overall benefits". Greenpeace also referred to Article 6 of the Basic Safety Standards Directive 80/836/EURATOM (as amended by 84/467/EURATOM), which provides that "... various types of activity resulting in an exposure

[53] [1994] 4 All E.R. 329.
[54] *R. v. Inspectorate of Pollution, ex p. Greenpeace* [1994] 4 All E.R. 321, C.A.

to ionising radiations shall have been justified in advance by the advantages which they produce ..."

Otton J. held that in considering the applications for variations to allow testing, HMIP and MAFF were under a general obligation "to consider the health and safety aspect and, in particular, whether the amount of radioactive waste to be discharged would pose a significant risk to the health or safety of the public" (p. 345). However, consideration was confined only to the testing process in respect of which the variations were sought and there was no need to consider the wider issues, which were already subject to public consultation under the main authorisations to operate the process permanently. The evidence of HMIP and MAFF stressed (a) the very low levels of discharges involved; and (b) that the discharges were below existing and authorised limits. These were, to the judge's mind, highly material considerations, which met the justification argument; it was not necessary to go into the wider social and economic issues arising out of the main application.

Otton J. concluded that, in any event, the justification requirements of the Guide and the Directive had, in fact, been satisfied. He based this conclusion on various circumstances: (a) the fact that justification issues had been considered by Ministers on the main application which there was "a distinct possibility" would be authorised; (b) the need for testing was established in terms of its overall benefit, the regulating bodies having taken into account the benefit which testing would produce; and (c) it was also relevant to bear in mind that the need for THORP had been considered at the planning stage in the Parker Report and the Parliamentary process, in Article 6 of the Directive, did not require re-evaluation.

The application of Greenpeace therefore failed. Whilst the comments of Otton J. are of interest, it must be remembered that they were made in the context of an application for variation of an existing authorisation to allow temporary testing only. The main authorisation process for THORP was the subject of a separate challenge, dealt with in the next paragraph.

Justification: *R. v. Secretary of State for the Environment, ex p.*   5–31 *Greenpeace Ltd*

In *R. v. Secretary of State for the Environment, ex p. Greenpeace Ltd*,[55] Greenpeace and Lancashire County Council sought judicial review to quash the decision of HMIP and MAFF to grant authorisations for the disposal of radioactive waste from the THORP plant.

The authorisations were granted on applications by BNFL, following two rounds of public consultation, dealing first with the terms of the proposed authorisation and secondly with wider policy issues raised during the first consultation, *i.e.* the justification for nuclear fuel reprocessing generally and for THORP in particular, and to the

---

[55] [1994] 4 All E.R. 352.

non-proliferation and security implications of an increasing stockpile of plutonium.

The substantive hearing (leave having been given by Laws J.) was before Potts J. who identified two separate issues relating to justification:

(a) whether there was a legal requirement to consider justification; and

(b) whether the finding that the relevant activities were justified was irrational.

WAS JUSTIFICATION REQUIRED? On the first of these issues, Greenpeace referred, as in *R. v. Inspectorate of Pollution, ex p. Greenpeace (No. 2)*, to the requirement for justification as set out in Directive 80/836/EURATOM and the Government's own *Guide to the Administration of the Radioactive Substances Act 1960*. They also referred to the place of justification in the system of radiological protection recommended by the ICRP in ICRP 60, and to *Re Ionising Radiation Protection, E.C. Commission v. Belgium*,[56] where Advocate General Jacobs stated in his opinion that the ICRP principles are reflected in the Directive.

However, the question for Potts J. was whether justification must be considered in exercising powers of authorisation and variation under the 1993 Act, which was silent on the issue. Potts J. held that the express language of the 1960 and 1993 Acts was not to be construed by reference to the guidance or policy statements in the DoE's Guide so as to require justification (p. 365). However, the argument that on principles of Community law the 1993 Act was to be construed consistently with the requirements of the directive was stronger. This raised difficult issues as to the relationship between Articles 6 and 13 of the Directive, whether the requirement of justification related to a type of activity in general rather than an activity at a specific site, and whether the principle is applicable to a group of sources of exposure (*e.g.* at Sellafield) or only to practices in the broader sense of generic human activities (see p. 367).

Potts J. felt some unease at dealing with issues that might be thought better dealt with by the European Court of Justice, but none of the parties had suggested the matter be referred (no doubt because of everyone's wish to have a speedy determination of the issues). Article 30 of the EURATOM Treaty, with its concentration on basic standards, sat uneasily in Potts J.'s view with Articles 6 and 13 of the directive. Nevertheless, he reached the following conclusions:

1. Article 13 of the directive makes reference to Article 6(a) and thus includes justification as a matter to which Member States must have regard; Article 13 would make no sense unless such a purposive approach were adopted (pp. 367–368).

2. The argument of Counsel for the Secretary of State and MAFF was

[56] Case C–376/90 [1993] 2 C.M.L.R. 513 at 524.

incorrect in suggesting that justification related to a type of activity in general rather than the carrying on of a particular activity at a specific site. What was required to be justified was thermal oxide reprocessing activities at Sellafield and the effect of that practice on particular individuals in particular circumstances. The inquiry and Report of Parker J. at the planning stage had not performed this exercise, nor had it purported to do so; likewise the ensuing Parliamentary process (p. 368). This aspect is contrary to the position taken by Otton J. (see para. 5–30) and, it is submitted, is the better view.

3.   Similarly, Potts J. rejected the submission by Counsel for BNFL that a distinction should be drawn between "source" and "practice" within the context of the Directive. The operation of THORP was the practice (*i.e.* the human activity) requiring to be justified; the "source" within that practice was Sellafield (p. 368).

4.   The Ionising Radiations Regulations 1985 did not (contrary to the submission of Counsel for BNFL) implement the whole of the Directive, which remained a relevant consideration in construing the 1993 Act.

On that basis Potts J. concluded that justification was a legal requirement and that the Ministers had erred in law in concluding that it was not. There was nothing in sections 13 and 16 of the 1993 Act to prevent those sections being construed so as to require the justification exercise to be carried out prior to the grant of authorisation.

WAS JUSTIFICATION PROPERLY CONSIDERED? The second question, having concluded that justification was required, was whether it had in fact been carried out, notwithstanding the Minister's error of law. Greenpeace argued that the Ministers had failed to consider all relevant information in that they had not checked BNFL's assertions as to future contracts and viability, they did not insist on seeing a report prepared for BNFL by Touche Ross on the future viability of THORP, and they did not consider possible alternatives to THORP such as dry storage. Consideration of these issues involved detailed examination of the authorisation process and the decision document. Effectively, the conclusion of the Ministers had been that THORP was a facility which BNFL's customers would use, that it would provide substantial financial benefit to BNFL and that it would provide a significant number of jobs for local people. Potts J. held that the Ministers were entitled to reach this conclusion on the material available to them; also they were entitled to reach that conclusion without seeing the Touche Ross report, which was an internal BNFL document. In terms of alternatives to reprocessing, the Ministers concluded that dry storage was not a true alternative, merely delaying the decision as to what to do with spent fuel. Potts J. found that they had approached the matter correctly.

The arguments of the other applicant, Lancashire County Council, centred on the health risk aspects of the emissions. Potts J. however concluded that the Ministers were entitled to follow the advice of the NRPB and ICRP in concluding that compliance with certain dose limits and targets (see further para. 5–53 below) would protect individual members of the public from unacceptable levels of risk, and to find that the discharge limits contained in the draft authorisations would result in doses well within those dose levels and targets.

Overall, Potts J. held that the Ministers had acted correctly in considering first the human issues related to acceptability of risk, and then examining the wider issues of justification as if they were relevant, even though the Ministers (incorrectly) believed they were not relevant:

> "The Ministers carried out a careful process of weighing the benefits against the detriments in reaching the conclusion that the balance came down on the side of justification. They were entitled to reach that conclusion."

5–32 Justification after the THORP litigation

The decision that justification is a legal requirement of the authorisation process (see para. 5–31 above) raises difficult practical issues for the regulatory bodies. The conclusions of the Government's *Review of Radioactive Waste Management Policy*[57] refers to the decision (para. 58) but barely hints at the difficulties. The Review's preliminary conclusions were that applicants should be encouraged to apply for authorisation at the early stages of a project (not just once the plant had been built, as in the past) so that justification could be considered before committing major capital investment. There was considerable support for this principle in responses to the consultation document; however, some environmental groups and local authorities were concerned that authorisation at too early a stage would not be adequate until full details of the plant's operations and waste arisings were known.

The Government's intention in the light of those responses is a "flexible system", to be fleshed out in a revised guide to the administration of the Radioactive Substances Act (see para. 62 of the Review). For major projects, it is expected that developers would make early application for authorisation at about the same time as seeking full planning permission for the project (although the applications would be determined separately). If authorisation were granted this could be on the basis of conditions to be met at specified stages, leading to approval to start operations in due course when plant is complete and commissioned. Parallels for this apply in the nuclear site licensing field (see para. 2–18 above). Should the design evolve then the authorisation may need to be varied, which may require public consultation in the event of significant changes.

[57] Cm. 2919 (1995).

The Government emphasises that no change to legislation is necessary to achieve this and that all applicants may apply for early authorisation if they are attracted by the greater certainty which would result (para. 62 of the Review).

Another aspect relates to the process of justification itself. It seems clear from the decision of Potts J. in *R. v. Secretary of State for the Environment, ex p. Greenpeace Ltd* that justification is a process specific to the particular activity in question, not a generic one. Therefore, the fact that one type of activity resulting in the emission of waste on one site is justified cannot mean that the same type of activity on another site can also be justified. Both the risks and the benefits may be different. Nonetheless, there may well be generic information which is common to both exercises and which is relevant to both, particularly where national considerations are an issue.

Local inquiries into authorisations   5–33

By subsection 24(2) of the 1993 Act, where an application for authorisation is referred to the Secretary of State for determination pursuant to a direction, the Secretary of State may cause a local inquiry to be held. In *R. v. Secretary of State for the Environment, ex p. Greenpeace Ltd* (see para. 5–31 above), the question arose as to how that discretion was to be exercised. Greenpeace put forward seven factors which it said were relevant to the decision whether to hold an inquiry:

(1) the importance of the decision, with a substantial increase in emissions of ionising radiations;
(2) the great public concern and the wish of many members of the public and of local authorities that there be an inquiry;
(3) the quality of the responses to consultation, raising scientific and technical issues that could only properly be tested in an inquiry;
(4) the failure of ministers to obtain certain information which could only properly be evaluated in an inquiry;
(5) the likelihood of public concern if Ministers decided issues where the Government had already indicated a policy and wish that THORP should proceed;
(6) the need to inform the public and allay their fears as to the effects of discharges; and
(7) the need to provide Ministers with the best possible information to allow the correct decision to be reached.

The Ministers decided not to hold a public inquiry, essentially on the basis that they were already adequately informed as a result of the consultation exercise; they were also conscious that a hearing or inquiry would cause considerable delay and likely adverse financial consequences for BNFL. Potts J. concluded that the Secretary of State had carried out the correct balancing exercise and had addressed the correct criteria in deciding not to hold an inquiry.[58] The Ministers had not given undue weight to the delay and financial consequences point; this was a collateral point, of which they were simply aware.

Greenpeace also argued that the Ministers had not referred to the need to allay the public's concerns and to better inform the public. Potts J. had some sympathy with these arguments, but essentially the Ministers had been entrusted by Parliament with a discretion as to whether to hold an inquiry and there was no evidence that they had exercised that discretion wrongly; the court might disagree with the Ministers as to the weight to be placed on matters such as public concerns, but it was not for the court to substitute its own judgment from that of the Ministers.

5–34 Environmental assessment

In *R. v. Secretary of State, ex p. Greenpeace Ltd*[59] (see para. 5–31 above) it was also argued by Greenpeace that no lawful authorisation for the disposal of radioactive waste from THORP could have been granted without an environmental assessment complying with Directive 85/337/EEC. This Directive identifies a number of projects involving radioactive materials as requiring environmental assessment. In particular, Annex I (projects subject to environmental assessment in all cases) refers to nuclear power stations and other nuclear reactors (except certain research reactors) and installations solely designed for the permanent storage, or final disposal of, radioactive waste. Annex II (projects subject to assessment where their characteristics so require) refers to installations for the production or enrichment of nuclear fuels and installations for the reprocessing of irradiated nuclear fuels. Various European countries, including Ireland, Denmark and The Netherlands had queried the legality of authorising THORP without going through formal environmental assessment procedures.[60]

The key issue in the THORP case was, however, whether the construction, operation and discharges from THORP were one single project, in which case its commencement predated the 1985 Directive, which would accordingly not apply to it.[61] Greenpeace argued that discharges of radioactive waste into the environment constituted another "intervention in the natural surroundings ..." within Article 1.2 of the Directive which was separate from the construction of the plant. This submission was rejected as a distortion of Article 1.2 and of the whole tenor of the Directive, which required environmental assessment at the earliest

---

[58] see: *Binney and Anscomb v. Secretary of State for the Environment* [1984] J.P.L. 871.
[59] [1994] 4 All E.R. 352.
[60] *Financial Times*, December 1, 1993.
[61] *Twyford Parish Council v. Secretary of State for the Environment* [1991] C.O.D. 210; [1992] 1 C.M.L.R. 276.

possible stage, prior to consent for the project being given. Possible alternatives to the process were to be considered at that stage, not at the stage of authorising emissions when the plant had already been built.

Environmental assessment will therefore generally be appropriate at the planning stage of a project involving the emission of ionising radiations, and will not require a separate formal environmental assessment at the stage of authorisation under sections 13 and 16 of the 1993 Act. Nonetheless, the process of considering the application and the related consultation will often in practice meet the requirements of the directive; certainly Potts J. was satisfied that this was so in the THORP case.

Revocation and variation of authorisations   5–35

By subsection 17(1), the Agency may at any time revoke an authorisation granted under section 13 or section 14, and by subsection 17(2) such an authorisation may be varied at any time. Before varying an authorisation in respect of the disposal of waste on or from a nuclear site, the Agency must consult the relevant Minister and the HSE (subs.17(2A); and see paras. 5–28 and 5–29 above). Notice of the revocation or variation must be given to the person to whom the authorisation was granted and a copy must be sent to any local or public authority to which a copy of the original authorisation was sent.

The scope of the variation power   5–36

In the case of the THORP plant at BNFL's Sellafield works, existing authorisations to discharge gaseous and liquid radioactive waste were varied so as to allow THORP to be tested prior to full authorisation. This variation was the subject of judicial review proceedings brought by Greenpeace,[62] referred to at para. 5–30 above. The variations had been made under subsection 8(7) of the Radioactive Substances Act 1960 (now subsection 17(2) of the 1993 Act) and Greenpeace argued that it was not a proper and lawful exercise of the variation power to include emissions from entirely new plant which did not exist at the time of the original authorisation. It was said by leading Counsel for Greenpeace that the power to vary conditions and limitations extended only to matters peripheral to the core activity, and did not permit variation of the core activity itself. Moreover, the effect of the variation route was to avoid public consultation which would have taken place on an application for a fresh authorisation.

The conclusion of Otton J. was that HMIP and MAFF had acted lawfully in granting the variations. The starting point was that subsection 6(1) of the 1960 Act (now subsection 13(1) of the 1993 Act) allowed BNFL to

---

[62] [1994] 4 All E.R. 321, C.A.

dispose of any radioactive waste, subject to complying with the terms of the authorisation, from "premises" used for the purpose of any undertaking carried on by BNFL. "Premises" meant land and would include the plant on site, but was not confined to the original plant; there was "... a wide power vested in BNFL to extend or contract their activities on site under their licence, in pursuance of their objectives but subject to strict independent regulatory control" (p. 342). In construing the authorisations, the term "premises" was to be given the same meaning as in the statute.[63] The wastes described in the authorisations were not limited to wastes arising from a particular type of plant but covered all wastes arising from BNFL's operations at Sellafield.

The power to vary the conditions of authorisations was "very wide but not unlimited"; any variation to the conditions or limitations of the authorisation could not have the effect of widening the general terms of the authorisation, but could be used to relax or make more stringent the conditions governing disposal of the relevant wastes. It was held that the variations granted had not extended the description of radioactive waste already included within the existing authorisations and, as mentioned above, it was irrelevant that waste would emanate from new plant. In the case of gaseous waste, one new discharge point would be added, and in the case of aqueous waste, the existing discharge point would be used. It was not illogical nor irrational to use the variation process with the effect that there would not be consultation; there was no mandatory requirement for public consultation either for authorisation or for variation.

The discussion provides useful clarification of the scope of the variation procedure, which can provide a flexible means of dealing with changes to plant provided that the scope of the authorisation in terms of the waste to be discharged is not enlarged. In view of the effect of subsection 13(1) in authorising discharges as outlined above, it will be worth considering carefully the terms of the authorisation to see whether any variation is necessary at all (see para. 5–39 below).

5–37 Review of authorisations

A regular process is followed for reviewing authorisations in the light of changed circumstances and past performance, including critical appraisal of the adequacy of existing conditions. These routine reviews are scheduled in the light of the individual circumstances of plant; for example, for Sizewell B the first routine review was scheduled to take place following the first major maintenance period or "statutory outage", during 1996.[64]

---

[63] Sections 11 and 21 of the Interpretation Act 1978.
[64] see *News Release H.M. 266* (August 1995).

248

No provision for transfer of authorisations 5–38

No provision is made in the 1993 Act for the transfer of authorisations and therefore where the relevant premises or undertaking changes hands, the appropriate course is re-authorisation by issuing a fresh authorisation to the incoming party, in the course of which the appropriate Agency may determine to change the conditions. This is significant in the context of privatisation of the nuclear power industry; changes in the operation of nuclear power stations involve the issue of new authorisations to the private sector companies involved. This process began in 1995 for the AGR and PWR stations, involving public consultation exercises and the justification process (see above) for continuation of the discharges.[65]

Effect of authorisation 5–39

In *R. v. Inspectorate of Pollution, ex p. Greenpeace Ltd (No. 2)*,[66] Otton J. considered the effect of authorisation pursuant to subsection 6(1) of the 1960 Act (now subsection 13(1) of the 1993 Act). He concluded that authorisation related to "premises" used for the purposes of the relevant undertaking, not to specific plant located at any given time on those premises. Thus, in the absence of any specific condition or limitation confirming the authorisation to the discharge of waste from a particular plant or materials, an authorisation is capable of accommodating the contraction or expansion of activities on site, subject to disposal remaining in accordance with its conditions (see para. 5–36 above).

Functions of local authorities—"special precautions" 5–40

The controversial practice of disposing of very low level waste with ordinary domestic refuse has been referred to (see para. 5–02 above).

Provisions exist under section 18 of the 1993 Act regarding circumstances in which the disposal of radioactive waste is likely to need special precautions to be taken by such bodies as local authorities, water bodies and other public authorities. Such consultation was intended to be additional to any other consultation requirements applying to applications generally.[67] Subsection 18(1) requires the Agency to consult with these bodies before granting the authorisation under section 13. If special precautions need to be taken by the appropriate consultee in

[65] *ibid.*
[66] [1994] 4 All E.R. 329.
[67] *Hansard*, H.L. Vol. 220, col. 1080.

disposing of radioactive waste, it is empowered to make a charge for taking those precautions. A further provision under subsection 18(3) imposes duties on a local authority in relation to radioactive waste that is brought to a local authority refuse site in compliance with an operator's authorisation under section 13. Under these circumstances, the local authority must accept that waste and deal with it in a manner required by the operator's authorisation.

The Government has no plans to extend section 18 to private site operators or local authority waste disposal companies' operating sites; since in England and Wales local authorities may no longer operate waste disposal sites, this latter aspect of section 18 is something of a dead letter, and the receipt of radioactive waste will therefore need to be agreed with the site operator.

Where radioactive waste is disposed of at a private or LAWDC-operated site, there is no obligation to consult local authorities. Nonetheless, in practice they are consulted, and the Government will issue formal guidance to the Environment Agencies on the need to consult local authorities about authorisations for controlled burial.[68]

5–41 Notices and records

The 1993 Act details a number of miscellaneous further obligations relating to a person authorised under section 13 or section 14. By section 19 any person to whom such an authorisation has been granted is under an obligation to cause copies of the certificate of authorisation to be kept posted on his premises in an appropriate fashion. Operators with an authorisation under section 13 or section 14 can be required to retain copies of site or disposal records for specified periods after the operator ceases to carry on the activities regulated by his authorisation. He may also be required under section 20 to provide copies of records where his authorisation is revoked or he ceases to carry on the authorised activities. For these purposes, "records" are defined as records required to be kept by virtue of the conditions attached to the authorisation relating to the activities regulated by that authorisation; "site records" relate to the condition of the relevant premises, while "disposal records" relate to the disposal of radioactive waste on or from the premises on which the relevant activities are carried on.

5–42 Enforcement and prohibition notices: sections 21 and 22

The same power to serve enforcement and prohibition notices as applies to registrations under sections 7 and 10 also applies to authorisations under sections 13 and 14 (see paras. 4–17 and 4–18 above).

---

[68] Cm. 2919 (1995), para. 119.

Secretary of State's powers   5–43

As with registrations under sections 7 and 10, the Secretary of State has powers under sections 23 and 24 to give directions to the appropriate Agency and to call in applications for his own determination (see para. 4–21 above). In relation to authorisations relating to the disposal of waste on or from nuclear sites in England, these functions are exercisable jointly by the Secretary of State for the Environment and the Minister of Agriculture, Fisheries and Food (subss.23(4A) and 24(4A)).

By section 25, the Secretary of State may direct that information contained in, or relating to, authorisations or applications under sections 13 and 14 be restricted on grounds of national security. However, in the case of authorisations under section 13 relating to nuclear sites, such directions may not detract from obligations to consult with, or provide information to, the relevant Minister (subs.25(3A); and see para. 5–27 above).

Appeals   5–44

The appeals mechanism in relation to sections 13 and 14 authorisations is to be found at sections 26 and 27 and is essentially the same as for sections 7 and 10 registrations (see para. 4–20 above). However, in relation to authorisations for the disposal of radioactive waste on or from nuclear sites in England, appeal is to the Secretary of State and the Minister of Agriculture, Fisheries and Food, acting jointly (subss.26(5A) and 27(7A)).

Powers of the Secretary of State relating to radioactive   5–45
waste: section 29

Section 29 confers residual powers on the Secretary of State, where it appears to him that adequate facilities are not available for the safe disposal or accumulation of radioactive waste, to provide such facilities or to arrange for their provision by such persons as the Secretary of State may think fit. In the context of the 1960 Act, this provision was described as carrying on an arrangement by which the Atomic Energy Authority had been acting as agent for the Minister in disposing of wastes which were particularly difficult, or in such large quantities as to require large-scale rather than local disposal.[69] By subsection 29(2), the Secretary of State is required to consult with any local authority in whose area it is proposed to provide such facilities, and such other public or local authorities (if any) as appear to him appropriate to be consulted. By subsection 29(3), reasonable charges may be made for the use of any facilities so provided.

[69] *Hansard*, H.L. Vol. 219, col. 879.

5–46 Power of appropriate Agency to dispose of radioactive waste: section 30

Residual powers are given to the Agencies by section 30 of the 1993 Act to deal with the situation where radioactive waste is present on premises and ought to be disposed of, but for various reasons is not likely to be. Those reasons are stated to include the premises being unoccupied or the occupier being absent or insolvent. In these unfortunate circumstances, the appropriate Agency has power to dispose of the waste as it thinks fit and to recover any expenses reasonably incurred from the occupier of the premises, or from the owner (a defined term) where the premises are unoccupied. Subsection 30(3) applies, in these circumstances, the provisions of section 294 of the Public Health Act 1936 so as to limit the liability of owners who are simply agents or trustees.

5–47 Rights of entry, inspection, etc.

The Agencies are provided with powers of entry, inspection and other enforcement powers by the general provisions of the Environment Act 1995. These are discussed at paras. 5–27 and 5–28 above.

5–48 Offences: sections 32–38

The various offences relating to authorisation are the same as for registration under sections 7 and 10, which are discussed at para. 4–07 above. Various prosecutions have been brought in recent years in relation to accumulation and disposal of radioactive waste. These include:

Nichols Institute Diagnostics who pleaded guilty to accumulating waste diagnostic kits without authorisation (as well as to holding radioactive materials exceeding registered limits) and were fined a total of £10,000 with £2,726 costs in April 1993;

Essex University who pleaded guilty to the incineration of radioactive waste in breach of an authorised limit on two successive occasions and were fined £800 with costs of £2,651 in 1993;

BNFL who pleaded guilty to unauthorised disposal of low-level solid radioactive waste and were fined £7,500 with costs of around £6,000 in September 1991;

Leicestershire Health Authority who pleaded guilty to failing to properly dispose of radioactive waste from Leicester Royal Infirmary and were fined £6,000 with costs of £5,000 in February 1992.

Sections 36 (liability of directors and similar officers) and 37 (offence due to another's fault) are applicable (see annotations to those sections at Appendix B). Section 38 restricts those who may bring criminal proceedings under the Act. Particulars of convictions are required to be made publicly available until spent.[70] The provisions of the Rehabilitation of Offenders Act 1974 relating to spent convictions do not apply to offences by corporations.[71]

Public access to information: section 39   5–49

When the Radioactive Substances Act 1960 was debated in Parliament, concerns were expressed as to the lack of provision for public registers to inform the public where radioactive wastes were stored and how they were to be disposed of.[72] However, it was not until the Environmental Protection Act 1990 amended the 1960 Act that such provision was made.

The issue of public access to information pursuant to section 39 of the 1993 Act is discussed generally in Chapter 4 (see para. 4–23 above). In relation to authorised discharges of radioactive waste, the Secretary of State has directed HMIP/the Agencies to send to all local authorities (*i.e.* county councils, district and London Borough Councils) copies of the annual reports of their monitoring programmes for such discharges, containing both summary data and an analysis of the main findings.[73] Such reports deal with analysis of effluent samples from nuclear sites, and environmental monitoring around such sites, relevant landfill sites, major registrations for keeping radioactive material, and other relevant industrial sources, *e.g.* certain smelters. Each local authority is required by subsections 39(2) and (5) to make copies of such documents available to the public.

In relation to local authorities consulted on applications relating to discharges from nuclear sites (and any others who request the information) the Ministry of Agriculture, Fisheries and Food sends copies of its annual Aquatic Monitoring Report and Terrestrial Radioactivity Monitoring Programme Report.[74] The mass of more detailed monitoring data available is not sent to local authorities, but is available subject to confidentiality constraints at regional offices of the Agency, or with the MAFF Directorate of Fisheries (aquatic data), or Food Safety Radiation Unit (terrestrial data). Similarly, monitoring data required by the terms of authorisations to be provided to the enforcing authority are sent to the relevant local authorities, and should generally then be made available to the public.[75] In the case of the Ministry of Defence and visiting forces, copies of the non-statutory authorisations which they hold are sent to the local authorities.[76]

---

[70] section 39(1)(d); and the Radioactive Substances (Records of Convictions) Regulations 1992 (S.I. 1992 No. 1685).
[71] see also *Circular 22/92*, paras. 11–12.
[72] *Hansard*, H.L. Vol. 220, cols. 798 and 802.
[73] section 23(4); and DoE Circular 22/92, para. 4 (Welsh Office: 43/92; FMP 1102).
[74] *Circular 22/92*, para. 5.
[75] *Circular 22/92*, para. 8.
[76] *Circular 22/92*, paras. 9–10.

5–50 Relationship of 1993 Act to other statutory provisions

The potential overlap between the legislation on radioactive substances and radioactive waste and other environmental controls is dealt with in various ways:

1. *Part I of the Environmental Protection Act 1990: prescribed processes*

Where activities comprising a process prescribed for the purposes of Part I of the 1990 Act are regulated both by an authorisation under Part I and an authorisation under the 1993 Act, if different obligations are imposed by the separate authorisations in respect to the same matter, the relevant Part I authorisation condition will be treated as not binding (EPA 1990, subs.28(2)).

2. *Part II of the Environmental Protection Act 1990: waste*

Except as may be provided by regulations (none have as yet been made) nothing in Part II of the 1990 Act applies to radioactive waste. Accordingly, radioactive waste is not subject to the duty of care under section 34 of the Act and the deposit or treatment of such waste does not require a waste management licence.

3. *Part IIA of the Environmental Protection Act 1990: contaminated land*

Radioactive substances and waste can result in very serious problems of contamination.[77] The new provisions on the identification and clean-up of contaminated land contained in Part IIA of the 1990 Act (inserted by section 57 of the Environment Act 1995) do not apply to land or water pollution attributable to radioactivity unless applied by regulations (s.78YC). The Government points out at para. 133 of its *Review of Radioactive Waste Management Policy*[78] that although the contaminated land provisions were not developed to deal specifically with land contaminated by radioactivity, they could provide a suitable overall framework subject to various issues being addressed. These issues include which enforcing authority is appropriate and what levels of radioactivity—bearing in mind that it is a natural phenomenon—should be regarded as a contaminant. The Government intends to issue draft regulations for comment during 1997, which will apply Part IIA with modifications.

4. *Disregard of radioactivity*

By section 40 of the 1993 Act, for the purposes of certain specified statutory provisions, no account is to be taken of any radioactivity possessed by any substance or article, or by any part of any premises. The objective of this provision was to avoid those who were becoming subject

---

[77] S. Tromans and R. Turrall-Clarke, *Contaminated Land* (Sweet & Maxwell, 1994), para. 5.36.
[78] Cm. 2919 (1995).

to the new regime of control remaining subject to other controls and the possibility of prosecution under those controls.[79] The section applies to all local enactments dealing with matters such as waste and nuisances (subsection 40(2)(b) and to the provisions listed at Schedule 3 to the 1993 Act). Among the more significant are Part III of the Environmental Protection Act 1990 (statutory nuisances), the provisions on pollution of controlled waters contained in the Water Resources Act 1991 and the Control of Pollution Act 1974, and Chapter III of Part IV of the Water Industry Act 1991 (discharge of trade effluent to sewer). Effectively, therefore, the enforcing or authorising bodies for these provisions must disregard any radioactivity of the relevant materials or effluent, leaving this as part of the matter to be regulated under the 1993 Act. Similarly, in any prosecution under these provisions (*e.g.* for causing polluting matter to enter controlled waters), the court would have to ignore the radioactivity.

5. *Special Waste*

Radioactive waste may have toxic or other qualities which would qualify it as special waste falling within the Special Waste Regulations 1996 (S.I. 1996 No. 972) (which as from September 1, 1996, replaced the Control of Pollution (Special Waste) Regulations 1980 (S.I. 1980 No. 1709)). Regulation 3 of the 1996 Regulations makes it clear (as did the 1980 Regulations) that radioactive waste may be special waste even though it is not controlled waste within Part II of the 1990 Act. Whereas the Radioactive Substances Act 1960, as amended, included the 1980 Regulations within the list of statutory provisions in respect of which radioactivity was to be disregarded (see sub-para. 4 above), reference to these Regulations is omitted from Schedule 3 to the 1993 Act.

Application to Crown: section 42 5–51

By subsection 42(1), the general position is that the 1993 Act binds the Crown, although contravention will not result in the Crown being criminally liable. However, by subsection 42(2), application of the Act to the Crown does not affect premises occupied on behalf of the Crown for naval, military or air force purposes, or for purposes of the Ministry of Defence, or by or for the purposes of a visiting force. Where premises are occupied by a contractor carrying out work for the armed forces or Ministry of Defence, the question will be whether those premises are being occupied on behalf of the Crown, or on the contractor's own behalf; this will depend on the nature of the arrangements between the parties.

In any event, even in relation to premises occupied by the armed forces or Ministry of Defence, the policy is to go through authorisation

---

[79] *Hansard*, H.L. Vol. 220, col. 1093.

procedures for the accumulation and disposal of radioactive waste. A good example is H.M. Naval Base at Devonport where applications for authorisation for the disposal of liquid radioactive waste to sewer, and very low-level waste to landfill, were in February 1996 subject to full public consultation by HMIP.[80] The practice is that, rather than being issued with registrations or authorisations, MoD and visiting forces are sent documents known as "certificates of notification", recording the holding of radioactive material, and "certificates of agreement" which approve the disposal of radioactive waste. Each will specify the conditions to be observed. The Government's policy is to send copies of such certificates to the relevant local authorities except where considerations of national security preclude it.

5–52 Effect on other rights and duties: section 46

By section 46 of the 1993 Act, nothing in the Act is to be construed as conferring a right of action in civil proceedings in respect of any contravention or as derogating from any right of action or other remedy, criminal or civil, in proceedings instituted otherwise than under the Act. This provision must be read subject to the provisions of section 40 on the disregard of radioactivity for certain purposes (see para. 5–50 above) and to the general provision of section 18 of the Interpretation Act 1978 on the avoidance of multiple punishments for offences arising from the same act or omission.

Although it is not stated expressly, it would appear to be the case that this provision should not be read as derogating from the express right to bring civil proceedings to enforce the legislation conferred by subsection 32(3) of the Act (added by the Environment Act 1995, s.120 and Sched. 22). Thus, where injury or damage is caused by radioactive waste accumulated on a site, or emitted or discharged from a site, the normal principles of civil law will govern liability. The position is, of course, different where nuclear installations are concerned, being governed by the special liability regime of the Nuclear Installations Act 1965 (see generally, Chapter 3).

PRACTICAL ISSUES IN AUTHORISATION

5–53 Dose limits for gaseous and liquid waste

As already mentioned (see para. 5–11 above), dose limits will be applied to authorisations for the discharge of liquid and gaseous waste. The current legal dose limits are those contained in the revised Basic Safety

---

[80] *HMIP Bulletin—No. 42* (February 1996).

Standards (BSS) Directive and incorporated in the Ionising Radiations Regulations 1985, *i.e.* 5 mSv/y for members of the public from all man-made sources other than medical exposure. The fact that in 1990 the ICRP recommended that this dose be reduced to 1mSv/y (ICRP 60) has been referred to previously. This new limit is reflected in the revised BSS Directive 96/29/EURATOM, and as such will need to be incorporated into United Kingdom law in due course.

In fact, limits based on the lower ICRP recommendation are already applied in the United Kingdom. Initially, this limit was applied on a flexible basis to allow yearly doses of up to 5 mSv provided the total dose did not exceed 70 mSv over a lifetime.[81] The Government, however, now accepts the advice of the NRPB in its *1993 Board Statement on the 1990 Recommendations of the ICRP* that this flexibility to average exposure over a number of years is no longer necessary and that the limit should be 1 mSv/y; also that assessments of dose should include the effects of past discharges.[82]

The practical question is then how to translate this overall dose limit into limits for individual authorisations. Given that exposure may arise from several sources, the Government's past policy has been to apply a target of 0.5 mSv/y in respect of assessed dose arising from radioactive waste discharges from any single nuclear site, irrespective of size or the number of installations. Discharge limits in authorisations were based on the totality of operations on site, and reflected the assumption that doses would be maintained at the maximum levels specified for the various radionuclides.[83] This issue was reconsidered by the Government as part of its review, within the principle of optimisation (keeping exposures as low as reasonably achievable). In the light of this review the Government now accepts[84] that a maximum constraint value of 0.3 mSv/y should replace the "target" of 0.5 mSv/y when determining applications for discharges from a single new source (*i.e.* a facility or groups of facilities which can be optimised as an integral whole in terms of radioactive waste disposals). The NRPB recommended to the Government that, in general, it should be possible for existing as well as new facilities to operate within the 0.3 mSv/y dose constraint. The NRPB has also recognised that in some cases a realistic assessment would suggest that an existing facility could not be operated within this figure and has recommended that, in such cases, the operator must demonstrate that doses deriving from the facility's continued operation are as low as reasonably achievable and within the legal dose limits. The Government accepts this advice.[85]

---

[81] Cmnd. 9852 (1986).
[82] Cm. 2919 (1995), para. 65.
[83] *ibid.*, para. 66.
[84] Cm. 2919 (1995), para. 68.
[85] *ibid.*, para. 69.

5–54 Use of the ALARA principle

A key concept in the principle of optimisation of releases is that exposure should be kept as low as is reasonably achievable (ALARA) (see para. 5–13 above). The relationship of the ALARA principle to other concepts, such as the use of best practicable means, has not always been clear. An article by Christopher E. Miller, *Economics v. Pragmatics: The Control of Radioactive Wastes*,[86] has pointed out that whilst the concept may appear to be explicitly economic in character, in practice the approach has tended to be pragmatic. The term "reasonably practicable" has been construed as implying a computation between the risk to be averted and the cost of measures involved.[87] Nuclear site licences and authorisations to discharge radioactive waste have frequently made use of terms such as "all reasonable steps" and "practicable" as well as the term "best practicable means". Such terminology was subject to consideration in the 1985 trial of BNFL for breach of authorisation conditions relating to its Sellafield works. The condition in question referred to a requirement to limit the amount of radioactive elements discharged and their rate of discharge such that radiation exposure was ALARA; another relevant condition referred to a requirement to take "reasonable steps" to minimise the exposure of persons to radiation. The judge (Rose J.) appears to have accepted the appropriateness of a weighing exercise of costs and other disadvantages against higher levels of risk.[88] Ultimately, the jury was directed to make "common sense" decisions, about what was reasonable in the circumstances, rather than detailed economic judgments; indeed, there was no detailed evidence before the jury on cost issues. An interesting sequel to the case is that Greenpeace attempted unsuccessfully to argue that illegality in failing to comply with the ALARA requirement constituted a defence against claims by BNFL for an injunction against Greenpeace in respect of trespass to BNFL's property.[89]

At about the same time as the BNFL case, the RWMAC argued in its Fifth Report[89a] that certificates of authorisation should be stated in terms of as low as reasonably practicable (ALARP) rather than ALARA; the RWMAC took this view apparently on the basis that actual exposures were outside the ultimate control of the operator, and that the ALARP concept would lend itself more readily to effective control over operations and processes leading to discharges, as well as enjoying the benefits of previous judicial interpretation.

However, there are grave doubts as to whether ALARP would achieve

---

[86] Dr C. Miller, *Economics v. Pragmatics: The Control of Radioactive Wastes* [1990] J.Env.L. (Vol. 2), No. 1, p. 65.

[87] *ibid.*, p. 68.

[88] *ibid.*, p. 73.

[89] *BNFL v. Greenpeace*, C.A., March 25, 1996, unreported.

[89a] HMSO, 1984.

any more effective protection than ALARA[90] or indeed, as discussed at the Sizewell B Inquiry, whether there is in fact any discernible difference between ALARA and ALARP.

A more fundamental issue is whether the use of concepts such as ALARA and BPM in authorisation wording (as distinct from criteria which result in the setting of quantitative limits) is a weakness in the regulatory system. This was certainly the view of the House of Commons Environment Committee in its report on *Radioactive Waste*.[91] Authorisations for discharges from Sellafield granted after the Report and after the BNFL trial appeared to move in the direction of specific emission limits for individual radionuclides and total alpha and beta activity, whilst using the ALARP requirement (rather than ALARA as in previous authorisations) for certain aspects and a defined version of BPM (excluding "grossly disproportionate" expenditure and embracing maintenance and manner of operation of plant) for a general limit on the activity of waste discharged.[92]

Dose limits and restructuring of the nuclear industry   5–55

The application of the principles set out in the previous paragraph are complicated by the restructuring of the nuclear industry whereby ownership of Magnox and other nuclear power stations is split between different companies. This may result in different sources within a single site being owned or operated by different organisations. To provide some reassurance that this will not cause standards to be relaxed (*i.e.* by aggregating the 0.3 mSv/y maximum applicable to the different facilities across the whole site), the Government proposes to use an additional "site constraint", equivalent to the previous site target of 0.5 mSv/y and applicable to the aggregate exposure from all sources with contiguous boundaries at a single location, whether operated by a single or separate organisations.[93]

Optimisation of discharges—lower threshold   5–56

The principle of constantly reducing risk under the concept of optimisation is constrained by the recognition that, at a certain level, risk becomes tolerable or acceptable. As part of its review of radioactive waste management policy, the Government proposes to introduce a lower limit for optimisation, applicable to radioactive waste discharges. This limit is based on the general principle that an annual risk of death of one in a

---

[90] Dr C. Miller, *Economics v. Pragmatics: The Control of Radioactive Wastes* [1990] J.Env.L. (Vol. 2), No. 1, p. 77.
[91] Session 1985–1986, Vol. 1, para. 153.
[92] Dr C. Miller, *Economics v. Pragmatics: The Control of Radioactive Wastes* [1990] J.Env.L. (Vol. 2), No. 1, p. 78.
[93] Cm. 2919 (1995), para. 70.

million (10^{-6}) is broadly acceptable; translated into an annual dose (and bearing in mind the inherent uncertainties in such an exercise) this produces a suggested figure of 0.02 mSv/y. The effect will be that if an operator can show the exposures involved to be below this level, the Agencies should not apply the optimisation principle to seek further reductions, "provided they are satisfied that the operator is using the best practicable means to limit discharges".[94]

As to the relationship with the BPM principle (which, in this context, probably relates to issues such as maintenance and manner of operation of plant) see para. 5–54 above.

5–57 Criteria for authorising solid waste disposal

A different approach to liquid and gaseous effluent is required when authorising facilities for the disposal of solid waste: the approach is essentially risk-based. Criteria set out in the 1984 "Green Book", *Disposal Facilities on Land of Low and Intermediate Level Radioactive Wastes: Principles for the Protection of the Human Environment* have been superseded both in terms of regulatory process and policy by the Government's *Review of Radioactive Wastes Management Policy*.[95]

The issue turns on the safety case provided by the developer or operator of the proposed facility as to the precautions to prevent radionuclides finding their way back to the human environment. As referred to in the *Review*, the question of what constitutes upper and lower levels of acceptability of risk for such facilities is a difficult one, which created a divergence of views within the RWMAC/ACSNI Study Group created by the Government to advise on the issue. The Government, faced with this difficulty, takes the view that the nature of such disposal systems makes them less amenable to the use of quantified, numerical risk criteria than facilities such as new nuclear power stations, and views it as inappropriate to rely on a specified risk limit alone to judge the acceptability of a proposed disposal facility. Instead, it proposes that a "risk target" of 10^{-6}/y of developing either a fatal cancer or serious hereditary defect should be used as an "objective" in the design process. Where the level of risk is below this figure, then no further reductions in this should be sought, provided the Agencies are satisfied that best practicable means to limit risks have been adopted. Above this figure, the regulators will need to be satisfied "not only that an appropriate level of safety is assured, but also that any further improvements in safety could be achieved only at disproportionate cost".[96]

The approach is to be set out more precisely in *Guidance on Requirements for Authorisation* to be prepared by the regulatory bodies, which will in turn take into account the more detailed recommendations made by the

[94] *ibid.*, para. 73.
[95] Cm. 2919 (1995), paras. 74–82.
[96] *ibid.*, para. 78.

RWMAC/ACSNI Study Group on environmental safety criteria (para. 82); the *Guidance* will also have to address the difficult issue of the timescale over which risk is to be assessed, although the Government's view is that the regulators should not prescribe any cut-off for this period (para. 81).

A draft of the Guidance was issued for consultation in October 1995. The draft Guidance suggests that early application for authorisation will be necessary, and considers what input the Agencies will have at the planning stage. It also deals with the principles which will be applied to safeguard the public and radiological and technical requirements.

Provision of data to the Commission   5–58

Article 37 of the EURATOM Treaty is described at para. 5–15 above, and may require data to be provided to the Commission to allow it to determine whether the implementation of any planned disposal of radioactive waste is liable to result in radioactive contamination of water, soil or airspace of another Member State. The authorisation for disposal cannot be issued until the Commission has given its Opinion on the matter, a process which normally requires a period of six months; the Commission will consult a group of experts before giving its Opinion.

Solid waste disposal—site selection generally   5–59

The method of selecting sites for solid waste disposal, whether waste of low, intermediate or high-level activity is a vexed and politically fraught issue, as history demonstrates (see paras. 5–03 to 5–05 above). The Government asked the RWMAC/ACSNI Study Group to advise on the general approach to be adopted, and the Group proposed a formal procedure involving 10 discrete phases, although the Group was not unanimous on the issue. The Government has expressed reservations to these recommendations[97] including concern to the "corporatist approach" which might weaken the responsibility of waste producers and create confusion between the role of existing agencies and authorities, and the "independent commission" proposed by the Group. Nevertheless, the Government recognises the valid concerns raised by the exercise as to the need for transparency of decision-making and public reassurance. It proposes to set out guidelines to developers of repositories about the need for transparency in site selection, publication of research results and the formulation of specific milestones in the process (para. 87). Essentially, the message is that responsibility for the adequacy of procedures rests with the developer, not the Government.

[97] Cm. 2919 (1995), para. 86.

5–60 Solid waste disposal—repository for intermediate-level waste

Whereas the process of site selection for high-level waste has not yet begun (allowing for the formulation of a clear site selection procedure), the process for intermediate-level waste has been under way for some time. Part of this process involves the planning application by Nirex to constitute a rock characterisation facility at Longlands Farm near Sellafield—effectively an underground laboratory to test the geology and hydrogeology of the site. Refusal of the application by Cumbria County Council led to a public inquiry held in 1995.[98]

The policy underlying the Nirex site selection process for a multi-purpose deep site for low and intermediate-level waste is described above (see para. 5–12 above). By 1989, Nirex had prepared a short list of 12 sites and the Government had accepted that a deep repository should be under land rather than sea and that the next steps should be to carry out detailed geological studies in the vicinity of Sellafield and Dounreay. This conclusion of Nirex was based on the view that it would be best to begin by exploring those sites where there was some measure of local support for civil nuclear activities. Following borehole tests, Nirex announced in 1991 that its preliminary conclusion was that either site could potentially support the safety case necessary for a deep site, but that it proposed to concentrate its investigations on Sellafield for logistical and transport reasons (most intermediate-level waste arising from BNFL's operations at Sellafield). If planning permission is obtained for the rock characterisation facility those investigations will proceed. Any planning application for construction of the repository itself will be subject to a public inquiry.[99]

This book ends as it began: the enormous potential of atomic energy for good or evil has also presented society, its legislators and decision-makers, with long-term, and still intractable, problems.

[98] *ibid.*, para. 105.
[99] Cm. 2919 (1995), para. 110.

APPENDIX A

NUCLEAR INSTALLATIONS ACT 1965
(c. 57)

265

Cover for compensation

Miscellaneous and general

Schedules

NUCLEAR INSTALLATIONS ACT 1965 (c. 57)

General Note

This Act came into force on December 1, 1965 (Nuclear Installations Act **A–01**
1965) (Commencement No.1) Order 1965 (S.I. 1965 No. 1880), and
repealed and replaced the Nuclear Installations (Licensing and
Insurance) Act 1959 and the Nuclear Installations (Amendment) Act
1965.

It has been amended by the Nuclear Installations Act 1969 and may be
cited together with that Act as the Nuclear Installations Acts 1965 and
1969 (s.5(2) of the 1969 Act). It has also been amended by the Congenital
Disabilities (Civil Liability) Act 1976; the Atomic Energy Authority Act
1971 and the Atomic Weapons Establishment Act 1991.

Extent

The Act applies to England, Scotland and Wales although the modifi- **A–02**
cations and extensions effected by the Congenital Disabilities (Civil
Liability) Act 1976 do not apply to Scotland. It applies to Northern
Ireland with modifications (s.27).

International obligations

Early appreciation of the potential benefits of the peaceful applications of **A–03**
nuclear power was rapidly tempered by the realisation of the catastrophic
trans-national consequences that could follow from failure to maintain
adequate levels of safety:

> "In atomic energy, because of its spectacularly military origins
> and the potential risk involved in its utilisation, the notion of
> calamities has taken high precedence over other considerations
> both in the public mind and in the minds of the lawyers and
> administrators concerned with the problems of regulation."[1]

This realisation led to the creation of a number of key international
organisations concerned to ensure adequate standards and safeguards.
First amongst these was the International Atomic Energy Agency, whose
Statute was approved on October 23, 1956 and came into force on July 29,

---

[1] Jerry L. Weinstein, *Progress in Nuclear Energy, Service X, Law and Administration* (1966), p. ix.

1957. The IAEA's objectives are to seek to accelerate and enlarge the contribution of atomic energy to peace, health and prosperity throughout the world. One of the functions of the Agency is to establish or adopt standards of safety for the protection of health and minimisation of danger to life and property; the work of the Agency in this sphere has resulted in numerous recommendations, principles, codes and safety standards on the use of nuclear energy, the safety of nuclear power plants, transportation of radioactive materials, radioactive waste management, notification of accidents and non-proliferation.[2] At the regional level the work of the IAEA was supplemented by agencies covering various parts of the globe, including the European Nuclear Energy Agency, the OECD Nuclear Energy Agency and the European Atomic Energy Community (the Treaty of which entered into force on January 1, 1958).

The thrust of the efforts of these agencies has been the prevention of nuclear accidents, but also securing adequate mechanisms for compensation in the event that such disasters do occur. As one commentator puts it:

> "The principal national and international legal problems posed by the development of the pacific uses of atomic energy have been on the one hand to keep calamities from happening and on the other to devise appropriate remedies so that damage to health and property may be compensated in the most humane, equitable and expeditious way possible."[3]

The issue of compensation is discussed below in the context of the Vienna, Paris and Brussels Conventions.

During the 1980s the realisation grew that nationally-orientated approaches to safety had led to potentially difficult discrepancies in technical requirements and standards: a view reinforced by the Chernobyl disaster in 1986. In 1992, the IAEA's Board of Governors authorised the process of preparing an international convention on nuclear safety, knowing that the future development of nuclear power was crucially dependent upon the public's perception as to its safety.[4]

International principles of civil liability

A–04 Several international conventions regulate liability for nuclear damage. One of these was the Vienna Convention on Civil Liability for Nuclear Damage of 1963, negotiated under the auspices of the IAEA. This Convention entered into force in 1978, but only 14 states are parties to it:

---

[2] see: El Baradei, Nwogugu and Rames (eds.), *The International Law of Nuclear Energy: Basic Documents* (1993), Parts 1 and 2.

[3] Jerry L. Weinstein, *Progress in Nuclear Energy, Service X, Law and Administration* (1996), p. ix.

[4] see: Louise de la Fayette, *International Environmental Law and the Problem of Nuclear Safety* [1993] J.E.L. (Vol. 5), No. 1., p. 31.

these do not include the United Kingdom or any other Western European country. Essentially the Convention makes the operator of a nuclear installation strictly liable for damage caused by a nuclear incident in the nuclear installation, or involving nuclear matter coming from or sent to the installation. Liability is by Article IV.1 expressed to be absolute, subject to only very limited defences; by Article V the liability of the operator may be limited by the State where the installation is located to a figure not less than U.S.$ 5 million for any one incident. The operator is required to maintain insurance or other financial security covering his liability for nuclear damage to such an amount as the relevant State shall specify; to the extent that the level of insurance or other financial security is inadequate to satisfy claims up to the financial limit established under Article V, the State is to provide the necessary funds to ensure payment.

The Paris Convention of 1960 is broadly comparable in effect; the Convention, concluded under the aegis of the OECD entered into force on April 1, 1968, and includes the United Kingdom amongst the state parties. By Article 3 of the Paris Convention the operator of a nuclear installation (as defined) is liable for damage to or loss of life of any person and for damage to or loss of any property other than the installation itself and connected on–site property; this régime is reflected closely in the provisions of the Nuclear Installations Act dealing with liability. As with the Vienna Convention, liability is subject to a maximum figure for each nuclear incident—though, unlike the Vienna Convention, the figure is specified in the Convention, subject to the ability of Contracting Parties to establish greater or lesser amounts, taking into account the availability of insurance or other financial security. By Article 10, the operator is required to have and maintain insurance or other financial security, providing sums which may be drawn on only for compensation for damage caused by a nuclear incident. Essentially, the Paris Convention may be said to rest on four fundamental principles:

(a) *Channelling of liability*: liability being absolute and falling exclusively on the operator of the installation.

(b) *Limitation of liability*: liability is limited to a finite amount, to be specified by each State Party within certain parameters.

(c) *Compulsory cover for liability*: liability up to the limit must be covered by financial security.

(d) *Single jurisdiction*: all proceedings in relation to the same nuclear incident must be dealt with in a single court, which must apply both the Convention and its national law. Judgements of the court are to be enforceable in the territory of all Contracting Parties, without the need for fresh inquiries into the facts.[5]

The main defect of the Paris Convention was the relative inadequacy of the compensation contemplated.

---

[5] see Rafaello Fornassier in Weinstein: note 1 *op. cit.*, p. 24.

It was duly supplemented by the Brussels Supplementary Convention of 1963, which entered into force on December 4, 1978. The Brussels Supplementary Convention marked what has been called "an unprecedented step in international collaboration and mutual confidence by agreeing to pay victims of nuclear incidents out of joint Government funds".[6] Based on a draft prepared by the then six EURATOM countries, 13 of the 16 signatories to the Paris Convention agreed that compensation in respect of damage from nuclear incidents should be provided up to 120 million units of account per incident; the first 5 million at least to be provided by insurance or other financial security, then up to 70 million from the relevant national public fund, then up to 120 million from public funds to be made available by the Contracting Parties according to a formula based on gross national product and the thermal power of the reactors located in the Contracting State concerned. Effectively, the Supplementary Convention creates a system of mutual assistance on the principle of financial solidarity.[7] This principle may "result in a Contracting Party being obliged to make available funds ... where the incident occurs in a foreign country and where the operator liable and all the victims are foreigners".[8]

An attempt to regulate the relationship between the Vienna and Paris Inventories (no State being a party to both) is provided by the Joint Protocol Relating to the Application of the Vienna Convention and the Paris Convention. This Protocol entered into force on April 27, 1992; the United Kingdom is not a party. It provides that either the Vienna or Paris Convention shall apply to a nuclear incident to the exclusion of the other; the determining factor being whether the nuclear installation involved is situated in the territory of a Party to the Vienna Convention or the Paris Convention.

Following the Chernobyl incident, efforts have been renewed to strengthen the liability régime. In 1990, the Board of Governors of the IAEA agreed to set up a Standing Committee on Liability for Nuclear Damage to consider, in particular, the relationship between civil and state liability. Preparatory work is being carried out for possible revision of the Vienna Convention, which may in due course lead to the convening of a revision conference. Similarly, a group of Governmental Experts, established by the OECD and Nuclear Energy Agency, is examining the practicalities of revision of the Paris and Brussels Supplementary Conventions in the light of the IAEA Standing Committee's current work.

The Nuclear Installations Acts

A–05 The Nuclear Installations Acts 1965 and 1969 provide a régime of control and liability in relation to the installation and operation of nuclear

---

[6] Jerry L. Weinstein, *Progress in Nuclear Energy, Service X, Law and Administration* (1996), p. x.
[7] see Rafaello Fornassier in Weinstein, p. 27.
[8] see Rafaello Fornassier in Weinstein, p. 29.

reactors (excluding those comprised in a means of transport)and other installations designed or adapted for the production or use of nuclear energy, for processes preparatory or ancillary to the production of such energy, and for the storage, processing and disposal of nuclear fuel and associated irradiated matter (s.1(1)). No site may be used for such purposes without a nuclear site licence granted by the Nuclear Installations Inspectorate (part of the Health and Safety Executive). In addition, certain operations involving the extraction of plutonium or uranium, or the treatment of uranium to increase the proportion of isotope 235, require a permit in writing granted by the United Kingdom Atomic Energy Authority or a Government department (s.2).

Sections 3–5 deal with the grant and variations of nuclear site licences, the attachment of conditions to such licences, and their revocation and surrender. A licence may be granted only to a body corporate and is not transferable (s.3(1)). Following surrender or revocation of a licence, the former licence-holder remains subject to certain obligations during a statutory "period of responsibility" (s.5(3)); this period extends until such time as the Inspectorate gives written notice that in its opinion there has ceased to be any danger from ionising radiations from anything on the site. The duty extends to the display of site notices and compliance with any directions given by the Inspectorate as to the prevention or warning of risk of injury or damage to property by ionising radiations from anything remaining on the site (s.5(2)). The Secretary of State for Energy is required to maintain a list showing every site in respect of which a nuclear site licence has been granted, together with a map showing the position and limits of each site; the list is not required to show sites in respect of which 30 years has elapsed since the expiration of the last licensee's period of responsibility (s.6).

Sections 7–10 impose a series of duties in relation to licensed sites, sites occupied by the UKAEA and the Crown, and nuclear matter in the course of carriage. These duties are essentially to avoid injury to persons or damage to property being caused by occurrences involving nuclear matter as a result of the radioactive properties (with or without toxic, explosive or other hazardous properties) of such matter. Breach of such duties gives rise to a right to compensation under section 12. A time limit (30 years) for bringing claims for compensation is provided by section 15 and a financial cap on such claims (£20 million) is provided by section 16.

The licensee of a nuclear site is required to make provision (by insurance or some other means) for funds to be made available to ensure that claims established for breach of the duty under section 7 are satisfied up to the required amount (s.19). By section 18, sums sufficient to satisfy statutory claims must be made available out of moneys provided by Parliament. Thus the basic features of the Acts, described more fully in the annotations to each individual section, mirror and reflect the requirements of the Paris and Supplementary Brussels Conventions, described above.

271

Control of certain nuclear installations and operations

Restriction of certain nuclear installations to licensed sites

A–06   1.—(1) Without prejudice to the requirements of any other Act, no person . . . shall use any site for the purpose of installing or operating—

(a) any nuclear reactor (other than such a reactor comprised in a means of transport, whether by land, water or air); or

(b) subject to subsection (2) of this section, any other installation of such class or description as may be prescribed, being an installation designed or adapted for—

(i) the production or use of atomic energy; or

(ii) the carrying out of any process which is preparatory or ancillary to the production or use of atomic energy and which involves or is capable of causing the emission of ionising radiations; or

(iii) the storage, processing or disposal of nuclear fuel or of bulk quantities of other radioactive matter, being matter which has been produced or irradiated in the course of the production or use of nuclear fuel,

unless a licence so to do (in this Act referred to as a "nuclear site licence") has been granted in respect of that site by the [Health and Safety Executive] and is for the time being in force.

(2) Regulations made by virtue of paragraph (b) of the foregoing subsection may exempt, or make provision for exempting, from the requirements of that subsection, either unconditionally or subject to prescribed conditions, any installation which the Minister is satisfied is not, or if the prescribed conditions were complied with would not be, a relevant installation.

(3) Any person who contravenes subsection (1) of this section shall be guilty of an offence . . .

Definitions

A–07   "atomic energy": section 26(1)
"contravenes": section 26(1)
"the Minister": section 26(1)
"nuclear reactor": section 26(1)
"nuclear site licence": section 26(1)
"prescribed": section 26(1)
"relevant installation": section 26(1)

Amendments

The references to the Health and Safety Executive in square brackets at **A–08** subsection (1) were substituted by the Nuclear Installations Act 1965, etc. (Repeals and Modifications) Regulations 1974 (S.I. 1974 No. 2056, reg. 2(1)(b) and Sched. 2). These Regulations were made in consequence of the establishment of the Health and Safety Executive and have the general effect of transferring licensing functions from the Minister to the HSE. The words omitted from subsection (1) ("other than the Authority") were deleted by the Nuclear Installations Act 1965 (Repeal and Modifications) Regulations 1990 (S.I. 1990 No. 1918).

The words at subsection (3) were repealed for England, Wales and Scotland but not for Northern Ireland by the Nuclear Installations Act 1965, etc. (Repeals and Modifications) Regulations 1974 (S.I. 1974 No. 2056).

General Note

Subsection (1) requires a nuclear site licence to be obtained from Her **A–09** Majesty's Nuclear Installations Inspectorate (part of the Health and Safety Executive) for the activities described in the subsection. These are:

(a) the use of a site for the purpose of operating or installing a nuclear reactor. This excludes the operation/installation of reactors comprised in a means of transport (*e.g.* a nuclear submarine); and

(b) the use of a site for operating prescribed installations which are designed or adapted for the production of nuclear energy, preparatory or ancillary processes capable of causing the emission of ionising radiations, and storage, processing or disposal of nuclear fuel or bulk quantities of other radioactive matter produced or irradiated in the course of producing or using nuclear fuel.

In relation to category (b), the relevant classes of installation are prescribed by the Nuclear Installations Regulations 1971 (S.I. 1971 No. 381), reg. 3, and include installations for the manufacture of nuclear fuel elements, enriched uranium or plutonium; nuclear assemblies designed for the production of neutrons; the processing of irradiated nuclear fuel; installations for the storage of fuel elements, irradiated nuclear fuel and bulk quantities of other radioactive matter irradiated in the course of producing or using nuclear fuel; and installations for treating irradiated matter to extract uranium or to treat uranium so as to increase the proportion of isotope 235. The Regulations are of a detailed nature and reference should be made to their wording for further clarification. By regulation 4 of the Regulations, pursuant to subsection (2), either the Secretary of State at the Department of Trade and Industry or the

Secretary of State for Scotland may by instrument in writing exempt from the requirements of subsection 1(1) of the Act any installation prescribed by regulation 3 which he is satisfied is not a relevant installation (*i.e.* not an installation to which a relevant international agreement applies: see subs.26(1)).

No definition is given of "bulk quantities" which can give rise to doubts in practice as to when a site licence is required.

The United Kingdom Atomic Energy Authority

A–10 As originally drafted, subsection (1) excluded the Authority from the requirement of a licence. That exclusion was removed by the Nuclear Installations Act 1965 (Repeal and Modifications) Regulations 1990 (S.I. 1990 No. 1918), with effect from October 31, 1990.

Offences: subsection (3)

A–11 In England, Scotland and Wales the maximum penalties are £5,000 on summary conviction and an unlimited fine or two years imprisonment or both on indictment.

In Northern Ireland for a body corporate convicted on indictment, the fine may be such amount as the court thinks just (subs.25(2)).

Prohibition of certain operations except under permit

A–12 2.—(1) Notwithstanding that a nuclear site licence is for the time being in force or is not for the time being required in respect thereof, no person other than the Authority shall use any site—
> (a) for any treatment of irradiated matter which involves the extraction therefrom of plutonium or uranium; or
> (b) for any treatment of uranium such as to increase the proportion of the isotope 235 contained therein,

except under, and in accordance with the terms of, a permit in writing . . . granted by the Authority or a government department [and for the time being in force]; and any fissile material produced under such a permit shall be disposed of only in such manner as may be approved by the authority by whom the permit was granted.

[(1A) A permit granted under this section, unless it is granted by the Minister, shall not authorise the use of a site as mentioned in paragraph (a) or paragraph (b) of the foregoing subsection otherwise than for purposes of research and development.

(1B) Where a permit granted under this section by the Minister to a body corporate authorises such a use of a site for purposes other than, or not limited to, research and development, the Minister may by order direct that the provisions set out in Schedule 1 to this Act shall have effect in relation to that body corporate.

(1C) Any power conferred by this section to make an order shall include power to vary or revoke the order by a subsequent order; and any such power shall be exercisable by statutory instrument, which shall be subject to annulment in pursuance of a resolution of either House of Parliament.

(1D) Any permit granted under this section by the Authority or by the Minister or any other government department may at any time be revoked by the Authority or by the Minister or that department, as the case may be, or may be surrendered by the person to whom it was granted.]

(2) Any person who contravenes [subsection (1) of this section] shall be guilty of an offence and be liable—

    (a) on summary conviction, to a fine not exceeding [the prescribed sum], or to imprisonment for a term not exceeding three months, or to both;

    (b) on conviction on indictment, to a fine . . . , or to imprisonment for a term not exceeding five years, or to both.

Definitions

"the Authority": section 26(1)   A–13
"contravenes": section 26(1)
"the Minister": section 26(1)
"nuclear site licence": section 26(1)

Amendments

Words omitted from subsection (1) were repealed by the Atomic Energy   A–14
Authority Act 1971, s.17(1). The words in square brackets were added by section 17(2) of the same Act.

    Subsections (1A)–(1D) were added by the Atomic Energy Authority Act 1971, section 17(1).

    At subsection (2) the words in square brackets were substituted by the Atomic Energy Authority Act 1971, s.17(2). The words in square brackets in paragraph (a) of subsection (2) were substituted by the Magistrates' Courts Act 1980, s.32(2) and the words omitted from paragraph (b) were removed by virtue of the Criminal Law Act 1977, s.32(1).

General Note

This section imposes additional requirements which relate to certain   A–15
activities even where a nuclear site licence under section 1 is in force or no such licence is required by virtue of an exemption. For these activities,

which have military as well as civil implications, a permit in writing is required from the Secretary of State or the United Kingdom Atomic Energy Authority. Specific approval is also required as to the manner in which fissile material produced by the process is to be dealt with. The Authority only has power to grant a permit for research and development purposes; otherwise Ministerial permission is needed (subs.(1A)). Additionally, where a permit is granted for purposes other than research and development, the Secretary of State may, by order, apply the provisions of Schedule 1 dealing with the security of the site, the role of Ministry of Defence police, and the safe-keeping of information and materials (subs.(1B)). The Nuclear Installations (Application of Security Provisions) Order 1971 (S.I. 1971 No. 569) has been made under subsection (1B) and applies to British Nuclear Fuels plc.

The amendments introduced into subsection (2) mean that the maximum fine for an offence under the section in the magistrates' court is £5,000 and unlimited in the Crown Court.

Nuclear site licences

Grant and variation of nuclear site licences

A–16   3.—(1) A nuclear site licence shall not be granted to any person other than a body corporate and shall not be transferable.

[(1A) The Health and Safety Executive shall consult the appropriate Agency before granting a nuclear site licence in respect of a site in Great Britain.]

(2) Two or more installations in the vicinity of one another may, if the [Health and Safety Executive] thinks fit, be treated for the purposes of the grant of a nuclear site licence as being on the same site.

(3) Subject to subsection (4) of this section, where it appears to the [Health and Safety Executive] appropriate so to do in the case of any application for a nuclear site licence in respect of any site, he may direct the applicant to serve on such bodies of any of the following descriptions as may be specified in the direction, that is to say—

   (a) any local authority;
   (b) ..., any water undertaker or any local fisheries committee;
   (c) any river purification board within the meaning of the Rivers (Prevention of Pollution) (Scotland) Act 1951, any district board constituted under the Salmon Fisheries (Scotland) Acts 1828 to 1868, the board of commissioners appointed under the Tweed Fisheries Act 1857, and any local water authority within the meaning of the Water (Scotland) Acts 1946 and 1949; and
   (d) any other body which is a public [or local] authority,

notice that the application has been made, giving such particulars as may be so specified with respect to the use proposed to be made of the site under the licence, and stating that representations with respect thereto may be made to the [Health and Safety Executive] by the body upon whom the notice is served at any time within three months of the date of service; and where such a direction has been given, the [Health and Safety Executive] shall not grant the licence unless he is satisfied that three months have elapsed since the service of the last of the notices required thereby nor until after he has considered any representations made in accordance with any of those notices.

(4) Subsection (3) of this section shall not apply in relation to an application in respect of a site for a generating station [where a consent under section 36 of the Electricity Act 1989 or article 39 of the Electricity Supply (Northern Ireland) Order 1992 is required for the operation of the station.]

(5) A nuclear site licence may include provision with respect to the time from which section 19(1) of this Act is to apply in relation to the licensed site, and where such provision is so included the said section 19(1) shall not apply until that time or the first occasion after the grant of the licence on which any person uses the site for the operation of a nuclear installation, whichever is the earlier [provided that no such provision shall be so included without the consent of the Secretary of State].

(6) The [Health and Safety Executive] may from time to time vary any nuclear site licence by excluding therefrom any part of the licensed site—

(a) which the licensee no longer needs for any use requiring such a licence; and

(b) with respect to which the [Health and Safety Executive] is satisfied that there is no danger from ionising radiations from anything on that part of the site.

[(6A) The Health and Safety Executive shall consult the appropriate Agency before varying a nuclear site licence in respect of a site in Great Britain, if the variation relates to or affects the creation, accumulation or disposal of radioactive waste, within the meaning of the Radioactive Substances Act 1993.]

Definitions

"appropriate Agency": section 26(1)   A–17
"licensed site": section 26(1)
"licensee": section 26(1)
"the Minister": section 26(1)
"nuclear installation": section 26(1)
"nuclear site licence": section 26(1)

Amendments

Apart from subsections (1A) and (6A), which are inserted by paragraph 7 **A–18** of Schedule 22 to the Environment Act 1995, the references in square brackets to the Health and Safety Executive were substituted by the

Nuclear Installations Act 1965, etc. (Repeals and Modifications) Regulations 1974 (S.I. 1974 No. 2056, reg. 2(1)(b) and Sched. 2).

Subsection (3)(b) was substituted by the Water Act 1989, s.190 and Sched. 25, although the words omitted have now ceased to have effect. The words in square brackets at subsection (3)(d) were repealed for Scotland by the Local Government (Scotland) Act 1973, Sched. 29.

The words in square brackets at subsection (4) were inserted by the Electricity Act 1989, s.112(1) and the Electricity Supply (Northern Ireland) Order 1992 (S.I. 1992 No. 231).

The words in square brackets at subsection (5) were added for England, Wales and Scotland (but not for Northern Ireland) by the Nuclear Installations Act 1965, etc. (Repeals and Modifications) Regulations 1974 (S.I. 1974 No. 2056).

General Note

A–19 This section deals with the grant and variation of nuclear site licences under section 1. The licence can only be granted to a body corporate and is personal and not transferable.

Whilst section 1 might indicate that each separate site of an installation requires a licence, subsection (3) allows a number of installations in the vicinity of each other to be treated as being on the same site for licensing purposes.

Consultation

A–20 There are no compulsory requirements of publicity or consultation in relation to applications for nuclear site licences. However, by subsection (3) the HSE may direct that the application be served on specified public bodies inviting representations with respect to it. This provision does not apply in relation to applications for generating stations covered by the consent system under the Electricity Act 1989, s.36 or the Northern Ireland equivalent.

Under section 36(1) of the 1989 Act, consent from the Secretary of State is typically required for the construction, extension or operation of a generating station. Provisions under section 36(4) empower the Secretary of State to exempt generating stations of a particular class or description. At February 1997, only the Off-shore Generating Station (Exemption) Order 1990 (S.I. 1990 No. 443) has been made under section 36(4), exempting certain off-shore installations from the need of a consent. Section 36(2) of the 1989 Act provides that a consent under section 36(1) is not required for generating stations that do not exceed a permitted capacity of 50 megawatts; the subsection also empowers the Secretary of State to make different provisions for different categories of generating stations. Further variation from the basic exemption under section 36(2) is provided by the powers given to the Secretary of State under section 36(3) whereby he may direct by order a change in the permitted capacity of generating stations under which no consent is required for construction, extension or operation. At February 1997, only an order relating to hydro-electric generating stations in Scotland of 1 megawatt permitted capacity has been made (The Electricity Act 1989 (Requirement of Consent for Hydro-electric Generating Stations) (Scotland) Order 1990 (S.I. 1990 No. 392)).

Variation

Variation of licences is provided for by subsections (6) and (6A). The **A–21** power to vary extends to the exclusion of all or any part of a site from the licence where it is no longer needed and where the HSE is satisfied that there is no danger from ionising radiations from the relevant land.

Financial cover

By subsection (5), with the Secretary of State's consent, special provision **A–22** may be made with respect to the time from which financial cover in respect of statutory liabilities is to be provided under section 19. The effect of the subsection is that the application of section 19 cannot be postponed beyond the first occasion after the grant of the licence on which any person uses the site for the operation of a nuclear installation.

Attachment of conditions to licences

4.—(1) The [Health and Safety Executive] by instrument in writing shall **A–23** on granting any nuclear site licence, and may from time to time thereafter, attach to the licence such conditions as may appear to the [Health and Safety Executive] to be necessary or desirable in the interests of safety, whether in normal circumstances or in the event of any accident or other emergency on the site, which conditions may in particular include provision—
- (a) for securing the maintenance of an efficient system for detecting and recording the presence and intensity of any ionising radiations from time to time emitted from anything on the site or from anything discharged on or from the site;
- (b) with respect to the design, siting, construction, installation, operation, modification and maintenance of any plant or other installation on, or to be installed on, the site;
- (c) with respect to preparations for dealing with, and measure to be taken on the happening of, any accident or other emergency on the site;

(d) without prejudice to sections [13 and 16] of the Radioactive Substances Act [1993], with respect to the discharge of any substance on or from the site.

(2) The [Health and Safety Executive] may at any time by instrument in writing attach to a nuclear site licence such conditions as the [Executive] may think fit with respect to the handling, treatment and disposal of nuclear matter.

(3) The [Health and Safety Executive] may at any time by a further instrument in writing vary or revoke any condition for the time being attached to a nuclear site licence by virtue of this section.

[(3A) The Health and Safety Executive shall consult the appropriate Agency:

(a) before attaching any condition to a nuclear site licence in respect of a site in Great Britain, or

(b) before varying or revoking any condition attached to such a nuclear site licence,

if the condition relates to or affects the creation, accumulation or disposal of radioactive waste within the meaning of the Radioactive Substances Act 1993.]

(4) While a nuclear site licence remains in force in respect of any site, the [Health and Safety Executive] shall consider any representations by any organisation representing persons having duties upon the site which may from time to time be made to him with a view to the exercise by him in relation to the site of any of his powers under the foregoing provisions of this section.

(5) At all times while a nuclear site licence remains in force, the licensee shall cause copies of any conditions for the time being in force under this section to be kept posted upon the site, and in particular on any part thereof which an inspector may direct, in such characters and in such positions as to be conveniently read by persons having duties upon the site which are or may be affected by those conditions.

(6) Any person who contravenes subsection (5) of this section, and, in the event of any contravention of any condition attached to a nuclear site licence by virtue of this section, the licensee and any person having duties upon the site in question by whom that contravention was committed, shall be guilty of an offence . . . and any person who without reasonable cause pulls down, injures or defaces any document posted in pursuance of the said subsection (5) shall be guilty of an offence and be liable on summary conviction to a fine not exceeding [level 2 on the standard scale: England, Wales and Scotland].

Definitions

A–24 "appropriate Agency": section 26(1)
"contravenes" and "contravention": section 26(1)
"inspector": section 26(1)

"licensee": section 26(1)
"nuclear matter": section 26(1)
"nuclear site licence": section 26(1)

Amendments

Apart from subsection (3A) which is inserted by paragraph 8 of Schedule **A–25**
22 to the Environment Act 1995, the references in square brackets to the
Health and Safety Executive were substituted by the Nuclear Installations
Act 1965, etc. (Repeals and Modifications) Regulations 1974 (S.I. 1974
No. 2056), reg. 2(1)(b) and Sched. 2.

The words in square brackets in paragraph (d) of subsection 1 were
substituted by the Radioactive Substances Act 1993, Sched. 4.

The words omitted in subsection (6) were repealed for England, Wales
and Scotland by the Nuclear Installations Act 1965 etc. (Repeals and
Modifications) Regulations 1974 (S.I. 1974 No. 2056), but remain in
force for Northern Ireland.

The words in square brackets in subsection 6 relating to standard scale
levels of fine were substituted by the Criminal Justice Act 1982, section 46
(for England and Wales) and the Criminal Procedure (Scotland) Act
1975, ss.289F and 289G (for Scotland).

General Note

The section requires the HSE to attach to the nuclear site licence such **A–26**
conditions as appear necessary or desirable in the interests of safety—
whether in normal, accidental or emergency circumstances.

Further, by subsection (2), conditions may be imposed in writing
relating to the handling, treatment and disposal of nuclear matter at and
from the site.

Variation and revocation of conditions

Site licence conditions may be varied or revoked in writing by the HSE **A–27**
(subs.(3)).

Representations by unions, etc.

By subsection (4) while a site licence remains in force, the HSE must **A–28**
consider any representations made by organisations representing persons having duties on the site (though not necessarily employed there)
with a view to the exercise of the HSE's powers to impose and modify
conditions.

Notice of conditions

A–29 Copies of the site licence conditions must be kept posted on the site in convenient positions and on such parts of the site as an HSE inspector may direct (subs.(5)).

Offences

A–30 Subsection (6) provides a number of offences, the most significant of which is contravention of site licence conditions; this offence may be committed by the licensee and by any person having duties on the site. On summary conviction, the maximum penalty is a fine of £100 or 3 months imprisonment or both. On indictment, the maximum penalty is a fine of £500 or five years imprisonment or both. An example of a prosecution for contravention of a licence condition is provided by the prosecution of British Nuclear Fuels Ltd as a result of an incident in November 1983 when a large quantity of slightly radioactive liquid was discharged into the Irish Sea from the pipeline at BNFL's Sellafield plant. Whilst the discharge did not violate the relevant certificate of authorisation under the Radioactive Substances Act 1960, it received considerable media attention as a result of protests by Greenpeace. BNFL were prosecuted under section 4(6) (and under section 13(1) of the Radioactive Substances Act). One of the counts was contravention of a licence condition requiring the licensee to "take all reasonable steps to minimise the exposure of persons to radiation". BNFL were fined £2,500 on that count.[9]

Revocation and surrender of licences

A–31 5.—(1) A nuclear site licence may at any time be revoked by the [Health and Safety Executive] or surrendered by the licensee.

[(1A) The Health and Safety Executive shall consult the appropriate Agency before revoking a nuclear site licence in respect of a site in Great Britain.]

(2) Where a nuclear site licence has been revoked or surrendered, the licensee shall, if so required by the [Health and Safety Executive] deliver up or account for the licence to such person as the [Health and Safety Executive] may direct, and shall during the remainder of the period of his responsibility cause to be kept posted upon the site such notices indicating the limits thereof in such positions as may be directed by an inspector; and the [Health and Safety Executive] may on revocation or surrender and from time to time thereafter until the expiration of the said period give to the licensee such other directions as the [Health and

---

[9] see: Dr C. E. Miller [1989] J.E.L. (Vol. 1), No. 1, p. 10.

Safety Executive] may think fit for preventing or giving warning of any risk of injury to any person or damage to any property by ionising radiations from anything remaining on the site.

(3) In this Act, the expression "period of responsibility" in relation to the licensee under a nuclear site licence means, as respects the site in question or any part thereof, the period beginning with the grant of the licence and ending with whichever of the following dates is the earlier, that is to say—

(a) the date when the [Health and Safety Executive] gives notice in writing to the licensee that in the opinion of the [Health and Safety Executive] there has ceased to be any danger from ionising radiations from anything on the site or, as the case may be, on that part thereof;

(b) the date when a new nuclear site licence in respect of a site comprising the site in question or, as the case may be, that part thereof is granted either to the same licensee or to some other person,

except that it does not include any period during which section 19(1) of this Act does not apply in relation to the site.

(4) If the licensee contravenes any direction for the time being in force under subsection (2) of this section, he shall be guilty of an offence ... and any person who without reasonable cause pulls down, injures or defaces any notice posted in pursuance of the said subsection (2) shall be guilty of an offence and be liable on summary conviction to a fine not exceeding [level 2 on the standard scale: England, Wales and Scotland].

(5) ...

Definitions

"appropriate Agency": section 26(1)   A–32
"contravenes": section 26(1)
"injury": section 26(1)
"inspector": section 26(1)
"licensee": section 26(1)
"the Minister": section 26(1)
"nuclear site licence": section 26(1)

Amendments

Apart from subsection (1A) which is inserted by paragraph 9 of Schedule A–33
22 to the Environment Act 1995, the reference in square brackets to the
Health and Safety Executive were substituted by the Nuclear Installations
Act 1965, etc. (Repeals and Modifications) Regulations 1974 (S.I. 1974
No. 2056), reg. 2(1)(b) and Sched. 2.

The words omitted in subsection (4) were repealed for England, Scotland and Wales by the Nuclear Installations Act 1965, etc. (Repeals and Modifications) Regulations 1974 (S.I. 1974 No. 2056), but remain in force for Northern Ireland.

The words in square brackets in subsection (4) relating to standard scale levels of fine were substituted by the Criminal Justice Act 1982, s.46 (for England and Wales) and the Criminal Procedure (Scotland) Act 1975, ss.289F and 289G (for Scotland).

Subsection (5) was repealed for England, Wales and Scotland by the Nuclear Installations Act 1965, etc. (Repeals and Modifications) Regulations 1974 (S.I. 1974 No. 2056), but remains in force for Northern Ireland.

General Note

A–34 This section deals with revocation and surrender of licences. A licence may at any time be revoked by the HSE; no statutory grounds of revocation are given. Similarly, there is no statutory restriction on the licensee's right to surrender the licence at any time.

However, there are provisions to ensure that the site of a revoked or surrendered licence does not present uncontrolled hazards (subs. (2)). In particular, the licensee is required to post site notices and to comply with directions of the HSE for the purpose of preventing or giving warning of risk of injury to persons or property by ionising radiations from anything remaining on the site. These obligations extend to the statutory period of responsibility defined in subsection (3), *i.e.* the date when the HSE notifies in writing its opinion that there is no further danger from ionising radiation or, if earlier, the date on which a new site licence is granted.

Maintenance of list of licensed sites

A–35   6.—(1) Subject to subsection (2) of this section, the Minister shall maintain a list showing every site in respect of which a nuclear site licence has been granted [by him] and including a map or maps showing the position and limits of each such site, and make arrangements for the list or a copy thereof to be available for inspection by the public; and he shall cause notice of those arrangements to be made public in such manner as may appear to him appropriate.

(2) The said list shall not be required to show any site or part of a site in the case of which—

(a) no nuclear site licence is for the time being in force; and

(b) thirty years have elapsed since the expiration of the last licensee's period of responsibility.

Definitions

"licensee": section 26(1).
"licence": section 26(1)
"the Minister": section 26(1)
"nuclear site licence": section 26(1)

A–36

Amendments

The words in square brackets at subsection (1) were repealed for England, Wales and Scotland by the Nuclear Installations Act 1965, etc. (Repeals and Modifications Regulations 1974 (S.I. 1974 No. 2056), reg. 2(1)(a) and Sched. 1, but remain in force for Northern Ireland.

A–37

General Note

The Secretaries of State are required to maintain a list and map showing sites in respect of which a nuclear site licence has been granted; where the site licence has been surrendered or revoked, details need only be kept for 30 years from expiry of the last licensee's period of responsibility (see s.5 above).

A–38

Duty of licensee, etc., in respect of nuclear occurrences

Duty of licensee of licensed site

7.—(1) [Subject to subsection (4) below, where] a nuclear site licence has been granted in respect of any site, it shall be the duty of the licensee to secure that—

    (a) no such occurrence involving nuclear matter as is mentioned in subsection (2) of this section causes injury to any person or damage to any property of any person other than the licensee, being injury or damage arising out of or resulting from the radioactive properties, or a combination of those and any toxic, explosive or other hazardous properties, of that nuclear matter; and

    (b) no ionising radiations emitted during the period of the licensee's responsibility—

        (i) from anything caused or suffered by the licensee to be on the site which is not nuclear matter; or

        (ii) from any waste discharged (in whatever form) on or from the site,

    cause injury to any person or damage to any property of any person other than the licensee.

A–39

(2) The occurrences referred to in subsection (1)(a) of this section are—

(a) any occurrence on the licensed site during the period of the licensee's responsibility, being an occurrence involving nuclear matter;

(b) any occurrence elsewhere than on the licensed site involving nuclear matter which is not excepted matter and which at the time of the occurrence—

    (i) is in the course of carriage on behalf of the licensee as licensee of that site; or

    (ii) is in the course of carriage to that site with the agreement of the licensee from a place outside the relevant territories; and

    (iii) in either case, is not on any other relevant site in the United Kingdom;

(c) any occurrence elsewhere than on the licensed site involving nuclear matter which is not excepted matter and which—

    (i) having been on the licensed site at any time during the period of the licensee's responsibility; or

    (ii) having been in the course of carriage on behalf of the licensee as licensee of that site,

has not subsequently been on any relevant site, or in the course of any relevant carriage, or (except in the course of relevant carriage) within the territorial limits of a country which is not a relevant territory.

(3) In determining the liability by virtue of subsection (1) of this section in respect of any occurrence of the licensee of a licensed site, any property which at the time of the occurrence is on that site, being—

(a) a nuclear installation; or

(b) other property which is on that site—

    (i) for the purpose of use in connection with the operation, or the cessation of the operation, by the licensee of a nuclear installation which is or has been on that site; or

    (ii) for the purpose of the construction of a nuclear installation on that site,

shall, notwithstanding that it is the property of some other person, be deemed to be the property of the licensee.

[(4) Section 8 of this Act shall apply in relation to sites occupied by the Authority.]

Definitions

A–40 "the Authority": section 26(1)
"excepted matter": section 26(1)
"injury": section 26(1)
"licensed site": section 26(1)
"licensee": section 26(1)

"nuclear installation": section 26(1)
"nuclear matter": section 26(1)
"period of responsibility": section 26(1)
"relevant carriage": section 26(1)
"relevant installation": section 26(1)
"relevant site": section 26(1)
"relevant territory": section 26(1)
"territorial limits": section 26(1)

Amendments

The words in square brackets at subsection (1) were added by the Nuclear A–41
Installations Act 1965 (Repeal and Modifications) Regulations 1990 (S.I.
1990 No. 1918), reg. 2. Subsection (4) was added by regulation 2.

General Note

This section imposes a statutory duty on the licensee of a licensed site to A–42
secure that specified occurrences or emissions do not cause injury to any
person, or damage to the property of any person other than the licensee.
Those occurrences are listed at subsection (2) and include both
occurrences on-site and certain off-site occurrences involving material in
the course of carriage and material which has been on site or which has
been in the course of carriage. In any event the injury or damage must
arise out of or result from the radioactive properties of the nuclear
matter, or from a combination of those and any toxic, explosive or other
hazardous properties of that matter.

The other limb of the duty relates to the emission of ionising radiations
during the period of the licensee's responsibility (see s.5) from anything
caused or suffered by the licensee to be on the site which is not nuclear
matter, or from any waste discharged on or from the site. The reference to
waste discharged from the site is potentially important, since radioactive
waste from nuclear sites is routinely discharged to the marine environment and to the atmosphere under authorisations under the Radioactive
Substances Act 1993.

Injury to persons

Section 7 refers to injury to any person and so applies to persons injured A–43
by occurrences or ionising radiation on-site and off-site, and whether
employees or third parties. An example of a claim under section 7 which
was settled out of court was that of Rudi Molinari, an employee of the

Ministry of Defence at Chatham Naval Dockyard from 1970–1983, during which time he was involved in the refitting of nuclear submarines. He contracted leukaemia in 1990. Liability was accepted by the Ministry of Defence (notwithstanding the inapplicability of section 1 of the Act to nuclear reactors comprised in means of transport) and an agreed award of damages was approved by the court.

Both BNFL and the UKAEA operate schemes which compensate the dependants of former and current employees who die of leukaemia or other forms of cancer; the scheme does not involve any concession of liability.

Property of the licensee

A–44 The duty does not extend to avoiding damage to property of the licensee. This term receives an extended definition by subsection (3) in the case of occurrences of the licensee; in this context it includes property which at the time of the occurrence is on the site and which is either a nuclear installation or is there for the purpose of using or ceasing to use the nuclear installation, or for the purpose of constructing a nuclear installation on the site.

Relationship of duty to site licence applications

A–45 The duty under section 7 does not impose a duty on an applicant for a nuclear site licence to prove absolutely that no such occurrence as is mentioned in section 7 can possibly occur.[10]   In that case, Friends of the Earth argued that subsection 7(1) had to be interpreted literally and required an assurance of absolute safety before a licence could be granted. The Court of Appeal (Gibson J.) rejected that contention which would effectively mean that the operator would have to demonstrate that no such event could possibly occur, "... even from an unforeseeable natural event, or from some combination, however wildly improbable, of equipment failure or human error or malice on the part of the operators. That would in effect mean that no licence could be granted."

Compensation for breach of duty

A–46 Compensation is payable under section 12 in respect of injury or damage caused in breach of the duty (see also annotations to s.12).

[10] see: *The Friends of the Earth* [1988] J.P.L. 93, C.A.; C. E. Miller [1989] J.E.L. (Vol. 1), No. 1, p. 10.

Nature of damage

Section 7 is not a duty to avoid risk to health or economic loss such as A–47
diminution in the value of property: *Merlin v. British Nuclear Fuels plc.*[11] In
that case, the plaintiffs claimed that their house had become contaminated by dust deposits from radioactive waste material discharged from
BNFL's Sellafield plant into the Irish Sea. A claim for damages based on
section 7 was made, alleging that by virtue of the contamination the house
had become worthless as a home; the plaintiffs moved from their
property because of fears as to health effects, and the property was sold at
a low price. Gatehouse J. held that the Act gave no remedy in respect of
economic loss. It was held that the Act provided, as was required by the
Vienna Convention, for damage to property, injury and loss of life; the
term property as used in the Convention was regarded as meaning
physical property, not incorporeal property or property rights, nor air
space within the walls of the house. Contamination *per se* was not to be
equated to damage to property. It was also held that the Act compensated
for proved personal injury, not the risk of future personal injury: if it were
otherwise a number of very difficult issues would arise in relation to
assessing risk. Further, Gatehouse J. was influenced by the current
position as to irrecoverability of economic loss under English law, the
undesirability of a construction which would result in an operator being
in continual breach of the statutory duty to a possibly very large number
of people, and the need for causation between the presence of ionising
radiations from the installation and some consequential damage.

United Kingdom Atomic Energy Authority

Section 7 applies with modifications to the Authority by section 8.   A–48

Merchant Shipping

The Merchant Shipping Act 1979, Sched. 4, Art. 3(c) excludes claims A–49
made for a breach of duty under section 7.

Duty of Authority

    8. Section 7 of this Act shall apply in relation to the Authority—   A–50
       (a) as if any premises which are or have been occupied by the
           Authority were a site in respect of which a nuclear site licence
           has been granted to the Authority; and
       (b) as if in relation to any such premises any reference to the period
           of the licensee's responsibility were a reference to any period
           during which the Authority is in occupation of those premises;

---

[11] [1990] 3 W.L.R. 383; [1991] J.E.L. (Vol. 3), No. 1, p. 122 with comment by R. Macrory.

[and section 7 shall so apply whether or not a nuclear site licence has been granted in respect of the premises in question.]

Definitions

A–51 "the Authority": section 26(1)
"licensee": section 26(1)
"nuclear site licence": section 26(1)
"period of responsibility": section 26(1)

Amendments

A–52 The words in square brackets were added by the Nuclear Installations Act 1965 (Repeal and Modifications) Regulations 1990 (S.I. 1990 No. 1918).

General Note

A–53 This section applies the duties of section 7 to the Authority as if a nuclear site licence had been granted in respect of the relevant site. The period of responsibility is equated to the period of occupation by the Authority.

The explanatory note to S.I. 1990 No. 1918, which adds words to section 8, states that the intention is to ensure that section 8 continues to apply to determine the duties of the Authority in respect of all premises which are or have been occupied by it, so that those duties apply whether or not a nuclear site licence has been granted in respect of the premises.

Merchant Shipping

A–54 Claims under section 8 are excluded by the Merchant Shipping Act 1979, Sched. 4, Art. 3(c).

Duty of Crown in respect of certain sites

A–55 9. If a government department uses any site for any purpose which, if section 1 of this Act applied to the Crown, would require the authority of a nuclear site licence in respect of that site, section 7 of this Act shall apply in like manner as if—

(a) the Crown were the licensee under a nuclear site licence in respect of that site; and
(b) any reference to the period of the licensee's responsibility were a reference to any period during which the department occupies the site.

Definitions

"licensee": section 26(1)   A–56
"nuclear site licence": section 26(1)
"period of responsibility": section 26(1)

General Note

The requirement of a nuclear site licence does not apply to the Crown, as A–57
a result of Crown Immunity. However, the duties imposed by section 7 are
applied to the Crown by this section, notwithstanding the absence of a site
licence. The period of responsibility for the purpose of section 7 is
equated to the period during which the relevant Government department occupies the site.

Merchant Shipping

Claims made under section 9 are excluded by the Merchant Shipping Act A–58
1979, Sched. 4, Art. 4(c).

Duty of certain foreign operators

10.—(1) In the case of any nuclear matter which is not excepted matter A–59
and which—
    (a) is—
        (i) in the course of carriage on behalf of a relevant foreign
           operator; or
        (ii) in the course of carriage to such an operator's relevant site
           with the agreement of that operator from a place outside
           the relevant territories,
        and is not for the time being on any relevant site in the United
        Kingdom; or
    (b) having been on such an operator's relevant site or in the course
        of carriage on behalf of such an operator, has not subsequently
        been on any relevant site or in the course of any relevant
        carriage or (except in the course of relevant carriage) within the
        territorial limits of a country which is not a relevant territory,
it shall be the duty of that operator to secure that no occurrence such as is
mentioned in subsection (2) of this section causes injury to any person or
damage to any property of any person other than that operator, being
injury or damage arising out of or resulting from the radioactive
properties, or a combination of those and any toxic, explosive or other
hazardous properties, of that nuclear matter.
    (2) The occurrences referred to in the foregoing subsection are—

(a) an occurrence taking place wholly or partly within the territorial limits of the United Kingdom; or

(b) an occurrence outside the said territorial limits which also involves nuclear matter in respect of which a duty is imposed on any person by section 7, 8 or 9 of this Act.

Definitions

A–60 "excepted matter": section 26(1)
"injury": section 26(1)
"nuclear matter": section 26(1)
"relevant carriage": section 26(1)
"relevant foreign operator": section 26(1)
"relevant site": section 26(1)
"relevant territory": section 26(1)
"territorial limits": section 26(1)

General Note

A–61 This section replaces the Nuclear Installations (Amendment) Act 1965, s.2. It sets out the duties of an operator based in a country outside the United Kingdom that is also a signatory to the Paris Convention, in relation to nuclear matter (other than excepted matter, as defined) under specified circumstances. The duty in subsection (1) is very similar to that in section 11 relating to the carriage of non-excepted nuclear material is the United Kingdom.

Merchant Shipping

A–62 Claims under section 10 are excluded by the Merchant Shipping Act 1979, Sched. 4, Art. 3(c).

Duty of persons causing nuclear matter to be carried

A–63 11. Where any nuclear matter, not being excepted matter, is in the course of carriage within the territorial limits of the United Kingdom on behalf of any person (hereafter in this section referred to as "the responsible party") and—

(a) the carriage is not relevant carriage; and

(b) the nuclear matter is not for the time being on any relevant site, it shall be the duty of the responsible party to secure that no occurrence involving that nuclear matter causes injury to any person or damage to any property of any person other than the responsible party, being injury or damage incurred within the said territorial limits and arising out of or resulting from the radioactive properties, or a combination of those and any toxic, explosive or other hazardous properties, of that nuclear matter.

Definitions

"carriage": section 26(2). A–64
"excepted matter": section 26(1).
"injury": section 26(1).
"nuclear matter": section 26(1).
"relevant carriage": section 26(1).
"relevant site": section 26(1).
"territorial limits": section 26(1).

General Note

The section replaces the Nuclear Installations (Amendment) Act 1965, A–65 section 3. It imposes duties on any persons on whose behalf nuclear matter is transported outside a relevant site, as defined, but within the territorial limits of the United Kingdom. The duty is expressed in terms that are very similar to the duty of foreign operators under subsection 10(1).

Merchant Shipping

Claims under section 11 are excluded by the Merchant Shipping Act A–66 1979, Sched. 4, Art. 3(c).

Right to compensation in respect of breach of duty

Right to compensation by virtue of sections 7 to 10

12.—(1) Where any injury or damage has been caused in breach of a A–67 duty imposed by section 7, 8, 9 or 10 of this Act—
    (a) subject to sections 13(1), (3) and (4), 15 and 17(1) of this Act, compensation in respect of that injury or damage shall be payable in accordance with section 16 of this Act wherever the injury or damage was incurred;
    (b) Subject to subsections (3) and (4) of this section and to section 21(2) of this Act, no other liability shall be incurred by any person in respect of that injury or damage.

(2) Subject to subsection (3) of this section, any injury or damage which, though not caused in breach of such a duty as aforesaid, is not reasonably separable from injury or damage so caused shall be deemed for the purposes of subsection (1) of this section to have been so caused.

(3) Where any injury or damage is caused partly in breach of such a duty as aforesaid and partly by an emission of ionising radiations which does not constitute such a breach, subsection (2) of this section shall not affect any liability of any person in respect of that emission apart from this Act, but a claimant shall not be entitled to recover compensation in respect of the same injury or damage both under this Act and otherwise than under this Act.

[(3A) Subject to subsection (4) of this section, where damage to any property has been caused which was not caused in breach of a duty imposed by section 7, 8, 9 or 10 of this Act but which would have been caused in breach of such a duty if in subsection (1)(a) or (b) of the said section 7 the words "other than the licensee" or in subsection (1) of the said section 10 the words "other than that operator" had not been enacted, no liability which, apart from this subsection, would have been incurred by any person in respect of that damage shall be so incurred except—

    (a) in pursuance of an agreement to incur liability in respect of such damage entered into in writing before the occurrence of the damage; or

    (b) where the damage was caused by an act or omission of that person done with intent to cause injury or damage.]

(4) Subject to section 13(5) of this Act, nothing in subsection (1)(b) [or in subsection (3A)] of this section shall affect—

    (a) ...

    (b) the operation of the Carriage by Air Act 1932, the Carriage by Air Act 1961 or the Carriage by Air (Supplementary Provisions) Act 1962 in relation to any international carriage to which a convention referred to in the Act in question applies; or

    (c) the operation of any Act which may be passed to give effect to the Convention on the Contract for the International Carriage of Goods by Road signed at Geneva on May 19, 1956.

Definition

A–68   "injury": section 26(1).

Amendments

A–69   Subsection (3A) and the words in square brackets in subsection (4) were inserted by the Nuclear Installations Act 1969, s.1. Subsection (4)(a) was repealed by the Carriage of Goods by Sea Act 1971, s.6(3)(b).

General Note

This section replaces the Nuclear Installations (Amendment) Act 1965, **A–70** s.4(1)–(4). It sets out the extent of the right to compensation of persons suffering injury or damage as a result of a breach of the duty imposed by sections 7–10. Such compensation is payable in accordance with the provisions of section 16 (see below). Compensation extends to injury or damage which is not reasonably separable from injury or damage caused by the breach. Where the injury or damage has been caused partly by a breach of duty under the Act and partly by a breach outside the Act and liability accrues outside as well as within the Act, a claimant cannot claim under the Act and outside the Act for the same injury or damage.

The operation of the section is unaffected by the Health and Safety at Work, etc. Act 1974 (see below).

Subsection (3A)

This subsection was added by the Nuclear Installations Act 1969, s.1. It **A–71** sets out the circumstances in which a licensee (under s.7) or a foreign operator (under s.10) may be liable following damage to their own property.

Health and Safety at Work

Section 47(1)(c) of the Health and Safety at Work, etc. Act 1974 **A–72** specifically exempts the compensation provisions of section 12 of the Nuclear Installations Act 1965 from the operation of Part I of the 1974 Act.

Exclusion, extension or reduction of compensation in certain cases

13.—(1) Subject to subsections (2) and (5) of this section, compen- **A–73** sation shall not be payable under this Act in respect of injury or damage caused by a breach of a duty imposed by section 7, 8, 9 or 10 thereof if the injury or damage—
    (a) was caused by such an occurrence as is mentioned in section 7(2)(b) or (c) or 10(2)(b) of this Act which is shown to have taken place wholly within the territorial limits of one, and one only, of the relevant territories other than the United Kingdom; or
    (b) was incurred within the territorial limits of a country which is not a relevant territory.
(2) In the case of a breach of a duty imposed by section 7, 8 or 9 of this Act, subsection (1)(b) of this section shall not apply to injury or damage

incurred by, or by persons or property on, a ship or aircraft registered in the United Kingdom.

(3) Compensation shall not be payable under this Act in respect of injury or damage caused by a breach of a duty imposed by section 10 of this Act in respect of such carriage as is referred to in subsection (1)(a)(ii) of that section unless the agreement so referred to was expressed in writing.

(4) The duty imposed by section 7, 8, 9, 10 or 11 of this Act—

(a) shall not impose any liability on the person subject to that duty with respect to injury or damage caused by an occurrence which constitutes a breach of that duty if the occurrence, or the causing thereby of the injury or damage, is attributable to hostile action in the course of any armed conflict, including any armed conflict within the United Kingdom; but

(b) shall impose such a liability where the occurrence, or the causing thereby of the injury or damage, is attributable to a natural disaster, notwithstanding that the disaster is of such an exceptional character that it could not reasonably have been foreseen.

(5) Where, in the case of an occurrence which constitutes a breach of a duty imposed by section 7, 8, 9 or 10 of this Act, a person other than the person subject to that duty makes any payment in respect of injury or damage caused by that occurrence and—

(a) the payment is made in pursuance of any of the international conventions referred to in the Acts mentioned in section 12(4) of this Act; or

(b) the occurrence took place [or the injury or damage was incurred] within the territorial limits of a country which is not a relevant territory, and the payment is made by virtue of a law of that country and by a person who has his principal place of business in a relevant territory or is acting on behalf of such a person,

the person making the payment may make the like claim under this Act for compensation of the like amount, if any, [subject to subsection (5A) of this section] as would have been available to him if—

(i) the injury in question had been suffered by him or, as the case may be, the property suffering the damage in question had been his; and

(ii) subsection (1) of this section had not been passed.

[(5A) The amount that a person may claim by virtue of subsection (5) of this section shall not exceed the amount of the payment made by him and, in the case of a claim made by virtue of paragraph (b) of that subsection, shall not exceed the amount applicable under section 16(1) or (2) of this Act to the person subject to the duty in question.]

(6) The amount of compensation payable to or in respect of any person under this Act in respect of any injury or damage caused in breach of a duty imposed by section 7, 8, 9 or 10 of this Act may be reduced by

reason of the fault of that person if, but only if, and to the extent that, the causing of that injury or damage is attributable to any act of that person committed with the intention of causing harm to any person or property or with reckless disregard for the consequences of his act.

Definitions

"injury": section 26(1).   A–74
"relevant territory": section 26(1).
"territorial limits": section 26(1).

Amendments

In subsection (5)(b) the first set of words in square brackets was added by   A–75
the Nuclear Installations Act 1969, s.3 and the second set of words was substituted by the Energy Act 1983, s.27(3). Subsection (5A) was also added by the Energy Act 1983, s.27(3). Subsection (6) has been modified for England, Wales and Northern Ireland but not for Scotland by the Congenital Disabilities (Civil Liability) Act 1976, s.3(4) (see below).

General Note

This section replaces the Nuclear Installations (Amendment) Act 1965,   A–76
section 5. It sets out those circumstances in which the right to compensation under section 12 may be excluded, extended or reduced. The section also provides for the circumstances in which a third party (not the person subject to the duty under ss.7–10) who makes payment in respect of injury or damage caused by an occurrence constituting a breach of the duty may be compensated; compensation paid out to such third parties is subject to a ceiling as set out in subsection (5A). Levels of compensation may be reduced where a person may have contributed to his own injury or damage by his own acts committed recklessly or by acts committed by him with the intention of causing harm to any person or property. This provision is modified in relation to children born with disabilities (see below).

Congenital Disabilities (Civil Liability) Act 1976

This Act makes provision as to civil liability in the case of children born   A–77
disabled in consequence of some person's fault. It extends the impact of the Nuclear Installations Act 1965 so that children born in a disabled condition in consequence of a breach of duty may claim compensation.

Section 3(4) modifies section 13(6) of the 1965 Act to the extent that in calculating the compensation to a child, the contributory fault due to one or other of the parents in suffering the consequences of a breach of duty under sections 7–10 is taken into account. Additionally, section 3(5) provides that compensation is not payable if the injury to the parent pre-dated the conception of the child and, at the time, either or both parents knew the risk of their child being born disabled.

Protection for ships and aircraft

A–78 14.—(1) A claim under this Act in respect of any occurrence such as is mentioned in section 7(2)(b) or (c), 10 or 11 of this Act which constitutes a breach of a person's duty under section 7, 8, 9, 10 or 11 of this Act shall not give rise to any lien or other right in respect of any ship or aircraft; and the following provisions of the Administration of Justice Act 1956 (which relate to the bringing of actions in rem against ships or aircraft in England and Wales, Scotland and Northern Ireland respectively), that is to say—
    (a) section 3(3) and (4);
    (b) section 47; and
    (c) paragraph 3(3) and (4) of Part I of Schedule 1
... shall not apply to that claim.
    (2) Subsection (1) of this section shall have effect in relation to any claim notwithstanding that by reason of section 16 of this Act no payment for the time being falls to be made in satisfaction of the claim.

Amendments

A–79 The words omitted in subsection (1) were repealed by the Merchant Shipping Act 1979, s.50(4) and Pt. I, Sched. 7.

General Note

A–80 This section replaces the Nuclear Installations (Licensing and Insurance) Act 1959, s.4(5) and Sched. 1, para. 2. It prevents a claimant for injury or damage caused by specified occurrences obtaining a lien or other right in respect of any ship or aircraft.

Administration of Justice Act 1956

A–81 Sections 3(3) and (4) and 47 and Schedule 1, Part I, paragraphs 3(3) and (4), which relate to the bringing of actions *in rem* against ships or aircraft do not apply for the purposes of this section. Section 3 of the 1956 Act was repealed by section 152(4) and Schedule 7 of the Supreme Court Act 1981 and replaced by section 21 of the 1981 Act.

Bringing and satisfaction of claims

Time for bringing claims under sections 7 to 11

15.—(1) Subject to subsection (2) of this section and to section 16(3) of A–82
this Act, but notwithstanding anything in any other enactment, a claim by
virtue of any of sections 7 to 11 of this Act may be made at any time
before, but shall not be entertained if made at any time after, the
expiration of thirty years from the relevant date, that is to say, the date of
the occurrence which gave rise to the claim or, where that occurrence
was a continuing one, or was one of a succession of occurrences all
attributable to a particular happening on a particular relevant site or to the
carrying out from time to time on a particular relevant site of a particular
operation, the date of the last event in the course of that occurrence or
succession of occurrences to which the claim relates.

(2) Notwithstanding anything in subsection (1) of this section, a claim
in respect of injury or damage caused by an occurrence involving nuclear
matter stolen from, or lost, jettisoned or abandoned by, the person whose
breach of a duty imposed by section 7, 8, 9 or 10 of this Act gave rise to
the claim shall not be entertained if the occurrence takes place after the
expiration of the period of twenty years beginning with the day when the
nuclear matter in question was so stolen, lost, jettisoned or abandoned.

Definitions

"injury": section 26(1).   A–83
"nuclear matter": section 26(1).   ·
"relevant site": section 26(1).

General Note

This section replaces the Nuclear Installations (Licensing and Insurance) A–84
Act 1959, s.4(4), and the Nuclear Installations (Amendment) Act 1965,
s.4(5) and (6) and Sched. 1, para. 1.

It imposes limits upon the length of time that may elapse from the date
of an occurrence up to the time that a claim under sections 7–11 is
brought. The time runs from the relevant date which is either the date of
the occurrence giving rise to the claim for injury or damage or for a
continuing occurrence, the date of the last event in a succession of
occurrences.

Where the nuclear matter is stolen from or lost or abandoned by the
person under a duty under sections 7–10 no claim for injury or damage
will be successful where it is made in relation to an occurrence that takes
place more than twenty years after the nuclear matter was stolen, lost or
abandoned.

Satisfaction of claims by virtue of sections 7 to 10

A–85   16.—(1) The liability of any person to pay compensation under this Act by virtue of a duty imposed on that person by section 7, 8 or 9 thereof shall not require him to make in respect of any one occurrence constituting a breach of that duty payments by way of such compensation exceeding in the aggregate, apart from payments in respect of interest or costs, [£140 million or, in the case of the licensees of such sites as may be prescribed, £10 million].

[(1A) The Secretary of State may with the approval of the Treasury by order increase or further increase either or both of the amounts specified in subsection (1) of this section; but an order under this subsection shall not affect liability in respect of any occurrence before (or beginning before) the order comes into force.]

(2) A relevant foreign operator shall not be required by virtue of section 10 of this Act to make any payment by way of compensation in respect of an occurrence—

    (a) if he would not have been required to make that payment if the occurrence had taken place in his home territory and the claim had been made by virtue of the relevant foreign law made for purposes corresponding to those of section 7, 8 or 9 of this Act; or

    (b) to the extent that the amount required for the satisfaction of the claim is not required to be available by the relevant foreign law made for purposes corresponding to those of section 19(1) of this Act and has not been made available under section 18 of this Act or by means of a relevant foreign contribution.

(3) Any claim by virtue of a duty imposed on any person by section 7, 8, 9 or 10 of this Act—

    (a) to the extent to which, by virtue of subsection (1) or (2) of this section, though duly established, it is not or would not be payable by that person; or

    (b) which is made after the expiration of the relevant period; or

    (c) which, being such a claim as is mentioned in section 15(2) of this Act, is made after the expiration of the period of twenty years so mentioned; or

    (d) which is a claim the full satisfaction of which out of funds otherwise required to be, or to be made, available for the purpose is prevented by section 21(1) of this Act,

shall be made to the appropriate authority, that is to say—

    (i) in the case of a claim by virtue of the said section 8, the Minister of Technology;

    (ii) in the case of a claim by virtue of the said section 9 (other than a claim in connection with a site used by a department of the Government of Northern Ireland), the Minister in charge of the government department concerned;

    (iii) in any other case, the Minister,

and, if established to the satisfaction of the appropriate authority, and to the extent to which it cannot be satisfied out of sums made available for the purpose under section 18 of this Act or by means of a relevant foreign contribution, shall be satisfied by the appropriate authority to such extent and out of funds provided by such means as Parliament may determine.

(4) Where in pursuance of subsection (3) of this section a claim has been made to the appropriate authority, any question affecting the establishment of the claim or as to the amount of any compensation in satisfaction of the claim may, if the authority thinks fit, be referred for decision to the appropriate court, that is to say, to whichever of the High Court, the Court of Session and the High Court of Justice in Northern Ireland would, but for the provisions of this section, have had jurisdiction in accordance with section 17(1) and (2) of this Act to determine the claim; and the claimant may appeal to that court from any decision of the authority on any such question which is not so referred; and on any such reference or appeal—

(a) the authority shall be entitled to appear and be heard; and
(b) notwithstanding anything in any Act, the decision of the court shall be final.

(5) In this section, the expression "the relevant period" means the period of ten years beginning with the relevant date within the meaning of section 15(1) of this Act.

Definitions

"home territory": section 26(1).   A–86
"the Minister": section 26(1).
"occurrence": section 26(1).
"prescribed": section 26(1).
"relevant foreign contribution": section 26(1).
"relevant foreign law": section 26(1).
"relevant foreign operator": section 26(1).

Amendments

In subsection (1) the words in square brackets were substituted, and A–87 subsection (1A) was inserted, by the Energy Act 1983, s.27(1).

The figure of £140 million and £10 million in subsection (1) were inserted by the Nuclear Installations (Increase in Operators' Limits of Liability) Order 1994 (S.I. 1994 No. 909).

General Note

This section replaces the Nuclear Installations (Amendment) Act 1965, A–88 s.6. The amendment in subsection (1) replaced a general £5 million limit to liability but did not affect liability relating to occurrences before (or beginning before) the commencement of the Energy Act 1983 (September 1, 1983). For occurrences after September 1, 1983, and before or beginning before April 1, 1994, the figure was £20 million, or in the case of prescribed sites, £5 million.

The section sets out the provisions regarding the satisfaction of claims resulting from a breach of duty under sections 7–10 of the Act. For prescribed sites there is currently an upper limit of £10 million for any individual occurrence and a ceiling of £140 million for licensees of non-prescribed sites. Both these figures may be altered by an order from the Secretary of State. The section also restricts the liability of foreign operators to pay compensation for breach of duty under section 10 under specified circumstances.

The section further sets out provisions to deal with circumstances in which claims against a breach of duty under sections 7–10 may not result in compensation being paid out by the person in breach. Such claims may be referred to an appropriate authority, *e.g.* the Secretary of State for Trade and Industry, who may refer them to the courts for settlement. Claimants may also appeal to the court from any decision of the relevant authority and the decision of the court will be final.

Prescribed Sites

A–89 Under subsection (1), as amended, a distinction is made between compensation payable by licensees of non-prescribed sites and that payable by licensees of prescribed sites. Sites are prescribed under the Nuclear Installations (Prescribed Sites) Regulations 1983 (S.I. 1983 No. 919), made by the Secretary of State under section 16(1).

Subsection (1A)

A–90 Inserted by the Energy Act 1983, s.27(1), this subsection gives the Secretary of State power by order to increase the figures for compensation expressed in subsection (1). The Nuclear Installations (Increase in Operators' Limits of Liability) Order 1994 (S.I. 1994 No. 909) has been made under this subsection and increased the limits to their current levels from April 1, 1994. The increase does not affect incidents where the occurrence took place before or began before that date.

Minister of Technology

A–91 The powers of the Minister of Technology were transferred to the Secretary of State for Trade and Industry by the Secretary of State for Trade and Industry Order 1970 (S.I. 1970 No. 1537), Art. 2(2).

Jurisdiction, shared liability and foreign judgments

17.—(1) No court in the United Kingdom or any part thereof shall have A–92 jurisdiction to determine any claim or question under this Act certified by the Minister to be a claim or question which, under any relevant international agreement, falls to be determined by a court of some other relevant territory or, as the case may be, of some other part of the United Kingdom; and any proceedings to enforce such a claim which are commenced in any court in the United Kingdom or, as the case may be, that part thereof shall be set aside.

(2) Where under the foregoing subsection the Minister certifies that any claim or question falls to be determined by a court in a particular part of the United Kingdom, that certificate shall be conclusive evidence of the jurisdiction of that court to determine that claim or question.

(3) Where by virtue of any one or more the following, that is to say, sections 7, 8, 9 and 10 of this Act and any relevant foreign law made for purposes corresponding to those of any of those sections, liability in respect of the same injury or damage is incurred by two or more persons, then, for the purposes of any proceedings in the United Kingdom relating to that injury or damage, including proceedings for the enforcement of a judgment registered under the Foreign Judgments (Reciprocal Enforcement) Act 1933—

    (a) both or all of those persons shall be treated as jointly and severally liable in respect of that injury or damage; and

    (b) until claims against each of those persons in respect of the occurrence by virtue of which the person in question is liable for that injury or damage have been satisfied—

        (i) in the case of a licensee, the Authority or the Crown, up to an aggregate amount [equal to that applicable to the person in question under section 16(1) of this Act]; or

        (ii) in the case of a relevant foreign operator, up to such aggregate amount, ... as may be provided for by the relevant foreign law made for purposes corresponding to those of section 19(1) of this Act,

no sums in excess of those required for the purposes of sub-paragraph (i) of this paragraph shall be required to be made available under section 18 of this Act for the purpose of paying compensation in respect of that injury or damage.

(4) Part I of the said Act of 1933 shall apply to any judgment given in a court of any foreign country which is certified by the Minister to be a relevant foreign judgment for the purposes of this Act, whether or not it would otherwise have so applied, and shall have effect in relation to any judgment so certified as if in section 4 of that Act subsections (1)(a)(ii), (2) and (3) were omitted.

(5) [Subject to subsection (5A) of this section], it shall be sufficient defence to proceedings in the United Kingdom against any person for the

recovery of a sum alleged to be payable under a judgment given in a country outside the United Kingdom for that person to show that—

    (a) the sum in question was awarded in respect of injury or damage of a description which is the subject of a relevant international agreement; and

    (b) the country in question is not a relevant territory; and

    (c) the sum in question was not awarded in pursuance of any of the international conventions referred to in the Acts mentioned in section 12(4) of this Act.

[(5A) Subsection (5) of this section shall not have effect where the judgment in question is enforceable in the United Kingdom in pursuance of an international agreement.]

(6) Where, in the case of any claim by virtue of section 10 of this Act, the relevant foreign operator is the government of a relevant territory, then, for the purposes of any proceedings brought in a court in the United Kingdom to enforce that claim, that government shall be deemed to have submitted to the jurisdiction of that court, and accordingly rules of court may provide for the manner in which any such action is to be commenced and carried on; but nothing in this subsection shall authorise the issue of execution, or in Scotland the execution of diligence, against the property of that government.

Definitions

A–93 "the Authority": section 26(1).
"injury": section 26(1).
"licensee": section 26(1).
"the Minister": section 26(1).
"occurrence": section 26(1).
"relevant foreign judgment": section 26(1).
"relevant foreign law": section 26(1).
"relevant foreign operators": section 26(1).
"relevant international agreement": section 26(1).
"relevant territory": section 26(1).

Amendments

A–94 In subsection (3)(b)(i), the words in square brackets were substituted and the words omitted from subsection (3)(b)(ii) were repealed, by the Energy Act 1983, ss.28(4), 36, and Pt. II, Sched. 4. These amendments do not have effect in relation to occurrences before (or beginning before) section 28 of the Energy Act 1983 was brought into force (September 1, 1983).

The words in square brackets in subsection (5) and the whole of subsection (5A) were inserted by the Energy Act 1983, s.31.

General Note

This section sets out the provisions concerning the jurisdiction of United A–95
Kingdom courts, relating to liability for claims for injury and damage; it
provides for circumstances where liability in respect of the same injury or
damage is incurred by two or more persons and the applicability of
foreign judgments in the United Kingdom. Subsection (6) which applies
to claims brought for breach of duty under section 10 against a
government of a relevant territory (as defined) in a United Kingdom
court, is specifically excluded from the operation of the State Immunity
Act 1978, Pt. I (see below).

Foreign Judgments

The Foreign Judgments (Reciprocal Enforcement) Act 1933 (as A–96
amended) makes provision for the enforcement in the United Kingdom
of judgments given in foreign countries which provide reciprocal
treatment to judgments given in the United Kingdom; this arrangement
facilitates the enforcement in foreign countries of judgments given in the
United Kingdom.

Subsections (5), (5A)

Both subsections entered into force on September 1, 1983 (the date of A–97
entry into force of the Energy Act 1983, s.31).

State Immunity

The State Immunity Act 1978 makes provisions relating to proceedings in A–98
the United Kingdom by or against other states and provides for the effect
of judgments given against the United Kingdom in the courts of states,
parties to the European Convention on State Immunity. Section 1 sets out
the general immunity of states from the jurisdiction of United Kingdom
courts and sections 2–11 list the exemptions. Section 16(3) of the Act
specifically exempts section 17(6) of the Nuclear Installations Act 1965
from the operation of Part I of the 1978 Act.

Cover for compensation

General cover for compensation by virtue of sections 7 to 10

18.—(1) In the case of any occurrence in respect of which one or more A–99
persons incur liability by virtue of section 7, 8, 9 or 10 of this Act or by
virtue of any relevant foreign law made for purposes corresponding to

those of any of those sections, but subject to subsections (2) [to (4B)] of this section and to sections 17(3)(b) and 21(1) of this Act, there shall be made available out of moneys provided by Parliament such sums as, when aggregated—

(a) with any funds required by, or by any relevant foreign law made for purposes corresponding to those of, section 19(1) of this Act to be available for the purpose of satisfying claims in respect of that occurrence against any licensee or relevant foreign operator; and

(b) in the case of a claim by virtue of any such foreign law, with any relevant foreign contributions towards the satisfaction of claims in respect of that occurrence[; and

(c) in the case of an occurrence in respect of which the Authority incurs liability, with any amounts payable under a contract of insurance or other arrangements for satisfying claims in respect of that occurrence against the Authority,]

may be necessary to ensure that all claims in respect of that occurrence made within the relevant period and duly established, excluding, but without prejudice to, any claim in respect of interest or costs, are satisfied up to [the aggregate amount specified in subsection (1A) of this section].

[(1A) The aggregate amount referred to in subsection (1) of this section is the equivalent in sterling of 300 million special drawing rights on—

(a) the day (or first day) of the occurrence in question, or

(b) if the Secretary of State certifies that another day has been fixed in relation to the occurrence in accordance with an international agreement, that other day.

(1B) The Secretary of State may with the approval of the Treasury by order increase or further increase the sum expressed in special drawing rights in subsection (1A) of this section; but an order under this subsection shall not have effect in respect of an occurrence before (or beginning before) the order comes into force.]

(2) Subsection (1) of this section shall not apply to any claim by virtue of such a relevant foreign law as is mentioned in that subsection in respect of injury or damage incurred within the territorial limits of a country which is not a relevant territory or to any claim such as is mentioned in section 15(2) of this Act which is not made within the period of twenty years so mentioned.

(3) Where any claim such as is mentioned in subsection (1) of this section is satisfied wholly or partly out of moneys provided by Parliament under that subsection, there shall also be made available out of moneys so provided such sums as are necessary to ensure the satisfaction of any claim in respect of interest or costs in connection with the first-mentioned claim.

(4) In relation to liability by virtue of any relevant foreign law [there shall be left out of account for the purposes of subsection (1) of this section any claim which, though made within the relevant period, was made after the

expiration of any period of limitation imposed by that law and permitted by a relevant international agreement.]

[(4A) Where—

(a) a relevant foreign law provides in pursuance of a relevant international agreement for sums additional to those referred to in subsection (1)(a) of this section to be made available out of public funds, but

(b) the maximum aggregate amount of compensation for which it provides in respect of an occurrence in pursuance of that agreement is less than that specified in subsection (1A) of this section,

then, in relation to liability by virtue of that law in respect of the occurrence, subsection (1) of this section shall have effect as if for the reference to the amount so specified there were substituted a reference to the maximum aggregate amount so provided.

(4B) Where a relevant foreign law does not make the provision mentioned in subsection (4A)(a) of this section, then in relation to liability by virtue of that law in respect of any occurrence—

(a) subsection (1) of this section shall not have effect unless the person (or one of the persons) liable is a licensee, the Authority or the Crown; and

(b) if a licensee, the Authority or the Crown is liable, subsection (1) shall have effect as if for the reference to the amount specified in subsection (1A) there were substituted a reference to the amount which would be applicable to that person under section 16(1) of this Act in respect of the occurrence (or, if more than one such person is liable, to the aggregate of the amounts which would be so applicable) if it had constituted a breach of duty under section 7, 8 or 9 of this Act.]

(5) Any sums received by the Minister by way of a relevant foreign contribution towards the satisfaction of any claim by virtue of section 7, 8, 9 or 10 of this Act shall be paid into the Exchequer.

(6) In this section, the expression "the relevant period" has the same meaning as in section 16 of this Act.

Definitions

"the Authority": section 26(1).
"injury": section 26(1).
"licensee": section 26(1).
"the Minister": section 26(1).
"occurrence": section 26(1).
"relevant foreign contribution": section 26(1).
"relevant foreign law": section 26(1).
"relevant foreign operator": section 26(1).
"relevant international agreement": section 26(1).

A–100

"relevant territory": section 26(1).
"territorial limits": section 26(1).

Amendments

A–101 In subsection (1), the first and third sets of words in square brackets were substituted by the Energy Act 1983, s.28(1) and the second set of words in square brackets was added by the Atomic Energy Act 1989, s.3.

Subsections (1A) and (1B) were inserted by the Energy Act 1983, s.28(2). Subsection (4) was substituted, and subsections (4A) and (4B) were inserted, by the Energy Act 1983, s.(3). Amendments introduced under the Energy Act 1983, s.28 only have effect in relation to an occurrence before (or beginning before) September 1, 1983 (when the Energy Act 1983, s.28 entered into force).

General Note

A–102 This section replaces the Nuclear Installations (Amendment) Act 1965, s.8. It sets out the provisions regarding general cover for compensation relating to claims made for breaches of duty under sections 7–10. It also sets out the circumstances in which the United Kingdom Parliament will make available funds to settle claims up to a specified ceiling.

Atomic Energy Act 1989

A–103 This Act introduces a number of amendments into nuclear legislation. Section 3 inserts paragraph (c) into subsection 18(1) of the Nuclear Installations Act 1965. For further amendments see the footnotes to sections 19 and 24.

Special cover for licensee's liability

A–104 19.—(1) Subject to section 3(5) of this Act and to subsection (3) of this section, where a nuclear site licence has been granted in respect of any site, the licensee shall make such provision (either by insurance or by some other means) as the Minister may with the consent of the Treasury approve for sufficient funds to be available at all times to ensure that any claims which have been or may be duly established against the licensee as licensee of that site by virtue of section 7 of this Act or any relevant foreign law made for purposes corresponding to those of section 10 of this Act (excluding, but without prejudice to, any claim in respect of interest or costs) are satisfied up to [the required amount] in respect of each severally of the following periods, that is to say—

(a) the current cover period, if any;

(b) any cover period which ended less than ten years before the time in question;

(c) any earlier cover period in respect of which a claim remains to be disposed of, being a claim made—

(i) within the relevant period within the meaning of section 16 of this Act; and

(ii) in the case of a claim such as is mentioned in section 15(2) of this Act, also within the period of twenty years so mentioned;

and for the purposes of this section the cover period in respect of which any claim is to be treated as being made shall be that in which the beginning of the relevant period aforesaid fell.

[(1A) In this section "the required amount", in relation to the provision to be made by a licensee in respect of a cover period, means an aggregate amount equal to the amount applicable under section 16(1) of this Act to the licensee, as licensee of the site in question, in respect of an occurrence within that period.]

(2) In this Act, the expression "cover period" means [subject to the following provisions of this section, the period of the licensee's responsibility; ...] and for the purposes of this definition the period of the licensee's responsibility shall be deemed to include any time after the expiration of that period during which it remains possible for the licensee to incur any liability by virtue of section 7(2)(b) or (c) of this Act, or by virtue of any relevant foreign law made for purposes corresponding to those of section 10 of this Act.

[(2A) When the amount applicable under section 16(1) of this Act to a licensee of a site changes as a result of—

(a) the coming into force of an order under section 16(1A) or of regulations made for the purposes of section 16(1), or

(b) an alteration relating to the site which brings it within, or takes it outside, the description prescribed by such regulations,

the current cover period relating to him as licensee of that site shall end and a new cover period shall begin.]

[(2B) The current cover period continues to run (and no new cover period begins) on the grant of a new nuclear site licence to the same licensee in respect of a site consisting of or including the site in respect of which his existing nuclear site licence is in force.]

(3) Where in the case of any licensed site the provision required by subsection (1) of this section is to be made otherwise than by insurance and, apart from this subsection, provision would also fall to be so made by the same person in respect of two or more other sites, the requirements of that subsection shall be deemed to be satisfied in respect of each of those sites if funds are available to meet such claims as are mentioned in that subsection in respect of all the sites collectively, and those funds would for the time being be sufficient to satisfy the requirements of that subsection in respect of those two of the sites in respect of which those requirements are highest:

Provided that the Minister may in any particular case at any time direct either that this subsection shall not apply or that the funds available as aforesaid shall be of such amount higher than that provided for by the foregoing provisions of this subsection, but lower than that necessary to satisfy the requirements of the said subsection (1) in respect of all the sites severally, as may be required by the direction.

(4) Where, by reason of the gravity of any occurrence which has resulted or may result in claims such as are mentioned in subsection (1) of this section against a licensee as licensee of a particular licensed site, or having regard to any previous occurrences which have resulted or may result in such claims against the licensee, the Minister thinks it proper so to do, he shall by notice in writing to the licensee direct that a new cover period for the purposes of the said subsection (1) shall begin in respect of that site on such date not earlier than two months after the date of the service of the notice as may be specified therein.

(5) If at any time while subsection (1) of this section applies in relation to any licensed site the provisions of that subsection are not complied with in respect of that site, the licensee shall be guilty of an offence and be liable—

    (a) on summary conviction to a fine not exceeding [the prescribed sum], or to imprisonment for a term not exceeding three months, or to both;

    (b) on conviction on indictment, to a fine . . ., or to imprisonment for a term not exceeding two years, or to both.

Definitions

A–105   "licensed site": section 26(1).
"licensee": section 26(1).
"the Minister": section 26(1).
"relevant foreign law": section 26(1).
"nuclear site licence": section 26(1).
"period of responsibility": section 26(1).

Amendments

A–106   In subsection (1), the words in square brackets were substituted by the Energy Act 1983, s.27(4). Subsections (1A) and (2A) were added by the Energy Act 1983, s.27(4). In subsection (2), the words in square brackets were substituted by the Atomic Energy Act 1989, s.4(1)(a) and subsection (2B) was added by the Atomic Energy Act 1989, s.4(1)(b). In subsection (5), the words in square brackets in paragraph (a) were substituted by virtue of the Magistrates' Courts Act 1980, s.32(2) and the words omitted in paragraph (b) ("not exceeding £500") are omitted by virtue of the Criminal Law Act 1977, s.32(1).

General Note

This section replaces the Nuclear Installations (Licensing and Insurance) **A–107**
Act 1959, subss.5(1), (2), (4), (6) and the Nuclear Installations (Amendment) Act 1965, ss.4(7), 9(1), 9(4), 12(3) and Sched. 1, para. 3. It
requires that a licensee makes sufficient financial provision, either by
insurance or some other means, to ensure that any claims established
against the licensee for breach of duty under section 7 or a foreign law
equivalent to section 10 of the Act can be satisfied up to a specified
required amount. The section sets out for how long the licensee requires
to hold such cover, the amount of cover to satisfy the section and the
special arrangements that apply for a licensee of more than one site.
Under subsection (5), the licensee is guilty of an offence when cover is
not provided; the penalty for conviction in the magistrates court may be a
fine of up to £5,000 or three months' imprisonment or both, while in the
Crown Court, the penalties consist of an unlimited fine or up to 2 years'
imprisonment or both.

Furnishing of information relating to licensee's cover

20.—(1) In the case of each licensed site, the licensee shall give notice **A–108**
in writing to the Minister forthwith upon its appearing to the licensee that
the aggregate amount of any claims such as are mentioned in section
19(1) of this Act made in respect of any cover period falling within
the period of the licensee's responsibility has reached [three-fifths of the
required amount within the meaning of section 19]; and where the
licensee has given such a notice, no payment by way of settlement of any
claim in respect of the cover period in question by agreement between
the licensee and the claimant shall be made except after consultation
with the Minister and in accordance with the terms of any direction which
the Minister may give to the licensee in writing with respect to any
particular claim.

(2) If in the case of any licensed site any cover period falling within the
period of the licensee's responsibility has ended, the licensee shall not
later than 31st January in each year send to the Minister in writing a
statement showing the date when that cover period ended and the
following particulars of any claims in respect of that cover period as at the
beginning and end respectively of the last preceding calendar year, that
is to say—
   (a) the aggregate number of claims received;
   (b) the aggregate number of claims established; and
   (c) the aggregate number and aggregate amount of claims
       satisfied.

(3) The Minister shall as soon as may be lay before each House of
Parliament a copy of any notice received by him under subsection (1) of
this section and a report (in such form as, having regard to section 16 of

this Act, he may consider appropriate) with respect to any statements received by him under subsection (2) of this section.

(4) Any person by whom any funds such as are mentioned in section 19(1) of this Act for the time being fall to be provided shall give to the Minister not less than two months notice in writing before ceasing to keep those funds available and, notwithstanding any such notice, so far as those funds relate to nuclear matter for the time being in the course of carriage, shall not so cease while that carriage continues.

Definitions

A–109   "carriage": section 26(2).
"cover period": section 26(1).
"licensed site": section 26(1).
"licensee": section 26(1).
"the Minister": section 26(1).
"nuclear matter": section 26(1).
"period of responsibility": section 26(1).
"required amount": section 19(1A).

Amendments

A–110   In subsection (1), the words in square brackets are substituted by the Energy Act 1983, s.27(6).

General Note

A–111   This section replaces the Nuclear Installations (Licensing and Insurance) Act 1959, s.5(3) and the Nuclear Installations (Amendment) Act 1965, subss.9(2), (3) and (5). It requires the licensee to give notice in writing to the Minister when the total value of claims against him for any cover period (as defined) has reached a specified amount. Thereafter, any settlement of claims by the licensee may only be accomplished after consultation with the Minister.

The licensee is further required to report the aggregate number of claims received and established and the aggregate number and aggregate amount of claims satisfied on a specified regular basis during the licensee's responsibility for the site.

The section also makes provision for circumstances in which a person holding funds for the purpose of satisfying section 19(1) may give the Minister notice in writing that he intends to cease keeping those funds available.

Supplementary provisions with respect to cover for compensation in respect of carriage

21.—(1) Where, in the case of an occurrence involving nuclear matter A–112 in the course of carriage, a claim in respect of damage to the means of transport being used for that carriage is duly established—
    (a) against any person by virtue of section 7, 8, 9 or 10 of this Act; or
    (b) against a licensee, the Authority or the Crown by virtue of any relevant foreign law made for purposes corresponding to those of the said section 10,

then, without prejudice to any right of the claimant to the satisfaction of that claim, no payment towards its satisfaction shall be made out of funds which are required to be available for the purpose by, or by any relevant foreign law made for purposes corresponding to those of, section 19(1) of this Act, or which have been made available for the purpose under section 18 of this Act or by means of a relevant foreign contribution, such as to prevent the satisfaction out of those funds up to an aggregate amount [which is the equivalent in sterling (on the day, or the first day, of that occurrence) of 5 million special drawing rights] of all claims which have been or may be duly established against the same person in respect of injury or damage caused by that occurrence other than damage to the said means of transport.

[(1A) The Secretary of State may with the approval of the Treasury by order increase or further increase the sum expressed in special drawing rights in subsection (1) of this section; but an order under this subsection shall not have effect in respect of any occurrence before (or beginning before) the order comes into force.]

(2) Where, in the case of an occurrence involving nuclear matter in the course of carriage, a claim in respect of damage to the means of transport being used for that carriage is duly established against a relevant foreign operator by virtue of section 10 of this Act, but by virtue of section 16(2)(a) thereof that operator is not required to make a payment in satisfaction of the claim, section 12(1)(b) of this Act shall not apply to any liability of that operator with respect to the damage in question apart from this Act.

(3) Where any nuclear matter is to be carried by, or on behalf or with the agreement of, a licensee, the Authority, a government department or a relevant foreign operator in such circumstances that, while the matter is in the course of that carriage, the licensee, the Authority, the Crown or the operator, as the case may be (in this and the next following subsection referred to as "the responsible party") may incur liability by virtue of section 7, 8, 9 or 10 of this Act or by virtue of any relevant foreign law made for purposes corresponding to those of the said section 10, the responsible party shall, before the carriage is begun, cause to be delivered to the person who is to carry that matter a document issued by or on behalf of the appropriate person mentioned in the next following subsection (in this subsection referred to as "the guarantor") which shall

contain such particulars as may be prescribed of the responsible party, of that nuclear matter and carriage, and of the funds available in pursuance of, or of the relevant foreign law made for purposes corresponding to those of, section 18 or 19(1) of this Act to satisfy any claim by virtue of that liability, and the guarantor shall be debarred from disputing in any court any of the particulars stated in that document; and if in any case there is a wilful failure to comply with this subsection, the responsible party (except where that party is the Crown), and also, if the carrier knew or ought to have known the matter carried to be such matter for carriage in such circumstances as aforesaid, the carrier, shall be guilty of an offence and liable on summary conviction to a fine not exceeding [level 3 on the standard scale].

(4) The person by whom or on whose behalf the document referred to in the last foregoing subsection is to be issued shall be—

    (a) where the responsible party is a licensee, the person by whom there fall to be provided the funds required by section 19(1) of this Act to be available to satisfy any claim in respect of the carriage in question;

    (b) where the responsible party is the Authority, the Minister of Technology;

    (c) where the responsible party is the Crown, the Minister in charge of the government department concerned;

    (d) where the responsible party is a relevant foreign operator, the person by whom there fall to be provided the funds required by the relevant foreign law made for purposes corresponding to those of section 18 or 19(1) of this Act to be made available to satisfy any claim in respect of the carriage in question.

[(4A) Subsection (3) of this section shall not apply where the carriage in question is wholly within the territorial limits of the United Kingdom.]

(5) The requirements of Part VI of the Road Traffic Act 1960 (which relates to compulsory insurance or security against third-party risks of users of motor vehicles) shall not apply in relation to any injury to any person for which any person is liable by virtue of section 7, 8, 9 or 10 of this Act.

Definitions

A–113   "the Authority": section 26(1).
"carriage": section 26(2).
"injury": section 26(1).
"licensee": section 26(1).
"the Minister": section 26(1).
"nuclear matter": section 26(1).
"relevant foreign contributions": section 26(1).
"relevant foreign law": section 26(1).

"relevant foreign operator": section 26(1).
"special drawing rights": section 25B.

Amendments

In subsection (1), the words in square brackets were substituted, and **A–114** subsections (1A) and (4A) were inserted, by the Energy Act 1983, subs.29(1), (2) and (3). Under section 29(4), these amendments have no effect in respect of any occurrence before (or beginning before) the commencement of section 29 (September 1, 1983).

In subsection (3), the words in square brackets were substituted by the Criminal Justice Act 1982, s.46.

General Note

This section replaces the Nuclear Installations (Amendment) Act 1965, **A–115** subs.10(1)–(4) and (6). It sets out the supplementary provisions with respect to cover for compensation available in relation to the carriage of nuclear matter. There are special provisions relating to the satisfaction of claims made in respect of damage to the means of transport. The section also sets out the procedure by which responsible parties (as defined) are required to deliver prior to carriage a certificate containing prescribed information to the person who is to carry the nuclear matter. This delivery is made on behalf of a guarantor of funds that will be needed to meet any claims through a breach of duty under sections 7–10 during carriage. The obligation on the responsible party is considerably diminished by the fact that it does not apply where the carriage occurs wholly within the territorial limits of the United Kingdom.

Compulsory insurance or security provisions against third-party risks of users of motor vehicles under Part VI of the Road Traffic Act 1960 (now replaced by Part VI of the Road Traffic Act 1972) do not apply in relation to injuries sustained by any person for which another person is liable for a breach of duty under sections 7–10 of the Nuclear Installations Act 1965.

Subsection (3)

The Nuclear Installations (Insurance Certificate) Regulations 1965 (S.I. **A–116** 1965 No. 1823) prescribe particulars to be contained in the certificate delivered to the person who is to carry the nuclear matter as set out in section 3. The particulars relate to the operator, the nuclear matter, the carriage and the funds available to satisfy any liability which may be incurred by the operator of a nuclear installation either under the Act or under corresponding foreign law. The 1965 Regulations have been amended by the Nuclear Installations (Insurance Certificate) (Amendment) Regulations 1969 (S.I. 1969 No. 64).

Minister of Technology

A–117 The powers of the Minister of Technology have been transferred to the Secretary of State for Trade and Industry by the Secretary of State for Trade and Industry Order 1970 (S.I. 1970 No. 1537), Art. 2(2).

Miscellaneous and general

Reporting of and inquiries into dangerous occurrences

A–118 22.—(1) The provisions of this section shall have effect on the happening of any occurrence of any such class or description as may be prescribed, being an occurrence—
   (a) on a licensed site; or
   (b) in the course of the carriage of nuclear matter on behalf of any person where a duty with respect to that carriage is imposed on that person by section 7, 10 or 11 of this Act.
   (2) The licensee or person aforesaid shall cause the occurrence to be reported forthwith in the prescribed manner to the [Health and Safety Executive] and to such other persons, if any, as may be prescribed in relation to occurrences of that class or description, and if the occurrence is not so reported the licensee or person aforesaid shall be guilty of an offence [and be liable on summary conviction—
   (a) in the case of a first offence under this subsection, to a fine not exceeding fifty pounds;
   (b) in the case of a second or subsequent offence under this subsection, to a fine not exceeding one hundred pounds, or to imprisonment for a term not exceeding three months, or to both.]
   [(3) For the purposes of subsection (2) of this section, a conviction under section 6(2) of the Act of 1959 shall be deemed to be a conviction under subsection (2) of this section.
   (4) The Minister may at any time direct an inspector to make a special report with respect to the occurrence, and the Minister may cause any such report, or so much thereof as it is not in his opinion inconsistent with the interest of national security to disclose, to be made public at such time and in such manner as he thinks fit.
   (5) The Minister may, where he thinks it expedient so to do, direct an inquiry to be held in accordance with the provision of [Schedule 2] to this Act into the occurrence and its causes, circumstances and effects; and any such inquiry shall be held in public except where or to the extent that it appears to the Minister expedient in the interests of national security to direct otherwise.]
   (6) . . .

Definitions

"carriage": section 26(2).
"inspector": section 26(1).
"licensed site": section 26(1).
"licensee": section 26(1).
"the Minister": section 26(1).
"nuclear matter": section 26(1).
"prescribed": section 26(1).

A–119

Amendments

In subsection (2), the words in the first set of square brackets were A–120 substituted by the Nuclear Installations Act 1965, etc. (Repeals and Modifications) Regulations 1974 (S.I. 1974 No. 2056), reg. 2(1)(b) and Sched. 2, para. 1. The words in the second set of square brackets in subsection (2) and the whole of subsections (3)–(6) were repealed for England, Wales and Scotland (but not for Northern Ireland) by regulation 2(1)(a) of, and Schedule 1 to, the same Regulations. The term "Schedule 2" in square brackets in subsection 5 was substituted by the Atomic Energy Authority Act 1971, s.17(3).

General Note

This section replaces the Nuclear Installations (Licensing and Insurance) A–121 Act 1959, subs.6(2)–(4) and (7) and the Nuclear Installations (Amendment) Act 1965, s.12(1). It makes provision for the reporting of prescribed occurrences to the Health and Safety Executive. Failure to make such a report is an offence under subsection (2). For Northern Ireland only, specified penalties are provided for summary convictions for offences under subsection (2) as are the powers of the Minister to obtain a special report of an occurrence and to set up an inquiry.

Prescribed Occurrences

Prescribed occurrences that need to be reported to the Health and Safety A–122 Executive are set out in the Nuclear Installations (Dangerous Occurrences) Regulations 1965 (S.I. 1965 No. 1824), reg. 3. Occurrences on a licensed site involving the emission of ionising radiations or the release of radioactive or toxic substances in specified circumstances, specified occurrences during the course of carriage of nuclear matter, explosions or outbreaks of fire on a licensed site and any uncontrolled criticality excursion are prescribed occurrences under the regulations. Regulation 4 sets out the manner in which occurrences are to be reported. This regulation is amended by the Nuclear Installations Act 1965, etc. (Repeals and Modifications) Regulations 1974 (S.I. 1974 No. 2056) in as much that any reference to the Minister is substituted by a reference to the Health and Safety Executive.

Registration in connection with certain occurrences

A–123   23.—(1) Without prejudice to any right of any person to claim against any person by virtue of any of sections 7 to 11 of this Act, the appropriate authority may, on the happening of any occurrence in respect of which liability may be incurred by virtue of any of those sections, by order make provision for enabling such particulars of any person shown to have been within such area during such period (being the period during which the occurrence took place) as may be specified in the order to be registered by or on behalf of that person in such manner as may be so specified, and any such registration in respect of any person shall be sufficient evidence of his presence within that area during that period unless the contrary is proved; and any such order shall be made by statutory instrument and be laid before Parliament after being made.

(2) In the foregoing subsection, the expression "the appropriate authority" means, in relation to any occurrence, the authority hereinafter specified in relation to the person against whom any claim in respect of that occurrence falls to be made, that is to say—

(a) where that person is the Authority, the Minister of Technology;
(b) where that person is the Crown, the Minister in charge of the government department concerned;
(c) in any other case, the Minister.

Definitions

A–124   "the Authority": section 26(1).
"the Minister": section 26(1).

General Note

A–125   This section replaces the Nuclear Installations (Licensing and Insurance) Act 1959, s.6(5), and the Nuclear Installations (Amendment) Act 1965, s.12(2). The section provides for a registration procedure by order of an appropriate authority, as defined in subsection (2), of persons in the area during the period of an occurrence. At February 1997, only local orders have been made under this section.

Minister of Technology

A–126   The powers of the Minister have been transferred to the Secretary of State for Trade and Industry by the Secretary of State for Trade and Industry Order 1970 (S.I. 1970 No. 1537, Art. 2(2)).

Inspectors (applicable to Northern Ireland only)

24.—(1) The Minister may appoint as inspectors to assist him in the A–127 execution of this Act such number of persons appearing to him to be qualified for the purpose as he may from time to time consider necessary or expedient, and may make to or in respect of any person so appointed such payments by way of remuneration, allowances or other payments as the Minister may with the approval of the Treasury determine.

(2) Any such inspector may, for the purposes of the execution of this Act, and subject to production, if so requested, of written evidence of his authority—

    (a) subject to subsection (3) of this section, enter—
        (i) at all reasonable times during the period of the licensee's responsibility, upon any premises comprised in any licensed site; or
        (ii) at all reasonable times, upon any premises comprised in any site which is being used for such purposes that, but for regulations made by virtue of section 1(2) of this Act, a nuclear site licence would be required in respect thereof,
    with such equipment, and carry out such tests and inspections, as the inspector may consider necessary or expedient;

    (b) require—
        (i) the licensee of any licensed site; or
        (ii) the person using any site as mentioned in paragraph (a)(ii) of this subsection; or
        (iii) any person with duties on or in connection with any licensed site or any site being used as aforesaid,
    to provide the inspector with such information, or to permit him to inspect such documents, relating to the use of the site as the inspector may specify;

    (c) enter any place, vehicle, vessel or aircraft involved in any such occurrence as is mentioned in section 22(1) of this Act with such equipment, and carry out such tests and inspections, as he may consider necessary or expedient;

    (d) require the licensee or other person referred to in the said section 22(1) concerned in any such occurrence and any other person with duties concerning the nuclear matter involved in the occurrence to provide him with such information, or to permit him to inspect such documents, relating to the nuclear matter as the inspector may specify.

(3) Before carrying out any test in pursuance of his powers under subsection (2)(a) of this section, the inspector shall consult with such persons having duties upon the site as may appear to him appropriate in order to secure that the carrying out of the test does not create any danger.

(4) Any person who obstructs an inspector in the exercise of his powers under subsection (2)(a) or (c) of this section or who refuses or

without reasonable excuse fails to provide any information or to permit any inspection reasonably required by the inspector under subsection (2)(b) or (d) thereof shall be guilty of an offence and be liable on summary conviction to a fine not exceeding [level 3 on the standard scale], or to imprisonment for a term not exceeding three months, or to both.

(5) Any person who, without the authority of the Minister, discloses any information obtained in the exercise of powers under this Act shall be guilty of an offence and be liable—

    (a) on summary conviction, to a fine not exceeding fifty pounds, or to imprisonment for a term not exceeding three months, or to both;

    (b) on conviction on indictment, to a fine not exceeding one hundred pounds, or to imprisonment for a term not exceeding two years, or to both.

(6) In such cases and to such extent as it may appear to the Minister, with the agreement of the Treasury, to be appropriate so to do, the Minister shall require a licensee to repay to the Minister such part as may appear to the Minister to be attributable to the nuclear installations in respect of which nuclear site licences have been granted to that licensee of—

    (a) any sums paid by the Minister under subsection (1) of this section; and

    (b) any expenses, being—

        (i) expenses incurred by the Minister; or

        (ii) expenses incurred by any other government department in connection with the Ministry of Power; or

        (iii) such sums as the Treasury may determine in respect of the use for the purposes of that Ministry of any premises belonging to the Crown,

    which the Minister may, with the consent of the Treasury, determine to be incurred in connection with the exercise by the Minister of his powers under the said subsection (1),

and the licensee shall comply with such requirement; and any sums so repaid to the Minister shall be paid into the Exchequer.

(7) Any liability of a licensee in respect of sums payable by him under subsection (6) of this section on account of pensions shall, if the Minister so determines, be satisfied by way of contributions, calculated, at such rate as may be determined by the Treasury, by reference to remuneration.

Definitions

A–128 "inspector": section 26(1).
       "licensed site": section 26(1).
       "licensee": section 26(1).
       "the Minister": section 26(1).

"nuclear installation": section 26(1).
"nuclear matter": section 26(1).
"nuclear site licence": section 26(1).
"occurrence": section 26(1).
"period of responsibility": section 26(1).

Amendments

This original section is now only applicable to Northern Ireland; it was **A–129**
substituted for England, Wales and Scotland by the Nuclear Installations
Act 1965, etc. (Repeals and Modifications) Regulations 1974 (S.I. 1974
No. 2056), reg. 2(1)(b) and Sched. 2, as amended by the Atomic Energy
Act 1989, s.6; see below, under "Substituted Section 24". In subsection
(4), the words in square brackets are substituted by the Criminal Justice
Act 1982, s.46.

General Note

The original section makes provision for the appointment of inspectors **A–130**
by the Minister and the powers of inspectors in carrying out their duties.
   Subsection (4) sets out the offences and penalties associated with the
failure of a person to co-operate with an inspector and subsection (5)
makes it an offence for an inspector to disclose information obtained as a
result of his activities under the Act, without the authority of the Minister.
The section also provides for payments by a licensee to the Minister under
specified circumstances.

[Inspectors

   24.—(1) The Secretary of State may appoint as inspectors for the **A–131**
purpose of assisting him in the execution of the provisions of this Act,
other than provisions which are mentioned in Schedule 1 to the Health
and Safety at Work, etc. Act 1974, such number of persons appearing to
him to be qualified for the purpose as he may from time to time consider
necessary or expedient, and may make to or in respect of any person so
appointed such payments by way of remuneration, allowances or other
payments as the Secretary of State may with the approval of the Minister
for the Civil Service determine.
   (2) Any such inspector may for that purpose exercise such of the
powers set out in section 20(2) of the Health and Safety at Work, etc. Act
1974 as are specified in his instrument of appointment and the provisions
of sections 28 (restrictions on disclosure of information), 33 (offences)
and 39 (prosecutions by inspectors) of that Act shall apply in the case of
inspectors so appointed as they apply in the case of inspectors
appointed under section 19 of that Act.

(3) In such cases and to such extent as it may appear to the Secretary of State, with the agreement of the Treasury, to be appropriate so to do, the Secretary of State shall require a licensee to repay to the Secretary of State such part as may appear to the Secretary of State to be attributable to the nuclear installations in respect of which nuclear site licences have been granted to that licensee of—

[(a) any sums paid by the Secretary of State under subsection (1) of this section;]; and

(b) any expenses ... being—

(i) expenses incurred by the Secretary of State; or

(ii) ... ; or

(iii) expenses incurred by any government department; or

(iv) such sums as the Treasury may determine in respect of the use of any premises belonging to the Crown,

which the Secretary of State may, with the consent of the Treasury, determine to be incurred in connection with the [exercise by the Secretary of State of his powers under the said subsection (1);],

and the licensee shall comply with such requirement; and any sums so repaid to the Secretary of State shall be paid into the Consolidated Fund ....

(4) Any liability of a licensee in respect of sums payable by him under subsection (3) of this section on account of pensions shall, if the Secretary of State so determines, be satisfied by way of contributions calculated, at such rate as may be determined by the Minister for the Civil Service, by reference to remuneration.]

Definitions

A–132 "inspector": section 26(1).
"licensed site": section 26(1).
"licensee": section 26(1).
"the Minister": section 26(1).
"nuclear installations": section 26(1).
"nuclear matter": section 26(1).
"nuclear site licence": section 26(1).
"period of responsibility": section 26(1).

Amendments

A–133 This substituted section is inserted by the Nuclear Installations Act 1965, etc. (Repeals and Modifications) Regulations 1974 (S.I. 1974 No. 2056), reg. 2(1)(b) and Sched. 2 for England, Wales and Scotland. In subsection (3), paragraph (a) has been substituted by the Atomic Energy Act 1989, s.6(1)(a); in paragraph (b) the omitted words were repealed, and the words in square brackets substituted, by the Atomic Energy Act 1989, s.6(1)(b) and (c).

General Note

The original section and the substituted section replace the Nuclear A-134 Installations (Licensing and Insurance) Act 1959, s.7 and the Nuclear Installations (Amendment) Act 1965, subs.12(4)–(6). The substituted section sets out the power of the Minister to appoint inspectors and the powers of inspectors in carrying out their duties. Provisions also exist for payments by the licensee to the Minister under specified circumstances.

Minister for the Civil Service

The functions of the Minister for the Civil Service are now exercised by A-135 the Treasury under the terms of the Transfer of Functions (Minister for the Civil Service and Treasury) Order 1981 (S.I. 1981 No. 1670), reg. 2.

[Recovery of expenses by Health and Safety Executive

24A.—(1) This section applies to any expenses incurred by the Health A-136 and Safety Executive ("the Executive") and any expenses incurred by the Health and Safety Commission ("the Commission"), which, in either case, the Executive may determine to be incurred wholly or partly in connection with—

(a) the carrying into effect of such of the provisions of this Act as are mentioned in Schedule 1 to the Health and Safety at Work, etc. Act 1974; or

(b) the carrying out of research into nuclear safety at the direction of the Commission.

(2) Without prejudice to the generality of subsection (1) of this section, the reference in that subsection to expenses incurred by the Executive includes any sums paid by it by way of remuneration, allowances or other payments to inspectors appointed under the Health and Safety at Work, etc. Act 1974.

(3) In such cases and to such extent as it may appear to the Executive appropriate to do so, the Executive shall require a person who has applied for a nuclear site licence to repay to it so much of any expenses to which this section applies as may appear to it to be attributable to dealing with the application.

(4) In such cases and to such extent as it may appear to the Executive to be appropriate to do so, the Executive shall require a person to whom a nuclear site licence has been granted to repay to it—

(a) so much of any expenses to which this section applies as may appear to it to be attributable to any nuclear installation in respect of which the licence has been granted; and

(b) so much of any expenses to which this section applies which are not otherwise recoverable under this section as it thinks fit.

(5) A person shall comply with any requirement made of him under this section.

(6) Any liability of a person in respect of sums payable by him under this section on account of pensions shall, if the Executive so determines, be satisfied by way of contributions calculated, at such rate as may be determined by the Treasury, by reference to remuneration.

(7) Where the Executive anticipates that a person who has applied for or has been granted a nuclear site licence will become subject to a liability under this section, it may require him to make to it a payment or payments on account of the liability.

(8) Where a person has made a payment under subsection (7) of this section on account of an anticipated liability, then—

(a) if he does not become subject to the liability, the Executive shall be liable to repay the payment to him; and

(b) if the amount of the liability to which he becomes subject is less than the amount paid under that subsection, the Executive shall be liable to repay the difference to him.]

Definitions

A–137 "nuclear installation": section 26(1).
"nuclear site licence": section 26(1).

Amendments

A–138 This section was inserted by the Atomic Energy Act 1989, s.2(1) and came into force on September 1, 1989, *i.e.* the date of entry into force of the 1989 Act.

General Note

A–139 The section makes provision for the recovery of expenses by the Health and Safety Executive and the Health and Safety Commission for their activities under the Act. These activities include the granting of nuclear site licences, inspection of sites and research into nuclear safety. The section also provides for circumstances in which the licensee may have overpaid the Executive.

Offences—general

A–140 25.—(1) Where a body corporate is guilty of an offence under [any of the provisions : Northern Ireland] [section 2(2) or 19(5): England Wales and Scotland] of this Act and that offence is proved to have been committed with the consent or connivance of, or to be attributable to any neglect on the part of, any director, manager, secretary or other similar officer of the body corporate or any person who was purporting to act in any such capacity, he, as well as the body corporate, shall be guilty of that offence and shall be liable to be proceeded against and punished accordingly; and where the body corporate was guilty of the offence in the capacity of licensee under a nuclear site licence, he shall be so liable as if he, as well as the body corporate, were the licensee.

In this subsection, the expression "director", in relation to a body corporate established by or under any enactment for the purpose of carrying on under national ownership any industry or part of an industry or undertaking, being a body corporate whose affairs are managed by its members, means a member of that body corporate.

(2) Where a body corporate is convicted on indictment of an offence under any of the following provisions of this Act, that is to say, sections [1(3)], 2(2), [4(6), 5(4)] and 19(5), so much of the provision in question as limits the amount of the fine which may be imposed shall not apply, and the body corporate shall be liable to a fine of such amount as the court thinks just.

(3) Proceedings in respect of any offence under [section 2(2) or 19(5) of] this Act shall not be instituted in England or Wales except by the Minister or by or with the consent of the Director of Public Prosecutions.

Definitions

"licensee": section 26(1).   **A–141**
"the Minister": section 26(1).
"nuclear site licence": section 26(1).

Amendments

The first set of words in square brackets in subsection (1) applies only to **A–142** Northern Ireland while the words in the second set of square brackets were substituted by the Nuclear Installations Act 1965, etc. (Repeals and Modifications) Regulations 1974 (S.I. 1974 No. 2056), reg. 2(1)(b) and Sched. 2 for England, Wales and Scotland. In subsection (2) the figures in both sets of square brackets were repealed by the same Regulations (reg. 2(1)(a) and Sched. 1) for England, Wales and Scotland, but not Northern Ireland, and the words in square brackets in subsection (3) were inserted by the same Regulations (reg. 2(1)(b) and Sched. 2).

General Note

This section replaces the Nuclear Installations (Licensing and Insurance) **A–143** Act 1959, s.8 and the Nuclear Installations (Amendment) Act 1965, s.10(5). It deals with general issues relating to offences including provisions for the prosecution of directors and officers where a body corporate is found guilty of an offence. It also sets the level of fine for a body corporate convicted on indictment for specified offences as an amount that the court thinks just. It further requires that proceedings relating to operating without a written permit under section 2(2) and the licensee's duty to provide the necessary cover for claims may only be instituted by the Minister or the Director of Public Prosecutions. In Northern Ireland, all proceedings for offences under the Act would need to be instituted in this way.

[Orders

A–144   25A. The power to make orders under section 16(1A), 18(1B) or 21(1A) of this Act shall be exercisable by statutory instrument; but no such order shall be made unless a draft of it has been laid before and approved by resolution of the House of Commons.]

General Note

A–145   This section was inserted by the Energy Act 1983, s.30.

[Special drawing rights

A–146   25B.—(1) In this Act "special drawing rights" means special drawing rights as defined by the International Monetary Fund; and for the purpose of determining the equivalent in sterling on any day of a sum expressed in special drawing rights, one special drawing right shall be treated as equal to such a sum in sterling as the International Monetary Fund have fixed as being the equivalent of one special drawing right—
    (a) for that day, or
    (b) if no sum has been so fixed for that day, for the last day before that day for which a sum has been so fixed.
  (2) A certificate given by or on behalf of the Treasury stating—
    (a) that a particular sum in sterling has been so fixed for a particular day, or
    (b) that no sum has been so fixed for a particular day and that a particular sum in sterling has been so fixed for a day which is the last day for which a sum has been so fixed before the particular day,
shall be conclusive evidence of those matters for the purposes of subsection (1) of this section; and a document purporting to be such a certificate shall in any proceedings be received in evidence and, unless the contrary is proved, be deemed to be such a certificate.
  (3) The Treasury may charge a reasonable fee for any certificate given

in pursuance of subsection (2) of this section and any fee received by the Treasury by virtue of this subsection shall be paid into the Consolidated Fund.]

General Note

This section was inserted by the Energy Act 1983, s.30 and sets out the A—147 means of calculating the sterling equivalent of a sum expressed in special drawing rights.

Interpretation

26.—(1) In this Act, except where the context otherwise requires, the A—148 following expressions have the following meanings respectively, that is to say—

"the Act of 1959" means the Nuclear Installations (Licensing and Insurance) Act 1959;

["the appropriate Agency" means:
(a) in the case of a site in England or Wales, the Environment Agency;
(b) in the case of a site in Scotland, the Scottish Environment Protection Agency;]

"atomic energy" has the meaning assigned by the Atomic Energy Act 1946;

"the Authority" means the United Kingdom Atomic Energy Authority;

"contravention", in relation to any enactment or to any condition imposed or direction given thereunder, includes a failure to comply with that enactment, condition or direction, and cognate expressions shall be construed accordingly;

"costs" in the application of this Act to Scotland, means expenses;

"cover period" has the meaning assigned by section 19(2) of this Act;

"excepted matter" means nuclear matter consisting only of one or more of the following, that is to say—
(a) isotopes prepared for use for industrial, commercial, agricultural, medical [scientific or educational] purposes;
(b) natural uranium;
(c) any uranium of which isotope 235 forms not more than 0.72 per cent;
(d) nuclear matter of such other description, if any, in such circumstances as may be prescribed (or, for the purposes of the application of this Act to a relevant foreign operator, as may be excluded from the operation of the relevant international agreement by the relevant foreign law);

"home territory", in relation to a relevant foreign operator, means the relevant territory in which, for the purposes of a relevant international agreement, he is the operator of a relevant installation;

"injury" means personal injury and includes loss of life;

"inspector" means an inspector appointed under section 24 of this Act: [Northern Ireland];

["inspector" in sections 4(5) and 5(2) of this Act means an inspector appointed by the Health and Safety Executive under section 24 of the Health and Safety at Work, etc. Act 1974: England, Wales and Scotland;]

"licensed site" means a site in respect of which a nuclear site licence has been granted, whether or not that licence remains in force;

"licensee" means a person to whom a nuclear site licence has been granted, whether or not that licence remains in force;

"the Minister" means—

(a) in the application of this Act to England and Wales, the Minister of Power;

(b) in the application of this Act to Scotland, the Secretary of State;

"nuclear installation" means a nuclear reactor or an installation such as is mentioned in section 1(1)(b) of this Act;

"nuclear matter" means, subject to any exceptions which may be prescribed—

(a) any fissile material in the form of uranium metal, alloy or chemical compound (including natural uranium), or of plutonium metal, alloy or chemical compound, and any other fissile material which may be prescribed; and

(b) any radioactive material produced in, or made radioactive by exposure to the radiation incidental to, the process of producing or utilising any such fissile material as aforesaid;

"nuclear reactor" means any plant (including any machinery, equipment or appliance, whether affixed to land or not) designed or adapted for the production of atomic energy by a fission process in which a controlled chain reaction can be maintained without an additional source of neutrons;

"nuclear site licence" has the meaning assigned by section 1(1) of this Act;

"occurrence" in sections 16(1) [and (1A)], 17(3) and 18 of this Act—

(a) in the case of a continuing occurrence, means the whole of that occurrence; and

(b) in the case of an occurrence which is one of a succession of occurrences all attributable to a particular happening on a particular relevant site or to the carrying out from time to time on a particular relevant site of a particular operation, means all those occurrences collectively;

"period of responsibility", in relation to a licensee, has the meaning assigned by section 5(3) of this Act;

"prescribed" means prescribed by regulations made by the Minister of Power and the Secretary of State acting jointly, which shall be made by statutory instrument and be subject to annulment in pursuance of a resolution of either House of Parliament;

"relevant carriage", in relation to nuclear matter, means carriage on behalf of—

(a) a licensee as the licensee of a particular licensed site; or

(b) the Authority; or

(c) a government department for the purposes of such use of a site by that department as is mentioned in section 9 of this Act; or

(d) a relevant foreign operator; or

(e) a person authorised to operate a nuclear reactor which is comprised in a means of transport and in which the nuclear matter in question is intended to be used;

"relevant foreign contribution", in relation to any claim, means any sums falling by virtue of any relevant international agreement to be paid by the government of any relevant territory other than the United Kingdom towards the satisfaction of that claim;

"relevant foreign judgment" means a judgment of a court of a relevant territory other than the United Kingdom which, under a relevant international agreement, is to be enforceable anywhere within the relevant territories;

"relevant foreign law" means the law of a relevant territory other than the United Kingdom or any part thereof regulating in accordance with a relevant international agreement matters falling to be so regulated and, in relation to a particular relevant foreign operator, means the law such as aforesaid of this home territory;

"relevant foreign operator" means a person who, for the purposes of a relevant international agreement, is the operator of a relevant installation in a relevant territory other than the United Kingdom;

"relevant installation" means an installation to which a relevant international agreement applies;

"relevant international agreement" means an international agreement with respect to third-party liability in the field of nuclear energy to which the United Kingdom or Her Majesty's Government therein are party, other than an agreement relating to liability in respect of nuclear reactors comprised in means of transport;

"relevant site" means any of the following, that is to say—

(a) a licensed site at any time during the period of the licensee's responsibility;

(b) any premises at any time when they are occupied by the Authority;

(c) any site at any time when it is occupied by a government department, if that site is being or has been used by that department as mentioned in section 9 of this Act;

(d) any site in a relevant territory other than the United Kingdom at any time when that site is being used for the operation of a relevant installation by a relevant foreign operator;

"relevant territory" means a country for the time being bound by a relevant international agreement;

"territorial limits" includes territorial waters.

(2) References in this Act to the carriage of nuclear matter shall be construed as including references to any storage incidental to the carriage of that matter before its delivery at its final destination.

(3) Any question arising under this Act as to whether—

(a) any person is a relevant foreign operator; or

(b) any law is the relevant foreign law with respect to any matter; or

(c) any country is for the time being a relevant territory,

shall be referred to and determined by the Minister.

(4) Save where the context otherwise requires, any reference in this Act to any enactment shall be construed as a reference to that enactment as amended, extended or applied by or under any other enactment.

Amendments

A–149 The definition of "the appropriate Agency" is inserted by paragraph 10 of Schedule 22 to the Environment Act 1995. The words in square brackets in paragraph (a) of the definition of "excepted matter", were inserted by the Energy Act 1983, s.32; the definition of "inspector" was substituted for England, Wales and Scotland by the Nuclear Installations Act 1965, etc. (Repeals and Modifications) Regulations 1974 (S.I. 1974 No. 2056), reg. 2(1)(b) and Sched. 2; the words in square brackets in the definition of "occurrence" were inserted by the Energy Act 1983, s.27(7).

Excepted Matter

A–150 The Nuclear Installations (Excepted Matter) Regulations 1978 (S.I. 1978 No. 1779) prescribe certain specified quantities and forms of nuclear matter as "excepted matter". They supersede the Nuclear Installations (Excepted Matter) Regulations 1965 (S.I. 1965 No. 1826) and bring the definition of excepted matter in those regulations into line with international decisions taken in relation to the Paris Convention on Third Party Liability in the Field of Nuclear Energy (see under general headnote to this Act).

If nuclear matter falls within the quantities and forms prescribed it is excluded from the provisions of the 1965 Act and so does not attract the strict liability for damage which is imposed on United Kingdom operators of nuclear installations under sections 7–9 of the Act or on responsible parties under section 11.

The quantities and forms prescribed correspond with those determined by the Steering Committee of the Nuclear Energy Agency of the OECD for the purposes of the Paris Convention. This Committee, on which all Contracting Parties to the Convention are represented, has power to exclude nuclear substances from the provisions of the Convention if in its view the small extent of the risk involved warrants it.

A further exemption applicable to "excepted matter" arises in relation to a certificate of insurance during carriage in the form prescribed by the Nuclear Installations (Insurance Certificate) Regulations 1965 (S.I. 1965 No. 1823). No such certificate is required during the carriage of excepted matter.

Injury

Although "injury" is defined in this section, the Congenital Disabilities A—151
(Civil Liability) Act 1976, subs.3(2) and (3) provide further clarification
of what might constitute injury for the purposes of the 1965 Act.

The Minister

Originally, the term referred to the Minister of Power but the Ministry was A—152
dissolved and its functions transferred to the Minister of Technology
whose powers in turn have been transferred to the Secretary of State for
Trade and Industry by the Secretary of State for Trade and Industry Order
1970 (S.I. 1970 No. 1537).

"Prescribed"

By operation of the Secretary of State for Trade and Industry Order 1970 A—153
(S.I. 1970 No. 1537), powers of the Minister of Power and the Secretary of
State to prescribe, are currently available to the Secretary of State for
Trade and Industry and the Secretary of State for Scotland.

Northern Ireland

27.—(1) In the application to Northern Ireland of the following A—154
provisions of this Act (hereafter in this section referred to as "the
designated provisions"), that is to say, sections 1 to 6 and 22 to [24A] and
[Schedules 1 and 2]—
  (a) any reference to the Minister shall be construed as a reference
      to the Minister of Commerce for Northern Ireland;

(b) the expression "prescribed" shall mean prescribed by regulations made by the said Minister of Commerce, which shall be subject to negative resolution within the meaning of section 41(6) of the Interpretation Act (Northern Ireland) 1954;

(c) any reference to the Treasury shall be construed as a reference to the Ministry of Finance for Northern Ireland;

(d) any reference to Parliament shall be construed as a reference to the Parliament of Northern Ireland;

[(dd) in section 2(1) and in section 2(1D) any reference to a government department shall be construed as including a reference to a department of the Government of Northern Ireland; and in section 2(1C), for the words from "'and any such power" onwards there shall be substituted the words "and any order under this section shall be subject to negative resolution within the meaning of section 41(6) of the Interpretation Act (Northern Ireland) 1954;]

(e) for section 3(3)(b) and (c) shall be substituted the following, that is to say—

"(b) any board of conservators for a fishery district constituted under the Fisheries Acts (Northern Ireland) 1842 to 1954 and any statutory water undertaking within the meaning of the Water Supplies and Sewerage Act (Northern Ireland) 1945";

(f) section 23(1) shall have effect as if the words "be made by statutory instrument and" were omitted;

(g) in section 24(6)—

(i) references to the Ministry of Power or to the Crown shall be construed as references respectively to the Ministry of Commerce for Northern Ireland or to the Crown in right of Her Majesty's Government in Northern Ireland;

(ii) for the words from "and any sums" onwards there shall be substituted the words "and any sums so repaid to the Ministry of Commerce shall be treated as part of the revenues of that Ministry";

[(gg) in section 24A—

(i) references to the Health and Safety Executive shall be construed as references to the Department of Economic Development;

(ii) references to the Health and Safety Commission shall be construed as references to the Health and Safety Agency for Northern Ireland;

(iii) references to the Health and Safety at Work, etc. Act 1974 shall be construed as references to the Health and Safety at Work (Northern Ireland) Order 1978; and

(iv) in subsection (1)(b), for the words "at the direction" there shall be substituted the words "on the recommendation";]

(h) in [Schedule 2], any reference to a master of the Supreme Court or to the High Court shall be construed respectively as a reference to the taxing master of the Supreme Court of Northern Ireland or to a judge of the High Court of Justice in Northern Ireland.

(2) In the application to Northern Ireland of any provision of this Act other than the designated provisions—

(a) any reference to the Minister shall be construed as a reference to the Minister of Power;

(b) any reference to an enactment of the Parliament of the United Kingdom shall be construed as a reference to that enactment as it applies in Northern Ireland.

(c) any reference to a government department shall be construed as including a reference to a department of the Government of Northern Ireland.

(3) In relation to a department of the Government of Northern Ireland using any site as mentioned in section 9 of this Act—

(a) references in this Act to the Crown shall be construed as references to the Crown in right of Her Majesty's Government in Northern Ireland;

(b) references in this Act to the Minister in charge of that department shall be construed as references to the Minister of the Government of Northern Ireland so in charge.

(4) In the application to Northern Ireland of section 21(5) of this Act, the reference to Part VI of the Road Traffic Act 1960 shall be construed as a reference to Part II of the Motor Vehicles and Road Traffic Act (Northern Ireland) 1930 as amended or re-enacted (with or without modification) by any subsequent enactment of the Parliament of Northern Ireland for the time being in force.

(5) Proceedings in respect of any offence under this Act shall not be instituted in Northern Ireland except—

(a) in the case of an offence under any of the designated provisions, by the said Minister of Commerce; or

(b) in the case of any other offence, by the Minister of Power; or

(c) in either case, by or with the consent of the Attorney-General for Northern Ireland.

(6) Nothing in this Act shall authorise any department of the Government of Northern Ireland to incur any expenses attributable to the provisions of this Act until provision has been made by the Parliament of Northern Ireland for those expenses to be defrayed out of moneys provided by that Parliament.

(7) . . . .

Amendments

In subsection (1), paragraph (dd) was inserted by the Atomic Energy A–155 Authority Act 1971, s.17(4)(b) and paragraph (gg) was inserted by the Atomic Energy Act 1989, s.6(2). The remaining words and figures in subsection (1) were substituted by the Atomic Energy Act 1971, section 17(4)(a) and (c) and the Atomic Energy Act 1989, section 6(2). Subsection (7) was repealed by the Northern Ireland Constitution Act 1973, s.41(1) and Pt. I, Sched. 6.

General Note

A–156 This section replaces the Nuclear Installations (Licensing and Insurance) Act 1959 subs.12(2)(c), (h) and (4) and the Nuclear Installations (Amendment) Act 1965, s.14.

Channel Islands, Isle of Man, etc.

A–157   28.—(1) Her Majesty may by Order in Council direct that any of the provisions of this Act specified in the Order shall extend, with such exceptions, adaptations and modifications as may be so specified, to any of the Channel Islands, to the Isle of Man or to any other territory outside the United Kingdom for the international relations of which Her Majesty's Government in the United Kingdom are responsible.

(2) Any Order in Council made by virtue of this section may be varied or revoked by any subsequent Order in Council so made.

General Note

A–158 This section replaces the Nuclear Installations (Licensing and Insurance) Act 1959, s.13 and the Nuclear Installations (Amendment) Act 1965, s.15.

Energy Act 1983

A–159 Section 33 of the 1983 Act extends the powers given by section 28 of the 1965 Act to include provisions in the 1965 Act as amended by the Energy Act 1983.

Congenital Disabilities (Civil Liability) Act 1976

A–160 Section 4(6) of the 1976 Act provides that section 3 of the 1976 Act is to be treated as if it were a provision of the 1965 Act for the purposes of section 28.

Orders under Subsection (1)

A–161 The following orders have be made:

- Nuclear Installations (Gibraltar) Order 1970 (S.I. 1970 No. 116), as amended by S.I. 1985 No. 752 extending sections 10, 17, 21, 26 and 30 of the 1965 Act with modifications and the Congenital Disabilities (Civil Liability) Act 1976, s.3 with modifications and adaptations, to Gibraltar;

- Nuclear Installations (Bahamas) Order 1972 (S.I. 1972 No. 121) extending sections 10–17, 21, 26 and 30 with modifications to the Bahama Islands;

- Nuclear Installations (British Solomon Islands Protectorate) Order 1972 (S.I. 1972 No. 122) extending sections 10–17, 21, 26 and 30 with modifications to the Protectorate;

- Nuclear Installations (Cayman Islands) Order 1972 (S.I. 1972 No. 123) extending sections 10–17, 21, 26 and 30 with modifications to the Cayman Islands;

- Nuclear Installations (Falkland Islands and Dependencies) Order 1972 (S.I. 1972 No. 124) extending sections 10–17, 21, 26 and 30 with modifications to the Falkland Islands and its Dependencies;

- Nuclear Installations (Gilbert and Ellice Islands) Order 1972 (S.I. 1972 No. 125) extending sections 10–17, 21, 26 and 30 with modifications to the Gilbert and Ellice Islands Colony;

- Nuclear Installations (Hong Kong) Order 1972 (S.I. 1972 No. 126) extending sections 10–17, 21, 26 and 30 with modifications to Hong Kong;

- Nuclear Installations (Montserrat) Order 1972 (S.I. 1972 No. 127) extending sections 10–17, 21, 26 and 30 with modifications to Montserrat;

- Nuclear Installations (St Helena) Order 1972 (S.I. 1972 No. 128) extending sections 10–17, 21, 26 and 30 with modifications to St. Helena;

- Nuclear Installations (Virgin Islands) Order 1973 (S.I. 1973 No. 235) extending sections 10–17, 21, 26 and 30 with modifications to the Virgin Islands;

- Nuclear Installations (Isle of Man) Order 1977 (S.I. 1977 No. 429) as varied by the Nuclear Installations (Isle of Man) (Variation) Order 1987 (S.I. 1987 No. 668) extending sections 10–17, 21, 26 and 30 with modifications to the Isle of Man;

- Nuclear Installations (Guernsey) Order 1978 (S.I. 1978 No. 1528) as amended by S.I. 1985 No. 1640 extending sections 10–17, 21–26, 30 and Sched. 2 and the Congenital Disabilities (Civil Liability) Act 1976, s.3, with modifications to Guernsey;

- Nuclear Installations (Jersey) Order 1980 (S.I. 1980 No. 1527) as varied by S.I. 1987 No. 2207 extending, with modifications, sections 10–17, 21–26 and 30 and Sched. 2, as amended, to the Bailiwick of Jersey;

- Nuclear Installations (Cayman Islands) Order 1983 (S.I. 1983 No. 1889), amending sections 17, 21 and 26, as extended to the Cayman Islands by (S.I. 1972 No. 123) and extending the Congenital Disabilities (Civil Liability) Act 1976, s.3, with adaptations and modifications, to the Cayman Islands;

- Nuclear Installations (Hong Kong) Order 1983 (S.I. 1983 No. 1890), as amended by (S.I. 1986 No. 2018), amending sections 17, 21 and 26, as extended to Hong Kong by (S.I. 1972 No. 126) and extending the Congenital Disabilities (Civil Liability) Act 1976, s.3, with adaptations and modifications to Hong Kong;

- Nuclear Installations (Montserrat) Order 1983 (S.I. 1983 No. 1891) amending sections 17, 21 and 26, as extended to Montserrat by (S.I. 1972 No. 127), and extending the Congenital Disabilities (Civil Liability) Act 1976, s.3, with adaptations and modifications, to Montserrat;

- Nuclear Installations (St Helena) Order 1983 (S.I. 1983 No. 1892) amending sections 17, 21 and 26, as extended to St Helena by (S.I. 1972 No. 128) and extending the Congenital Disabilities (Civil Liability) Act 1976, s.3 with adaptations and modifications, to St Helena;

- Nuclear Installations (Virgin Islands) Order 1983 (S.I. 1983 No. 1893) amending sections 17, 21 and 26, as extended to the Virgin Islands by (S.I. 1973 No. 235) and extending the Congenital Disabilities (Civil Liability) Act 1976, s.3, with modifications and adaptations, to the Virgin Islands.

Repeals and savings

A–162   29.—(1) ...

(2) Anything done under or by virtue of any enactment repealed by this Act shall be deemed for the purposes of this Act to have been done under or by virtue of the corresponding provision of this Act, and anything begun under any of the enactments so repealed may be continued under the corresponding provision of this Act.

(3) So much of any enactment or document as refers expressly or by implication to any enactment repealed by this Act shall, if and so far as the context permits, be construed as a reference to this Act or the corresponding enactment therein.

(4) Nothing in this section shall be construed as affecting the general application of section 38 of the Interpretation Act 1889 with respect to the effect of repeals.

Amendment

A–163   Subsection (1) was repealed by the Statute Law (Repeals) Act 1974, Pt. XI.

Short title and commencement

30.—(1) This Act may be cited as the Nuclear Installations Act 1965. A–164
(2) This Act shall come into force on such day as Her Majesty may by Order in Council appoint; and a later day may be appointed for the purposes of section 17(5) than that appointed for the purposes of the other provisions of this Act.

General Note

Subsection (2) replaces the Nuclear Installations (Amendment) Act A–165
1965, s.18(3).

Apart from section 17(5), the Nuclear Installations Act 1965 (Commencement No. 1) Order 1965 (S.I. 1965 No. 1880) brought the Act into force on December 1, 1965. Section 17(5) came into force on the date on which the Energy Act 1983, s.35 came into force (September 1, 1983).

Amendments and additions to the Act made by the Atomic Energy Act 1989 entered into force on September 1, 1989.

SCHEDULE 1

Section 2

SECURITY PROVISIONS APPLICABLE BY ORDER UNDER SECTION 2

1. In this Schedule "the specified body corporate", in relation to an A–166
order made under section 2 of this Act, means the body corporate specified in that order, as being a body to whom the Minister has granted a permit as mentioned in subsection (1B) of that section, and "site to which a permit applies" means a site in respect of which a permit so granted to the specified body corporate is for the time being in force.
2. . . .
3.—(1) Every site to which a permit applies shall, for the purposes of section 3(c) of the Official Secrets Act 1911 (which provides that places belonging to or used for the purposes of Her Majesty may be declared by order of the Secretary of State to be prohibited places for the purposes of that Act), be deemed to be a place belonging to or used for the purposes of Her Majesty.
(2) No person other than—
    (a) a constable acting in the execution of his duty as such, or
    (b) an officer of customs and excise or inland revenue, acting in the execution of his duty as such, or
  [(bb) a person designated as an inspector of the International Atomic Energy Agency under article 85 of the Agreement made on

September 6, 1976 for the application of Safeguards in the United Kingdom in connection with the Treaty on the Non-Proliferation of Nuclear Weapons (Cmnd. 6730), or]

(c) an inspector appointed under section 24 of this Act, or

[(cc) an inspector appointed under section 19 of the Health and Safety at Work, etc. Act 1974 and specially authorised in that behalf by or on behalf of a Minister of the Crown, or]

(d) an officer of any government department specially authorised in that behalf by or on behalf of a Minister of the Crown,

shall, except with the consent of the specified body corporate and in accordance with any conditions imposed by them, be entitled to exercise any right of entry (whether arising by virtue of any statutory provision or otherwise) upon any site which is for the time being declared to be a prohibited place by virtue of an order made under the said section 3(c) as extended by the preceding subparagraph:

Provided that any person aggrieved by a refusal of the specified body corporate to consent to, or by conditions imposed by that body on, the exercise of any such right of entry may apply to the Minister who may, if he thinks fit, himself authorise the exercise of the right subject to such conditions, if any, as he may think fit to impose.

4.—[(1) Section 3 of the Special Constables Act 1923 shall have effect as if all premises in the occupation or under the control of the specified body corporate were under the control of the Authority.]

(2) . . .

[(3) For the purposes of section 2 of the Metropolitan Police Act 1860 (which limits the use of the powers of special constables to property of the Crown in certain circumstances) any property of the specified body corporate shall be deemed to be property of the Crown; and in this subparagraph property of the specified body corporate includes property which (though not owned by them) is in their possession or under their control, and property which has been unlawfully removed from their possession or control.]

5.—(1) The specified body corporate shall comply with any directions which the Minister may give to them for the purpose of safeguarding information in the interests of national security; and a direction under this subparagraph may in particular require the specified body corporate to terminate the employment of any person specified in the direction who is an officer of, or employed by, that body or may require that body not to appoint a person so specified to be an officer of, or to any employment under, that body.

(2) The specified body corporate shall also comply with any directions given to them by the Minister with respect to the safekeeping of material of any description specified in the directions, whether in the interests of national security or of safety.

(3) The Minister may with the approval of the Treasury make grants out of moneys provided by Parliament for reimbursing to the specified body corporate, in whole or in part, any expenses incurred by that body in

complying with any directions given under subparagraph (1) of this paragraph and any directions given under subparagraph (2) of this paragraph with respect to the safe-keeping of material in the interests of national security.

6.—(1) Except with the consent of the Minister the specified body corporate shall not terminate on security grounds the employment of any person employed by them.

(2) In this paragraph "security grounds" means grounds which are grounds for dismissal from the civil service of Her Majesty, in accordance with any arrangements for the time being in force relating to dismissals from that service for reasons of national security.

7. In the application of this Schedule to Northern Ireland—

    (a) in paragraph 3(2)(d) the reference to a government department shall be construed as including a reference to a department of the Government of Northern Ireland; and

    (b) in paragraph 4(1), for the reference to section 3 of the Special Constables Act 1923 there shall be substituted a reference to paragraph 1(2) of Schedule 2 to the Emergency Laws (Miscellaneous Provisions) Act 1947.]

Definitions

"the Authority": section 26(1).   A–167
"the Minister": section 26(1).

Amendments

This schedule was inserted by the Atomic Energy Authority Act 1971, A–168
s.17(6).

Paragraph 2 was repealed by the Official Secrets Act 1989, s.16(4) and Sched. 2; and paragraph 4(2) was repealed by the Ministry of Defence Police Act 1987, s.7(4).

Paragraph 3(2)(bb) was inserted by the Nuclear Safeguards and Electricity (Finance) Act 1978, s.2(3)(b); and paragraph 3(2)(cc) was added by the Nuclear Installations Act 1965, etc. (Repeals and Modifications) Regulations 1974 (S.I. 1974 No. 2056), reg. 2(1)(b) and Sched. 2, for England, Wales and Scotland, but not for Northern Ireland.

Paragraph 4(1) was substituted and paragraph 4(2) was repealed by the Ministry of Defence Police Act 1987, s.7(4), and paragraph 4(3) was inserted for England, Wales and Scotland, but not for Northern Ireland, by the Atomic Energy Authority (Special Constables) Act 1976, s.2(1).

SCHEDULE 2

Section 22(5)

[Applicable to Northern Ireland only]

INQUIRIES UNDER SECTION 22(5)

A–169   1. An inquiry in pursuance of a direction under section 22(5) of this Act with respect to any occurrence shall be held by a competent person appointed by the Minister, and that person may conduct the inquiry either alone or with the assistance of an assessor or assessors so appointed.

2. The Minister may pay to the person appointed to hold the inquiry and to any assessor appointed to assist him such remuneration and allowances as the Minister may, with the approval of the Treasury, determine.

3. The person appointed to hold the inquiry (hereafter in this Schedule referred to as "the court") shall hold the inquiry in such manner and under such conditions as the court thinks most effectual for ascertaining the causes, circumstances and effects of the occurrence and for enabling the court to make the report hereafter in this Schedule mentioned.

4. The court shall, for the purposes of the inquiry, have power—

    (a) to enter and inspect any place or building the entry or inspection whereof appears to the court requisite for the said purposes;

    (b) by summons signed by the court to require any person to attend, at such time and place as is specified in the summons, to give evidence or produce any documents in his custody or under his control which the court considers it necessary for the purposes of the inquiry to examine;

    (c) to require a person appearing at the inquiry to furnish to any other person appearing thereat, on payment of such fee, if any, as the court thinks fit, a copy of any document offered, or proposed to be offered, in evidence by the first mentioned person;

    (d) to take evidence on oath, and for that purpose to administer oaths, or, instead of administering an oath, to require the person examined to make and subscribe a declaration of the truth of the matter respecting which he is examined;

    (e) to adjourn the inquiry from time to time; and

    (f) subject to the foregoing subparagraphs, to regulate the procedure of the court.

5. A person attending as a witness before the court shall be entitled to be paid by the Minister such expenses as would be allowed to a witness attending on subpoena before a court of record, and any dispute as to the amount to be so allowed shall be referred by the court to a master of the

Supreme Court who, on request signed by the court, shall ascertain and certify the proper amount of the expenses.

6. The court shall make a report to the Minister stating the causes, circumstances and effects of the occurrence, adding any observations which the court thinks it right to make, and the Minister shall cause copies of the report, or so much thereof as it is not in his opinion inconsistent with the interests of national security to disclose, to be laid before Parliament.

7. If any person—
   (a) without reasonable excuse (proof whereof shall lie on him), and after having the expenses (if any) to which he is entitled tendered to him, fails to comply with any summons or requisition of the court; or
   (b) does any other thing which would, if the court had been a court of law having power to commit for contempt, have been contempt of that court,

the court may, by instrument signed by the court, certify the offence of that person to the High Court, or in Scotland, the Court of Session, and the High Court or Court of Session may thereupon inquire into the alleged offence and after hearing any witnesses who may be produced against or on behalf of the person charged with the offence, and after hearing any statement that may be offered in defence, punish or take steps for the punishment of that person in like manner as if he had been guilty of contempt of the High Court or, as the case may be, the Court of Session.

8. In the application of this Schedule to Scotland, for references to a master of the Supreme Court, to a witness attending on subpoena before a court of record, and to a summons there shall be respectively substituted references to the Auditor of the Court of Session, to a witness attending on citation the High Court of Justiciary, and to an order.

Amendments

This schedule was renumbered by the Atomic Energy Authority Act 1971, **A–170** s.17(6). It was repealed by the Nuclear Installations Act 1965, etc. (Repeals and Modifications) Regulations 1974 (S.I. 1974 No. 2056) for England, Wales and Scotland, but not for Northern Ireland.

APPENDIX B

RADIOACTIVE SUBSTANCES ACT 1993
(c. 12)

Further obligations relating to registration or authorisation

Enforcement notices and prohibition notices

Powers of Secretary of State in relation to applications, etc.

Appeals

Further powers of Secretary of State in relation to radioactive waste

Offences

Public access to documents and records

Operation of other statutory provisions

General

Schedules

RADIOACTIVE SUBSTANCES ACT 1993 (c. 12)

Introduction and General Note

The Radioactive Substances Act 1993 (called "the Act" in this introduc- **B–01** tion) regulates the keeping and use of radioactive materials, and the emotive matters of the disposal and accumulation of radioactive waste. The Act is a consolidation of the Radioactive Substances Act 1960 (the RSA 1960) which has been heavily amended in recent years, in particular by the Environmental Protection Act 1990. The Department of the Environment felt that the previously existing structure of radioactive substances regulation would benefit from re-arrangement and also from the repeal of the now defunct provisions of the Radioactive Substances Act 1948.

The purpose of the consolidation effected by the Act in 1993 was to restructure the provisions of the RSA 1960 in such a way as to reflect and clarify Parliament's original intentions more accurately. As with all consolidations, no substantive amendments were introduced during the passage of the Act through the Houses of Parliament and therefore no change to the law on radioactive substances was made at that time: see below, however, in relation to the Environment Act 1995.

A Guide to the Administration of the Radioactive Substances Act 1960

The original motivation to introduce the RSA 1960 came out of a **B–02** report produced by an expert panel appointed by the Radioactive Substances Advisory Committee in 1956. That report, which concluded that the then existing laws on the regulation of radioactive substances would benefit from strengthening, was published in a White Paper "*The Control of Radioactive Waste*".[1] It is the recommendations and objectives set out in that report that led to the RSA 1960 and, together with additions made by the Environmental Protection Act 1990 and the Environment Act 1995, are now contained in the Act.

Scheme of the Act and the 1995 Act amendments

The scheme and content of the Act are as follows: the registration of **B–03** premises; registration of mobile radioactive apparatus; the cancellation or variation of registrations; the authorisation of the disposal or

---

[1] *The Control of Radioactive Waste*, Cmnd. 884 (1959).

accumulation of radioactive waste; central control as a feature of radioactive substances control; the requirement to display certificates of registration or authorisation; enforcement and prohibition notices; criminal offences of non-compliance; regulators' powers of entry, etc.; record keeping and public access to information; application of the Act to the Crown; other controls; and finally an account of the complex series of judicial reviews in relation to the THORP reprocessing plant in 1993 and 1994. Changes were effected to the Act by the Environment Act 1995 ("the 1995 Act") and these and other key issues in the Act are summarised below.

Changes effected in Great Britain by the 1995 Act

B–04 *(1) Enforcing authority*—this key change reflects the central purpose of the 1995 Act: **to create a unified Environment Agency** ("the Agency") **for England and Wales, and** its Scottish counterpart **the Scottish Environment Protection Agency** ("SEPA"). These bodies took over from the "chief inspector" his role as the Act's regulatory authority in Great Britain. Administratively the chief inspector's role was carried out in England and Wales by H.M. Inspectorate of Pollution ("HMIP") and in Scotland by H.M. Industrial Pollution Inspectorate ("HMIPI"), and those bodies now form part of the Agency and SEPA respectively.

The implications of this development can be seen as falling into four categories:

- throughout the Act, references to the chief inspector became references to "the appropriate Agency": primarily by virtue of the 1995 Act, Sched. 22, para. 200.
- a number of sections of the Act were revoked, either because they were no longer applicable (ss.4 and 5, dealing with the appointment of the inspectors) or because they were superseded by 1995 Act provisions (ss.31, 35 and 43 and Sched. 2, dealing with rights of entry, etc., fees and charges and the offence of obstructing inspectors in the execution of their duty);
- the expanded role in environmental protection generally envisaged for the Agencies (as reflected, for example, in section 38 of the 1995 Act which anticipates the delegation to them of further Ministerial responsibilities) prompted the devolution to them of certain, previously Ministerial functions: see the revocation of section 16(3) in relation to authorisations for the disposal of radioactive waste on or from nuclear sites; section 28 was consequently revoked (and see further *(3)* below);
- the range of duties, powers, aims and objectives provided for the Agencies by the 1995 Act now apply to the exercise of their regulatory functions under this Act. These include the duty in some circumstances to take account of costs and benefits (1995 Act, s.39) and the

duty to exercise their functions under this Act for the purpose of preventing or minimising, or remedying or mitigating the effects of, pollution of the environment (1995 Act, ss.5 and 33).

(2) *Injunctions*—an express power for the Agencies to take civil proceedings, where they believe that a criminal prosecution would afford an inadequate remedy for non-compliance with enforcement or prohibition notices, was inserted as section 32(3). If an injunction is obtained in this way, non-compliance is contempt of court which may result in imprisonment or sequestration of assets.

(3) *Role of Ministers*—As noted in *(1)* above, the Ministerial function (s.16(3)) of granting certain authorisations jointly with the regulatory authority (at that time the chief inspector) in England, Wales and Northern Ireland was revoked by the 1995 Act. (The authorisations in question are issued under section 13(1) in relation to the disposal of radioactive waste on or from nuclear sites in England, Wales and Northern Ireland.) This had ramifications elsewhere in the Act in four ways:

- the provision (s.28) for representations to be made to the Secretary of State (in the absence of the right of appeal) where authorisation was refused, was consequently removed—there is now a right of appeal under section 26;
- the Ministerial role in jointly revoking or varying the relevant authorisations was similarly removed (revocation of s.17(4));
- those Ministers must now be consulted (as now must the Health and Safety Executive) before a s.13(1) authorisation is granted (s.16(4A)) or varied (s.17(2A)). In addition, the Minister of Agriculture, Fisheries and Food, as one of the Ministers affected by the revocation of section 16(3), has also had his role enhanced elsewhere: in relation to section 13(1) authorisations, the power to give directions to the Agency (s.23), the power to take over the authorising role in relation to an application (s.24), and the role as appellate authority (s.26) is now exercisable in England jointly by the Secretary of State and the Minister of Agriculture, Fisheries and Food (ss.23(4A), 24(4A) and 26(5A) respectively: see also s.27(7A)).

(4) *Public access to information*—Amendments to section 25 make it clear that only such information in or relating to a registration as affects national security may be subject to a direction that knowledge of it be restricted.

(5) *False and misleading statements*—A new section 34A, inserted by the 1995 Act, Sched. 19, created new offences of (a) knowingly or recklessly making statements to the Agencies which are false or misleading in a

material particular; or (b) making false entries in a record kept pursuant to a condition in a registration or authorisation.

(6) *Contaminated land*—Section 40 provides that radioactivity is to be disregarded for the purposes of certain other specified statutory regimes. The contaminated land provisions introduced by the 1995 Act as Part IIA of the Environmental Protection Act 1990 have not been added to that list, but, by section 78YC of the 1990 Act similar provision is made. However, that section goes on to provide that those provisions may be modified so as to deal with radioactivity in the context of contaminated land, and that regulations may modify this Act (presumably again to take account of contaminated land issues).

Registration of Premises

B–05 Any person who keeps or uses radioactive material on premises used for the purpose of an undertaking must either be registered or subject to an exemption under the provisions of the Act (s.6).

Applications for registration are made to the Environment Agency ("the Agency") in England and Wales, the Scottish Environment Protection Agency ("SEPA") in Scotland and the Alkali and Radiochemical Inspectorate in Northern Ireland). Applications have to contain specified information and be accompanied by a fee (s.7). Copies of applications are sent to the local authority except to the extent that the circulation of an application has been restricted by the Secretary of State for reasons of national security (s.25).

The appropriate Agency (the Agency or SEPA) may either refuse to grant a registration or grant one (which may be subject to conditions and limitations). When setting conditions or limitations the appropriate Agency must have regard only "to the amount and character of the radioactive waste likely to arise from the keeping or use of the radioactive material on the premises in question" (s.7(7)). Conditions and limitations may deal with the structure of premises, apparatus, equipment or appliances relating to the use of radioactive material. Additional conditions may require the production of information on the movement of radioactive material, or prohibit the sale of incorrectly labelled radioactive material.

Premises covered by available exemptions include those which are subject to a nuclear site licence (granted under the Nuclear Installations Act 1965) and premises on which clocks and watches which contain radioactive material are kept or used (although this exemption does not extend to premises on which clocks or watches are manufactured or repaired by processes involving the use of luminous material). Other exemptions are set out in statutory orders and include a wide range of low activity radioactive materials (see the General Note to s.7).

Registration of Mobile Radioactive Apparatus

In addition to the registration of premises, special provisions are made for **B–06** mobile radioactive apparatus. Such equipment must be registered wherever it might be kept, used, lent or hired, unless such use is covered by an exemption (s.9).

The provisions on applications and the powers of the appropriate Agency for registering mobile radioactive apparatus are very similar to those for registering premises, although the appropriate Agency may impose any limitations and conditions that it thinks fit (s.10). A registration of mobile radioactive apparatus is only valid in the country in which it was issued.

Exemptions from the necessity to register mobile radioactive apparatus are to be set out in statutory instruments of which to date there have been none (s.11).

Cancellation or Variation of Registrations

The appropriate Agency may cancel or vary any registrations made in **B–07** respect of premises or mobile radioactive apparatus at any time (s.12). Such cancellation or variation is subject to a right of appeal (s.26).

Authorisation of Disposal or Accumulation

It is useful to distinguish between "disposal" and "storage". Storage of **B–08** radioactive waste applies where the material is placed at a facility (either engineered or natural) with the intention of action being taken at a later time for its subsequent disposal. That later action may involve the retrieval of the substances, their in-site treatment or a declaration that no further action is needed and that the storage has, in the event, become disposal.

The disposal of radioactive waste is the dispersal of the waste into an environmental medium or placement in a facility with the intention of taking no further action apart from some possible monitoring for technical or reassurance purposes. In the United Kingdom it is only low-level waste such as that produced by hospitals, research facilities and industry, etc. *e.g.* contaminated packaging, gloves, rags, glass, small tools, paper, filters and effluents which may be disposed of by way of incineration, landfill or discharge to sewers. Intermediate level waste and high level waste are generally accumulated in long-term storage facilities.

The disposal (on land, into water or by discharge into the atmosphere) of any radioactive waste, on or from any premises is prohibited unless it is authorised by the appropriate Agency or is exempted under the Act (s.13). A similar prohibition applies to the disposal of any radioactive waste from mobile radioactive apparatus or the receipt of radioactive waste for the purpose of its disposal.

353

The exemptions cover premises with a nuclear site licence, the disposal of any radioactive waste arising from clocks and watches (on a similar basis as for registration) and others as set out in statutory instruments (see the General Note to s.15).

The accumulation of radioactive waste with a view to its subsequent disposal is also prohibited unless it is either authorised by the appropriate Agency or exempted (s.14). The exemption provisions are as for the disposal of radioactive waste (s.15).

Applications for both authorisation for disposal or accumulation are made to the appropriate Agency (s.16). Where the disposal of radioactive waste is on or from a nuclear site in Great Britain, the appropriate Agency is under a duty to consult a number of bodies before granting an authorisation (subss.16(4A) and (5)). (For England, the relevant Minister is the Minister of Agriculture, Fisheries and Food and for Wales and Scotland the responsibility is solely that of the Secretaries of State for Wales and Scotland respectively; in Northern Ireland it is the Departments of Environment and Agriculture jointly.)

As for registration, authorisation may be granted or refused. Where granted, the authorisation may be subject to such limitations or conditions as the appropriate Agency may think fit. Copies of applications will be sent to the local authority except in so far as there are overriding issues of national security (s.25).

The appropriate Agency may (subject to consultation requirements in the case of variation: s.17(2A)) vary or revoke any authorisations for disposal or accumulation of radioactive waste in which case there is a right of appeal (s.17).

Central Control

B–09 The control of radioactive materials is a central and not a local government function. Local bodies are specifically prohibited from taking account of radioactivity when exercising their functions under public health and clean air legislation as enumerated in Schedule 3, or local enactments dealing with nuisances, pollution and waste discharges (s.40: see, however, the General Note to that section in relation to contaminated land). However, the non-radioactive aspects of any substances must still be dealt with. For example, the discharge of waste effluent to a sewer must be properly authorised under the Water Industry Act 1991, regardless of whether or not it is radioactive. Where the appropriate Agency believes that the disposal of radioactive waste is likely to involve the need for any special precautions to be taken, it is to consult with relevant authorities, including any local or water authorities, who would have to take special precautions, before the authority to dispose of waste is granted (s.18). Only where special precautions are necessary can the authority in question make a charge (s.18(2)).

The Secretary of State can give directions to the appropriate Agency on the conduct of any applications or in relation to any registration or authorisation granted (ss.23 and 24).

Where the Secretary of State believes that inadequate provisions for the safe disposal or accumulation of radioactive waste are not available, he may arrange for such facilities to be provided. The Secretary of State may then make a charge for their use. The site at Drigg near Sellafield has been provided pursuant to this power. That site is owned and operated by British Nuclear Fuels Limited and is used for the disposal of solid waste which is considered unsuitable for special precaution tipping at a landfill site (s.29).

On a similar note, the appropriate Agency has the power to dispose of radioactive waste where the premises on which it is are unoccupied or the occupier is absent or insolvent or for some other reason, and it is unlikely that the waste would be lawfully disposed of unless it intervened. The appropriate Agency may recover its reasonably incurred costs from the occupier or owner of the premises (s.30).

Posting of Certificates

A copy of any certificate of registration or authorisation must be displayed **B–10** at the appropriate premises (s.19). The appropriate Agency can require a registered or authorised person to retain various documents following cessation of the regulated activity, and produce on request, site and disposal documentation (s.20).

Enforcement and Prohibition Notices

The appropriate Agency can issue enforcement or prohibition notices in **B–11** relation to any registration or authorisation (ss.21 and 22).

The appropriate Agency may consider issuing an enforcement notice where there is either an actual or likely failure to comply with the conditions and limitations of an authorisation or registration.

Prohibition notices may be used where the continuation of an authorised or registered activity involves an imminent risk of pollution of the environment or of harm to human health. There is no pre-requisite to the service of such notice of non-compliance with any limitation or conditions. Prohibition notices are likely to be used in the event of an unauthorised or unusual happening.

When either a prohibition or enforcement notice has been served there is a right to appeal unless the Secretary of State himself has directed that it be served (subss.26(2) and (3)).

Criminal Offences

B–12 The Act creates a number of criminal offences arising out of failure to obtain authorisations and registrations as well as other provisions of the Act (ss.32–37). Section 35 was revoked by the Environment Act 1995, and offence provisions in relation to obstructing an Agency/SEPA officer are now found in section 10(1) of the 1995 Act.

Rights of Entry

B–13 Extensive rights of entry to premises at any reasonable time for testing and inspecting, etc., for the purposes of enforcing a registration or authorisation, together with other powers, are provided for in sections 108 and 109 of the 1995 Act.

Record Keeping and Access to Information

B–14 The appropriate Agency must keep copies of applications as well as documents issued by it or sent by it to local authorities. Additionally, it must keep a record of convictions under the Act (s.39). The appropriate Agency must send documentation to local authorities (ss.16(6) and (9) and 23(4)) and those documents are to be made available to the public (s.39), subject to requirements of confidentiality or national security (s.39(1)). Local authorities are also under a duty to make available to the public copies of documents, unless the appropriate Agency directs otherwise (s.39).

The Crown

B–15 The Crown is generally bound by the Act (s.42). However, the Act does not apply to premises occupied for military or defence purposes, by Her Majesty in her personal capacity or by visiting forces. The Crown cannot be held criminally liable for an infringement of the Act, but the courts can make a declaration that an unlawful act or omission has been committed.

Other Controls

B–16 Nuclear installations in the United Kingdom are regulated by the Nuclear Installations Act 1965. This Act provides for their licensing and inspection with a view to ensuring the maximum possible safety in their construction and operation. That Act also provides that where injury or damage results from the emission of ionising radiation from or in connection with a nuclear site, compensation is available.

The transportation by road of radioactive material is regulated by the Radioactive Material (Road Transport) Act 1991 and its attendant Regulations (see Appendix C for the text of the 1991 Act). That Act makes provision for the consignment and carriage of radioactive material in Great Britain and imposes strict requirements on the transportation of such material which are enforced by criminal sanctions, etc.

Radioactive Substances and Judicial Review

The three decisions in the judicial review proceedings brought by B–17 Greenpeace against HMIP (the predecessor to the Environment Agency as regulator under the Act) in respect of the thermal oxide reprocessing plant (THORP) at BNFL's Sellafield site deal with a range of important issues both on the interpretation of the Act and on the use of judicial review in relation to environmental matters. The complex facts surrounding this litigation may be summarised briefly as follows. BNFL held authorisations under section 6(1) of the RSA 1960 to discharge liquid and gaseous radioactive waste from its nuclear reprocessing facility at Sellafield. In 1992, BNFL applied for new authorisations, to include the proposed operation of THORP. HMIP and the Minister of Agriculture for Fisheries and Foods as the joint responsible authority prepared to draft authorisations on which an extensive public consultation exercise, generating 84,000 responses, took place. Pending the grant of the new authorisation, BNFL applied for and obtained a variation of its existing authorisation to enable it to test new plant before it became fully operational.

Greenpeace applied by judicial review to quash the decision to vary the existing authorisations and for a stay of the implementation of the authorisations as varied, which would have the effect of halting the proposed testing of THORP. Greenpeace was successful in its application to Brooke J. (September 1, 1993) for leave to apply for judicial review, but unsuccessful in that Brooke J. refused to grant the stay of the variation decision. Greenpeace appealed unsuccessfully against that aspect of the decision.[2]

The substantive application for judicial review was heard by Otton J. on September 14, 15, 16 and 29, 1993.[3] Otton J. dealt with the issue of whether Greenpeace had standing to bring the application, finding in favour of Greenpeace. However, the substantive attack by Greenpeace on the variation procedure failed. The decision to vary was also attacked, unsuccessfully, for failure adequately to justify the releases as required by Article 6 of Directive 80/836/EURATOM (the Basic Safety Standards Directive).

In terms of the application for the new authorisation for THORP, following the consultation exercise the authorities concluded that no new

---

[2] R. v. Inspectorate of Pollution, ex p. Greenpeace [1994] 4 All E.R. 321, C.A.
[3] R. v. Inspectorate of Pollution, ex p. Greenpeace (No. 2) [1994] 4 All E.R. 329.

matters had been raised to cause them to reconsider the terms of the draft authorisations, though further consultation took place in 1993 to consider wider policy issues (justification and proposed emissions of radioactivity in terms of overall benefits and the non-proliferation implications of an increasing stock pile of plutonium). The new authorisations were granted in December 1993 pursuant to sections 13 and 16 of the Act. Greenpeace, together with Lancashire County Council, applied for judicial review of this decision and the earlier decision by the Secretary of State not to exercise his discretion under section 24(2) of the Act to call in the applications for authorisations and hold a local inquiry. This challenge was also unsuccessful.[4]

R. v. Inspectorate of Pollution, ex p. Greenpeace Limited: the Interlocutory relief issue

B–18   The Court of Appeal held on this issue that if the real purpose of interlocutory relief in a judicial review case was to prevent executive action by a third party being carried out pursuant to the decision attacked, the most suitable procedure would be to have the third party in question joined and then to seek an interlocutory injunction, rather than a stay of the decision. Where the application for a stay was pursued, then the courts should look to the substance rather than the form, and apply the same principles to the application as would have been appropriate had the application been for an interlocutory injunction. It followed that Brooke J. had applied the correct principles on the issue of balance of convenience in refusing the stay, given the very heavy costs which would be incurred by BNFL (estimated at £250,000 a day) if the testing process was held up, and in the absence of any cross-undertaking in damages by Greenpeace.

R. v. Inspectorate of Pollution, ex p. Greenpeace Limited (No. 2): the standing issue

B–19   It was held that in deciding whether an applicant for judicial review had sufficient interest in the matter, the court should take into account in exercising its discretion the nature of the applicant, the extent of its interest in the issues raised, and the nature of the relief which was sought. Relevant factors here were that Greenpeace was a responsible and respected body with a genuine interest in the issues raised, that it had 2,500 members in the area where the plant was situated, that Greenpeace had been actively involved in the consultation process relating to the application for operation of the new plant, and that the primary relief

[4] *R. v. Secretary of State for the Environment, ex p. Greenpeace* [1994] 4 All E.R. 325 *per* Potts J.

sought was the less stringent remedy of *certiorari*, which would still leave the issue of an injunction to stop the testing at the discretion of the court. The court also found it relevant that if standing to Greenpeace were denied, then those it represented might not have an effective way to bring the issues before the court, given that individual local residents would not be able to command the expertise and resources at the disposal of Greenpeace. Otton J. emphasised that in exercising his discretion to grant Greenpeace standing, the question of what relief was appropriate would remain a separate discretionary issue, and that although he would probably have granted an order of *certiorari* if they had been successful on substantive grounds, it would have been a matter of considerable arguments and further representation before he would have taken the step of granting an injunction to end the testing.

R. v. Inspectorate of Pollution, ex p. Greenpeace Limited (No. 2): the variation issue

Otton J. held that the authorities had acted lawfully in varying the existing **B–20** authorisations of BNFL to dispose of any radioactive waste from their Sellafield premises used for the purposes of their undertaking. The testing of THORP fell within that definition and there was "a wide power vested in BNFL to extend or contract their activities on site under their licence, in pursuance of their objectives but subject to strict independent regulatory control". The variations were not such as to take the discharges covered outside the scope of the authorisations; they did not extend the description of radioactive waste or authorised disposals of new descriptions of radioactive waste and it was irrelevant in this respect that the relevant waste would emanate from new plant.

R. v. Inspectorate of Pollution, ex p. Greenpeace Limited (No. 2): the justification point

Greenpeace argued that, when considering the application for variation, **B–21** the authorising bodies were obliged to ensure that the process for which the variation was sought was justified. Greenpeace relied on the 1982 Guide to the Administration of the Radioactive Substances Act (at that time, of course, the RSA 1960) issued by the Department of the Environment and more specifically on Article 6 of Council Directive 80/836/EURATOM, as amended. Otton J. held that whilst the authorising bodies were under a general obligation to consider health and safety aspects and whether the amount of radioactive waste to be discharged by the testing process would impose a significant risk to the health and safety of the public, there was no obligation on the authorising bodies to consider the wider issues which were already under consideration

359

through the consultation process relating to the main application. The court regarded it as material that the amount of radioactivity released from the testing process would be very low, and in view of this it was not necessary to go into social and economic issues arising out of the main authorisation. In any event, the court was satisfied that the regulating bodies did in fact take account of wider issues of justification before reaching their decision: those bodies had taken into account the advantages which the testing process would produce. Also, the court thought it was relevant to bear in mind that the need for THORP was considered at the planning stage (the Parker Report on the Windscale Inquiry) and during the Parliamentary process before approval was given; thus in a broad sense THORP had already been "justified in advance".

R. v. Secretary of State for the Environment, ex p. Greenpeace Limited: the justification issue

B–22   Potts J. held that there was no reason why the relevant sections of the Act should not be construed to accord with Directive 80/836/EURATOM, requiring the issue of justification in terms of net benefit to be considered prior to the grant of any authorisation for the discharge for radioactive waste. It was held that Directive 80/836/EURATOM was concerned with the justification of particular practices in particular circumstances. The Ministers had erred in concluding that justification was not relevant in the context of the exercise by them of their functions under the Act.

However, although the Ministers had erred in law in this respect, their general approach to justification could not be faulted. They had in fact carried out a careful process of weighing the benefits against the detriments in reaching their conclusion that the balance came down on the side of justification. In particular, they were entitled to conclude that there was a good economic case for proceeding with THORP and that compliance with international and national standards on radiation limits would protect the public from unacceptable health risks.

R. v. Secretary of State for the Environment, ex p. Greenpeace Limited: environmental assessment

B–23   On the applicability of Directive 85/337/EEC on environmental assessment, it was held that the construction of THORP, the commissioning of the plant and the consequent discharges were a single project which predated the directive. Accordingly, the directive did not apply to the project and there was no legal duty to provide an environmental impact assessment. In any event, although no formal assessment had taken place, it was clear that the information provided and made available in fact met the subsequent requirements of the directive.

R. v. Secretary of State for the Environment, ex p. Greenpeace Limited: call-in and local inquiry

Greenpeace attacked the Secretary of State's decision not to exercise his   B–24
discretion under section 24(2) of the Act to direct a public local inquiry.
The court held that the Secretary of State had adequately and properly
addressed the matters relevant to his decision. There had been extensive
consultation, and the Secretary of State could properly be satisfied that he
was in a position to take account of the representations arising from that
exercise in deciding whether to grant the authorisations. Greenpeace
also referred to the scale of public concern about THORP and argued
that the Secretary of State should have taken into account the need to
inform the public and to allay fears or concerns about the issue. It was
held that, whilst it might be thought that a Minister, sensible to the scale
of representations and the desirability of putting to rest public anxiety,
would have directed an inquiry, that was not an issue for the court. The
central question was whether the Secretary of State in refusing to direct
an inquiry was acting lawfully within the powers conferred upon him by
Parliament, applying the relevant principles of administrative law, and
whether the Secretary of State had applied his mind genuinely and
rationally to the issue.

Abbreviations

the Act	The Radioactive Substances Act 1993	B–25
the 1995 Act	The Environment Act 1995	
the appropriate Agency ...	the Environment Agency, in England and Wales, and Scottish Environment Protection Agency	
DoE	Department of the Environment	
EPA 1990	Environmental Protection Act 1990	
HMIP	Her Majesty's Inspectorate of Pollution	
HMIPI	Her Majesty's Industrial Pollution Inspectorate	
MAFF	Ministry of Agriculture, Fisheries and Food	
MOD	Ministry of Defence	
RSA 1960	Radioactive Substances Act 1960	

Preliminary

Meaning of "radioactive material"

1.—(1) In this Act "radioactive material" means anything which, not   B–26
being waste, is either a substance to which this subsection applies or an
article made wholly or partly from, or incorporating, such a substance.

(2) Subsection (1) applies to any substance falling within either or both of the following descriptions, that is to say,—

(a) a substance containing an element specified in the first column of Schedule 1, in such a proportion that the number of becquerels of that element contained in the substance, divided by the number of grams which the substance weighs, is a number greater than that specified in relation to that element in the appropriate column of that Schedule;

(b) a substance possessing radioactivity which is wholly or partly attributable to a process of nuclear fission or other process of subjecting a substance to bombardment by neutrons or to ionising radiations, not being a process occurring in the course of nature, or in consequence of the disposal of radioactive waste, or by way of contamination in the course of the application of a process to some other substance.

(3) In subsection (2)(a) "the appropriate column"—

(a) in relation to a solid substance, means the second column,

(b) in relation to a liquid substance, means the third column, and

(c) in relation to a substance which is a gas or vapour, means the fourth column.

(4) For the purposes of subsection (2)(b), a substance shall not be treated as radioactive material if the level of radioactivity is less than such level as may be prescribed for substances of that description.

(5) The Secretary of State may by order vary the provisions of Schedule 1, either by adding further entries to any column of that Schedule or by altering or deleting any entry for the time being contained in any column.

(6) In the application of this section to Northern Ireland, the reference in subsection (5) to the Secretary of State shall have effect as a reference to the Department of the Environment for Northern Ireland.

Definitions

B–27 "article": section 47(1).
"contamination": section 47(1).
"prescribed": section 47(1).
"substance": section 47(1).
"waste": section 47(1).

General Note

B–28 This definition of "radioactive material" is a reproduction of the definition within section 18 of the RSA 1960 except that the original reference to microcuries has been replaced with its metric equivalent, the *becquerel*. The Units of Measurement Regulations 1986 (S.I. 1986 No.

1082) were made pursuant to section 2(2) of the European Communities Act 1972 and set out the becquerel as the only unit of radiation authorised for any "specified circumstances" as defined in those Regulations. There are 37,000 becquerels to a microcurie and the amount set out in Schedule 1 represents an exact conversion.

Subsection (4)

This subsection originated in the EPA 1990 and no substances have yet been prescribed as exempt from the definition of "radioactive material". **B–29**

Subsection (5)

The Secretary of State is empowered to change the parameters of what constitutes radioactive material, but to date no order has been made. **B–30**

Meaning of "radioactive waste"

2. In this Act "radioactive waste" means waste which consists wholly or partly of— **B–31**
    (a) a substance or article which, if it were not waste, would be radioactive material, or
    (b) a substance or article which has been contaminated in the course of the production, keeping or use of radioactive material, or by contact with or proximity to other waste falling within paragraph (a) or this paragraph.

Definitions

"article": section 47(1). **B–32**
"contaminated": section 47(5).
"radioactive material": section 1.
"substance": section 47(1).
"waste": section 47(1) and (4).

General Note

The meaning of "waste" is considered in section 47(1) where it is given a wide interpretation and is to be judged largely by reference to the state of mind of the person discarding or wishing to dispose of the substance or material. In the case of *Berridge Incinerators v. Nottinghamshire County Council High Court* (1987), unreported, but cited at paragraph 2.7 of *DoE* **B–33**

Circular 13/88 on the *Collection and Disposal of Waste Regulations*, it was held by Deputy Judge P.J. Crawford, Q.C. that:

> "It is of course, a truism that one man's waste is another man's raw material. The fact that a price is paid by the collector of the material to its originator is, no doubt, relevant, but I do not regard it as crucial. If I have an old fireplace to dispose of to a passing rag and bone man, its character as waste is not affected by whether or not I can persuade the latter to pay me 50p for it. In my judgment, the correct approach is to regard the material from the point of view of the person who produces it. Is it something which is produced as a product, or even as a by-product of his business, or is it something to be disposed of as useless?"

Circular 13/88 goes on to suggest, in the light of this decision, that disposal authorities may find it helpful to consider four questions from the point of view of the person producing the material when considering whether any particular material is waste: (a) Is it what would ordinarily be described as waste?; (b) Is the substance a scrap metal, effluent or other unwanted surplus?; (c) Is the substance or article required to be disposed of broken, worn out, contaminated or otherwise spoiled?; (d) Is the material being discarded or dealt with as if it were waste?

The Circular suggests that an answer of "yes" to any of these questions indicates that the material is waste. This is consistent with the definition of "waste" given at section 47(1) and (4).

Thus "radioactive waste" includes any scrap, surplus or spoiled "radioactive material" (as defined in s.1) and any other waste substance or article which has become radioactive or has acquired an increased concentration of radioactivity by any means.

It should also be noted that radioactive waste is not a "controlled" waste for the purposes of Part II of the EPA 1990 (s.75 of that Act).

Meaning of "mobile radioactive apparatus"

B–34   3. In this Act "mobile radioactive apparatus" means any apparatus, equipment, appliance or other thing which is radioactive material and—
    (a) is constructed or adapted for being transported from place to place, or
    (b) is portable and designed or intended to be used for releasing radioactive material into the environment or introducing it into organisms.

Definition

B–35   "radioactive material": section 1.

General Note

This definition was recast by the EPA 1990 to include plant which is **B–36** mobile and designed for either releasing radioactive material into the environment or introducing it into organisms.

Constructed or adapted

It is generally considered that this phrase "constructed or adapted" means **B–37** originally constructed or subsequently altered so as to make suitable.[5]

Inspectors and Chief Inspector appointed by Secretary of State

4. ...   **B–38**

General Note

This section was revoked in relation to Great Britain by the Environment Act 1995, Sched. 24.

Appointment of inspectors by Minister of Agriculture, Fisheries and Food

5. ...   **B–39**

General Note

This section was revoked in relation to Great Britain by the Environment Act 1995, Sched. 24.

Registration relating to use of radioactive material and mobile radioactive apparatus

Prohibition of use of radioactive material without registration

6. No person shall, on any premises which are used for the purposes of **B–40** an undertaking carried on by him, keep or use, or cause or permit to be kept or used, radioactive material of any description, knowing or having

---

[5] see: *Hubbard v. Messenger* (1938) 1 K.B. 300; and *Davison v. Birmingham Industrial Co-operative Society* [1920] LJKB 206.

reasonable grounds for believing it to be radioactive material, unless either—

    (a) he is registered under section 7 in respect of those premises and in respect of the keeping and use on those premises of radioactive material of that description, or

    (b) he is exempted from registration under that section in respect of those premises and in respect of the keeping and use on those premises of radioactive material of that description, or

    (c) the radioactive material in question consists of mobile radioactive apparatus in respect of which a person is registered under section 10 or is exempted from registration under that section.

Definitions

B–41 "mobile radioactive apparatus": section 3.
"premises": section 47(1).
"radioactive material": section 1.
"undertaking": section 47(1).

General Note

B–42 It is a criminal offence, pursuant to section 32, to keep or use radioactive material without registration. It is also an offence to cause or permit the keeping or using of radioactive material without the requisite registration. In all cases the offence must be committed with knowledge that or reasonable grounds for belief that the substance in question is radioactive material.

The different wording implies distinct offences and to "permit" an offence is generally considered to be a "looser and vaguer" term than to "cause".[6]

Meaning of "cause" and "permit"

B–43 The term "cause" does not imply any intention or negligence and its meaning should be approached in an everyday common sense way.[7] However, in another water pollution case, *Wychavon District Council v. National Rivers Authority*,[8] it was held that a person could only "cause" a discharge of sewage into controlled waters if he took some positive or deliberate act to bring it about. Failure to prevent a discharge was not enough to sustain a conviction. The facts of that case were such that had

---

[6] see: *McCleod v. Buchanan* [1940] 2 All E.R. 179, 187, *per* Lord Wright; and on the differences between "cause" and "permit" see *Shave v. Rosner* [1954] 2 Q.B. 113.
[7] see: *Alphacell v. Woodward* [1972] A.C. 824.
[8] [1993] 1 W.L.R. 125.

the appellants been charged with "knowingly permitting" they may well have been convicted. That case was further discussed and approved in *National Rivers Authority v. Welsh Development Agency.*[9] The House of Lords has held (although its remarks did not form part of the deciding principle of the case in question) that the *Wychavon* line of cases does not establish a general principle, they were simply the application by the Courts of the common sense term "cause" to the facts before them.[10]

However, to "cause" the keeping or use of radioactive material does imply some degree of positive participation or control on the defendant's part (see *McCleod v. Buchanan*).[11] Somebody can only be said to have "caused" another person to have undertaken a particular course of action when he either knew or deliberately chose not to know what was being done.[12]

To "permit" an act to occur imports the necessity for either express or implied permission for that act. Some knowledge of the facts constituting the offence is necessary to establish permitting, although turning a blind eye to events or allowing a course of action to occur during which the commission of an offence would be likely and not caring whether an offence occurs or not is sufficient.[13]

Knowing or having reasonable grounds for believing

To make out an offence under this section, the prosecution must prove **B–43A** that the defendant had the requisite knowledge.[14] As mentioned above, knowledge may be imputed to the person who turns a blind eye to the obvious.[15] Additionally, where a person deliberately does not make enquiries, the results of which he does not wish to know, there is authority that this may constitute in law, actual knowledge of the facts in question.[16]

With regard to the terms "reasonable grounds for believing" case law implies that these words require not only that the person in question has reasonable grounds for believing but that he does also actually believe.[17]

For a corporation to be found guilty of an offence under this provision the prosecution must prove that an individual for whom the corporation could be found responsible caused or permitted the commission of an offence with the requisite knowledge. A corporation is generally only liable for senior personnel who could be said to be exercising a "directing mind" over the company's affairs.[18]

---

[9] (1994) 158 J.P. 506.
[10] see: *NRA v. Yorkshire Water Services* [1994] 3 W.L.R. 1202.
[11] see: *McCleod v. Buchanan* [1940] 2 All E.R. 179.
[12] see: *James & Son v. Smee; Green v. Burnett* [1955] 1 Q.B. 78.
[13] see note **12** *ibid.*
[14] see: *Gaumont British Distributors Ltd v. Henry* [1939] 2 K.B. 711.
[15] *James & Son v. Smee; Green v. Burnett* [1955] 1 Q.B. 78.
[16] see: *Knox v. Boyd* [1941] J.C. 82; and *Mallon v. Allon* [1964] 1 Q.B. 385, 394.
[17] see: *R. v. Banks* [1916] 2 K.B. 621.
[18] see: *Tesco Supermarkets v. Natrass* [1972] A.C. 153; and *R. v. Boal (Francis)* [1992] 1 Q.B. 591.

Person

B–44 Under section 5 of the Interpretation Act 1978, unless a contrary intention is apparent, the term "person" includes a body of persons corporate or unincorporate.

Registration of users of radioactive material

B–45 7.—(1) Any application for registration under this section shall be made to the [appropriate Agency] and shall—
   (a) specify the particulars mentioned in subsection (2),
   (b) contain such other information as may be prescribed, and
   (c) be accompanied by the [charge prescribed for the purpose by a charging scheme under section 41 of the Environment Act 1995].
   (2) The particulars referred to in subsection (1)(a) are—
   (a) the premises to which the application relates,
   (b) the undertaking for the purposes of which those premises are used,
   (c) the description or descriptions of radioactive material proposed to be kept or used on the premises, and the maximum quantity of radioactive material of each such description likely to be kept or used on the premises at any one time, and
   (d) the manner (if any) in which radioactive material is proposed to be used on the premises.
   (3) On any application being made under this section, the [appropriate Agency] shall, subject to directions under section 25, send a copy of the application to each local authority in whose area the premises are situated.
   (4) Subject to the following provisions of this section, where an application is made to the [appropriate Agency] for registration under this section in respect of any premises, [the appropriate Agency] may either—
   (a) register the applicant in respect of those premises and in respect of the keeping and use on those premises of radioactive material of the description to which the application relates, or
   (b) if the application relates to two or more descriptions of radioactive material, register the applicant in respect of those premises and in respect of the keeping and use on those premises of such one or more of those descriptions of radioactive material as may be specified in the registration, or
   (c) refuse the application.
   (5) An application for registration under this section which is duly made to [the appropriate Agency] may be treated by the applicant as having been refused if it is not determined within the prescribed period for

determinations or within such longer period as may be agreed with the applicant.

(6) Any registration under this section in respect of any premises may (subject to subsection (7)) be effected subject to such limitations or conditions as [the appropriate Agency] thinks fit, and in particular (but without prejudice to the generality of this subsection) may be effected subject to conditions of any of the following descriptions—

(a) conditions imposing requirements (including, if the [appropriate Agency] thinks fit, requirements involving structural or other alterations) in respect of any part of the premises, or in respect of any apparatus, equipment or appliance used or to be used on any part of the premises for the purposes of any use of radioactive material from which radioactive waste is likely to arise,

(b) conditions requiring the person to whom the registration relates, at such times and in such manner as may be specified in the registration, to furnish [the appropriate Agency] with information as to the removal of radioactive material from those premises to any other premises, and

(c) conditions prohibiting radioactive material from being sold or otherwise supplied from those premises unless it (or the container in which it is supplied) bears a label or other mark—

(i) indicating that it is radioactive material, or

(ii) if the conditions so require, indicating the description of radioactive material to which it belongs,

and (in either case) complying with any relevant requirements specified in the conditions.

(7) In the exercise of any power conferred on [it] by subsection (4) or (6), the [appropriate Agency], except in determining whether to impose any conditions falling within paragraph (b) or (c) of subsection (6), shall have regard exclusively to the amount and character of the radioactive waste likely to arise from the keeping or use of radioactive material on the premises in question.

(8) On registering a person under this section in respect of any premises, the [appropriate Agency]—

(a) shall furnish him with a certificate containing all material particulars of the registration, and

(b) subject to directions under section 25, shall send a copy of the certificate to each local authority in whose area the premises are situated.

Definitions

"appropriate Agency": section 47(1).
"local authority": section 47(1).
"prescribed": section 47(1).

B–46

"prescribed period for determination": section 47(1) and (2).
"radioactive material": section 1.
"radioactive waste": section 2.

Amendment

B–47 Subsection 7(1)(c) was amended by the Environment Act 1995, Sched. 22, para. 202.

General Note

B–48 This section sets out the provisions relating to the making of an application for a registration for the use or keeping of radioactive material.

Subsection (1)

B–49 To date no statutory instrument has been made prescribing further information to be furnished to the appropriate Agency on an application for registration.

   As to the provisions relating to fees payable, see the Environment Act 1995, s.41.

Subsections (3) and (8)

B–50 Under section 25, the Secretary of State may, on the grounds of national security, restrict the copying of the relevant parts of applications and certificates of registration to local authorities.

Subsection (4)

B–51 The provisions on appeals against a refusal to grant a registration, a deemed refusal to grant a registration or the imposition of conditions or limitations on a registration are set out in sections 26 and 27.

Subsection (5)

B–52 The prescribed period is four months beginning with the day on which the application was received.

Subsection (6)

The types of conditions that the appropriate Agency may impose upon a   B–53
registration are set out at subsection (6). It should also be noted that
under section 44(2) the Secretary of State is empowered to make
regulations setting out general limitations or conditions applicable to
registrations. Breach of limitations or conditions is a criminal offence
under section 32. The statutory controls imposed by Part I of the EPA
1990 on integrated pollution control and local air pollution control raise
issues of duplication and overlap with other areas of control such as this
Act. Section 28(2) of the EPA 1990 deals with this potential problem by
providing that where activities within a prescribed process are regulated
by both an authorisation under the EPA 1990 and by registration under
this Act, then if different obligations are imposed as respects the same
matter those obligations imposed by the EPA 1990 are not binding. Thus,
any conditions imposed by this Act are superior to those imposed under
the EPA 1990.

Subsection (8)

The certificate provided to the applicant must be posted at the registered   B–54
premises pursuant to section 19.

Exemptions from registration under section 7

    8.—(1) At any time while a nuclear site licence is in force in respect of a   B–55
site, and at any time after the revocation or surrender of such a licence
but before the period of responsibility of the licensee has come to an end,
the licensee (subject to subsection (2)) is exempted from registration
under section 7 in respect of any premises situated on that site and in
respect of the keeping and use on those premises of radioactive material
of every description.

    (2) Where, in the case of any such premises as are mentioned in
subsection (1), it appears to the [appropriate Agency] that, if the licensee
had been required to apply for registration under section 7 in respect of
those premises, the [appropriate Agency] would have imposed conditions such as are mentioned in paragraph (b) or (c) of subsection (6) of
that section, the [appropriate Agency] may direct that the exemption
conferred by subsection (1) of this section shall have effect subject to
such conditions (being conditions which in the opinion of the [appropriate
Agency] correspond to those which [it] would so have imposed) as may
be specified in the direction.

    (3) On giving a direction under subsection (2) in respect of any
premises, the [appropriate Agency] shall furnish the licensee with a copy
of the direction.

(4) Except as provided by subsection (5), in respect of all premises all persons are exempted from registration under section 7 in respect of the keeping and use on the premises of clocks and watches which are radioactive material.

(5) Subsection (4) does not exempt from registration under section 7 any premises on which clocks or watches are manufactured or repaired by processes involving the use of luminous material.

(6) The Secretary of State may by order grant further exemptions from registration under section 7, by reference to such classes of premises and undertakings, and such descriptions of radioactive material, as may be specified in the order.

(7) Any exemption granted by an order under subsection (6) may be granted subject to such limitations or conditions as may be specified in the order.

(8) In the application of this section to Northern Ireland, the reference in subsection (6) to the Secretary of State shall have effect as a reference to the Department of the Environment for Northern Ireland.

Definitions

B–56 "appropriate Agency": section 47(1).
"licensee": section 47(1).
"nuclear site licence": section 47(1).
"period of responsibility": section 47(1).
"premises": section 47(1).
"prescribed": section 47(1).
"radioactive material": section 1.
"undertaking": section 47(1).

General Note

Subsection (1)

B–57 Premises for which there is an operative nuclear site licence are exempted from the necessity for registration under section 7. Nuclear site licences are issued under the Nuclear Installations Act 1965.

Subsection (2)

B–58 If it appears to the appropriate Agency that section 7 type conditions are necessary at a nuclear site, it may impose such conditions by direction notwithstanding the exemption. Failure to comply with such limitations or conditions imposed is a criminal offence under section 32.

Subsection (6)

The following statutory instruments granting exemptions under this **B–59** section are currently in force for England and Wales:

- The Radioactive Substances (Exhibitions) Exemption Order 1962 (S.I. 1962 No. 2645).

- The Radioactive Substances (Storage in Transit) Exemption Order 1962 (S.I. 1962 No. 2646).

- The Radioactive Substances (Phosphatic Substances, Rare Earths, etc.) Exemption Order 1962 (S.I. 1962 No. 2648).

- The Radioactive Substances (Lead) Exemption Order 1962 (S.I. 1962 No. 2649).

- The Radioactive Substances (Uranium and Thorium) Exemption Order 1962 (S.I. 1962 No. 2710).

- The Radioactive Substances (Prepared Uranium and Thorium Compounds) Exemption Order 1962 (S.I. 1962 No. 2711).

- The Radioactive Substances (Geological Specimens) Exemption Order 1962 (S.I. 1962 No. 2712).

- The Radioactive Substances (Schools, etc.) Exemption Order 1963 (S.I. 1963 No. 1832).

- The Radioactive Substances (Precipitated Phosphate) Exemption Order 1963 (S.I. 1963 No. 1836).

- The Radioactive Substances (Electronic Valves) Exemption Order 1967 (S.I. 1967 No. 1797).

- The Radioactive Substances (Smoke Detectors) Exemption Order 1980 (S.I. 1980 No. 953) as amended by (S.I. 1991 No. 477).

- The Radioactive Substances (Gaseous Tritium Light Devices) Exemption Order 1985 (S.I. 1985 No. 1047).

- The Radioactive Substances (Luminous Articles) Exemption Order 1985 (S.I. 1985 No. 1048).

- The Radioactive Substances (Testing Instruments) Exemption Order 1985 (S.I. 1985 No. 1049).

- The Radioactive Substances (Substances of Low Activity) Exemption Order 1986 (S.I. 1986 No. 1002) as amended by (S.I. 1992 No. 647).

- The Radioactive Substances (Hospitals) Exemption Order 1990 (S.I. 1990 No. 2512), as amended by (S.I. 1995 No. 2395).

Similar orders have been issued in relation to exemptions for Scotland and Northern Ireland.

Prohibition of use of mobile radioactive apparatus without registration

B–60   9.—(1) No person shall, for the purpose of any activities to which this section applies—

(a) keep, use, lend or let on hire mobile radioactive apparatus of any description, or

(b) cause or permit mobile radioactive apparatus of any description to be kept, used, lent or let on hire,

unless he is registered under section 10 in respect of that apparatus or is exempted from registration under that section in respect of mobile radioactive apparatus of that description.

(2) This section applies to activities involving the use of the apparatus concerned for—

(a) testing, measuring or otherwise investigating any of the characteristics of substances or articles, or

(b) releasing quantities of radioactive material into the environment or introducing such material into organisms.

Definitions

B–61   "article": section 47(1).
"mobile radioactive apparatus": section 3.
"substance": section 47(1).

General Note

Subsection (1)

B–62   Breach of this provision is a criminal offence pursuant to section 32. For further discussion on the components of the crime, see the General Note to section 6.

Person

B–63   As regards the meaning of "person" see further the General Note to section 6.

For revocation and variation of registrations see section 12. For exemptions from registration see section 11.

Registration of mobile radioactive apparatus

B–64   10.—(1) Any application for registration under this section shall be made to the [appropriate Agency] and—

(a) shall specify—

      (i) the apparatus to which the application relates, and

      (ii) the manner in which it is proposed to use the apparatus,

    (b) shall contain such other information as may be prescribed, and

    (c) shall be accompanied by the [charge prescribed for the purpose by a charging scheme under section 41 of the Environment Act 1995].

(2) Where an application is made to the [appropriate Agency] for registration under this section in respect of any apparatus, the [appropriate Agency] may register the applicant in respect of that apparatus, either unconditionally or subject to such limitations or conditions as the [appropriate Agency] thinks fit, or may refuse the application.

(3) On any application being made the [appropriate Agency] shall, subject to directions under section 25, send a copy of the application to each local authority in whose area it appears to the [appropriate Agency] the apparatus will be kept or will be used for releasing radioactive material into the environment.

(4) An application for registration under this section which is duly made to the [appropriate Agency] may be treated by the applicant as having been refused if it is not determined within the prescribed period for determinations or within such longer period as may be agreed with the applicant.

(5) On registering a person under this section in respect of any mobile radioactive apparatus, the [appropriate Agency]—

    (a) shall furnish him with a certificate containing all material particulars of the registration, and

    (b) shall, subject to directions under section 25, send a copy of the certificate to each local authority in whose area it appears to the [appropriate Agency] the apparatus will be kept or will be used for releasing radioactive material into the environment.

Definitions

"appropriate Agency": section 47(1).   B–65
"mobile radioactive apparatus": section 3.
"prescribed": section 47(1).
"prescribed period for determinations": section 47(1) and (2).

General Note

This section sets out the provisions relating to applications for regis- B–66
tration of mobile radioactive apparatus.

Amendment

Section 10(1)(c) was amended by the Environment Act 1995, Sched. 22, B–67
para. 204.

Subsection (1)

B–68 To date no statutory instrument has been made prescribing further information to be furnished to the appropriate Agency on an application for a registration.

As to the provision relating to fees payable, see the 1995 Act, s.41.

Subsection (2)

B–69 The appropriate Agency has the power to impose conditions on a registration. Additionally, under section 44(2) the Secretary of State is empowered to make regulations setting out general limitations or conditions applicable to registrations. Breaches of limitations or conditions is a criminal offence pursuant to section 32.

The provisions on appeals against a refusal to grant a registration, a deemed refusal to grant a registration or the imposition of limitations or conditions to a registration are set out in sections 26 and 27.

Subsections (3) and (5)

B–70 Under section 25, the Secretary of State may, on the grounds of national security, direct that the relevant parts of an application for registration or a certificate of registration should not be copied to the local authorities.

Subsection (4)

B–71 The prescribed period is four months beginning with the day on which the application was received.

Exemptions from registration under section 10

B–72   11.—(1) The Secretary of State may by order grant exemptions from registration under section 10, by reference to such classes of persons, and such descriptions of mobile radioactive apparatus, as may be specified in the order.

(2) Any exemption granted by an order under subsection (1) may be granted subject to such limitations or conditions as may be specified in the order.

(3) In the application of this section to Northern Ireland, the reference to the Secretary of State shall have effect as a reference to the Department of the Environment for Northern Ireland.

Definition

B–73 "mobile radioactive apparatus": section 3.

General Note

Subsection (1)

This provision sets out the exemption from the necessity to register **B–74** mobile radioactive apparatus as required by section 9.

Person

See the General Note to section 6.   **B–75**

Subsection (2)

Non-compliance with a limitation or condition is a criminal offence **B–76** under section 32.

The orders currently in force under this section are:

- The Radioactive Substances (Electronic Valves) Exemption Order 1967 (S.I. 1967 No. 1797).

- The Radioactive Substances (Testing Instruments) Exemption Order 1985 (S.I. 1985 No. 1049).

Similar orders have been issued in relation to exemptions for Scotland and Northern Ireland.

Cancellation and variation of registration

12.—(1) Where any person is for the time being registered under **B–77** section 7 or 10, the [appropriate Agency] may at any time cancel the registration, or may vary it—
  (a) where the registration has effect without limitations or con-
      ditions, by attaching limitations or conditions to it, or
  (b) where the registration has effect subject to limitations or
      conditions, by revoking or varying any of those limitations or
      conditions or by attaching further limitations or conditions to the
      registration.
  (2) On cancelling or varying a registration by virtue of this section, the
[appropriate Agency] shall—
  (a) give notice of the cancellation or variation to the person to
      whom the registration relates, and
  (b) if a copy of the certificate was sent to a local authority in
      accordance with section 7(8) or 10(5), send a copy of the notice
      to that local authority.

Definitions

B–78 "appropriate Agency": section 47(1).
"local authority": section 47(1).

General Note

B–79 This provision empowers the appropriate Agency to cancel or vary any registration of premises or mobile apparatus made under section 7 or section 10 respectively. Where a registration has been varied or cancelled the registered person may appeal to the Secretary of State pursuant to the provisions of sections 26 and 27.

Limitation or Condition

B–80 To breach a limitation or condition attached to a registration is a criminal offence pursuant to section 32.

Authorisation of disposal and accumulation of radioactive waste

Disposal of radioactive waste

B–81   13.—(1) Subject to section 15, no person shall, except in accordance with an authorisation granted in that behalf under this subsection, dispose of any radioactive waste on or from any premises which are used for the purposes of any undertaking carried on by him, or cause or permit any radioactive waste to be so disposed of, if (in any such case) he knows or has reasonable grounds for believing it to be radioactive waste.

(2) Where any person keeps any mobile radioactive apparatus for the purpose of its being used in activities to which section 9 applies, he shall not dispose of any radioactive waste arising from any such apparatus so kept by him, or cause or permit any such radioactive waste to be disposed of, except in accordance with an authorisation granted in that behalf under this subsection.

(3) Subject to subsection (4) and to section 15, where any person, in the course of the carrying on by him of an undertaking, receives any radioactive waste for the purpose of its being disposed of by him, he shall not, except in accordance with an authorisation granted in that behalf under this subsection, dispose of that waste, or cause or permit it to be disposed of, knowing or having reasonable grounds for believing it to be radioactive waste.

(4) The disposal of any radioactive waste does not require an authorisation under subsection (3) if it is waste which falls within the

provisions of an authorisation granted under subsection (1) or (2), and it is disposed of in accordance with the authorisation so granted.

(5) In relation to any premises which—

(a) are situated on a nuclear site, but

(b) have ceased to be used for the purposes of an undertaking carried on by the licensee,

subsection (1) shall apply (subject to section 15) as if the premises were used for the purposes of an undertaking carried on by the licensee.

Definitions

"disposal": section 47(1).   B–82

"dispose of": section 47(1).

"mobile radioactive apparatus": section 3.

"nuclear site": section 47(1).

"period of responsibility": section 47(1).

"premises": section 47(1).

"radioactive waste": section 2.

"undertaking": section 47(1).

General Note

The disposal of radioactive waste is prohibited unless it is in accordance B–83
with an authorisation issued under this section, or exempted by section 15. The disposal of waste includes the removal of the waste from the premises as well as its deposit or destruction (s.47(1)). Pursuant to subsection (5) this section also applies to redundant nuclear sites.

Subsection (2)

The use of mobile radioactive apparatus to dispose of any radioactive B–84
waste arising from the use of such apparatus is prohibited unless authorised under this subsection or exempted pursuant to section 11.

Subsection (3)

The receipt of waste for disposal is prohibited unless authorised under B–85
this subsection or exempted under section 15. However, there is no need for an authorisation under this subsection where the disposal is already the subject of authorisation under subsections (1) or (2).

The provisions in relation to the application for an authorisation are set out at section 16.

Subsection (5)

B–86   The power of the Health and Safety Executive to attach conditions to nuclear site licences with regard to the discharge of substances on or from such a site is without prejudice to this subsection (Nuclear Installations Act 1965, s.4).

For the provisions on appeals in relation to applications for authorisations see sections 26 and 27.

For the right to a hearing before an application is refused or limitation or condition attached to it see section 28.

Contravention of this section is a criminal offence under section 32.

For discussion of the elements of the criminal offences, see the General Note to section 6.

Person

B–87   For the meaning of "person" see the General Note to section 6.

Accumulation of radioactive waste

B–88   14.—(1) Subject to the provisions of this section and section 15, no person shall, except in accordance with an authorisation granted in that behalf under this section, accumulate any radioactive waste (with a view to its subsequent disposal) on any premises which are used for the purposes of an undertaking carried on by him, or cause or permit any radioactive waste to be so accumulated, if (in any such case) he knows or has reasonable grounds for believing it to be radioactive waste.

(2) Where the disposal of any radioactive waste has been authorised under section 13, and in accordance with that authorisation the waste is required or permitted to be accumulated with a view to its subsequent disposal, no further authorisation under this section shall be required to enable the waste to be accumulated in accordance with the authorisation granted under section 13.

(3) Subsection (1) shall not apply to the accumulation of radioactive waste on any premises situated on a nuclear site.

(4) For the purposes of this section, where radioactive material is produced, kept or used on any premises, and any substance arising from the production, keeping or use of that material is accumulated in a part of the premises appropriated for the purpose, and is retained there for a period of not less than three months, that substance shall, unless the contrary is proved, be presumed—

(a) to be radioactive waste, and
(b) to be accumulated on the premises with a view to the subsequent disposal of the substance.

Definitions

"disposal": section 47(1).   **B–89**
"nuclear site": section 47(1).
"premises": section 47(1).
"radioactive material": section 1.
"radioactive waste": section 2.
"substance": section 47(1).
"undertaking": section 47(1).
"waste": section 47(1) and (4).

General Note

Subsection (1)

The accumulation of radioactive waste with a view to its subsequent **B–90** disposal is prohibited unless it is in accordance with an authorisation granted under this subsection.

Subsection (2)

Where the accumulation of radioactive waste is in accordance with an **B–91** authorisation granted under section 13, then no additional authorisation is required.

Subsection (3)

The accumulation of radioactive waste on a nuclear site does not require **B–92** authorisation under this subsection.

Subsection (4)

For the purposes of this section the accumulation of radioactive material **B–93** for at least three months is presumed to be both radioactive waste and accumulated with a view to its disposal and so must be authorised under subsection (1). For these purposes "months" means "calendar months", see section 5 of the Interpretation Act 1978. The provisions for appeals in relation to applications for authorisations are contained in sections 26

381

and 27. It is a criminal offence pursuant to section 32 to fail to comply with the requirements of this section. For a discussion on the elements and the criminal offences see the General Note to section 6.

Person

B–94   For the meaning of the word "person", see the General Note to section 6.

Further exemptions from sections 13 and 14

B–95   15.—(1) Sections 13(1) and (3) and 14(1) shall not apply to the disposal or accumulation of any radioactive waste arising from clocks or watches, but this subsection does not affect the operation of section 13(1) or section 14(1) in relation to the disposal or accumulation of radioactive waste arising from clocks or watches on or from premises which, by virtue of subsection (5) of section 8, are excluded from the operation of subsection (4) of that section.

(2) Without prejudice to subsection (1), the Secretary of State may by order exclude particular descriptions of radioactive waste from any of the provisions of section 13 or 14, either absolutely or subject to limitations or conditions; and accordingly such of those provisions as may be specified in an order under this subsection shall not apply to a disposal or accumulation of radioactive waste if it is radioactive waste of a description so specified, and (where the exclusion is subject to limitations or conditions) the limitations or conditions specified in the order are complied with.

(3) In the application of this section to Northern Ireland, the reference to the Secretary of State shall have effect as a reference to the Department of the Environment for Northern Ireland.

Definitions

B–96   "disposal": section 47(1).
"radioactive waste": section 2.

General Note

B–97   This section contains the exemptions from the necessity to hold an authorisation for disposing of or accumulating radioactive waste as required by sections 13 and 14.

Subsection (1)

The disposal or accumulation of radioactive waste from watches or clocks   **B–98**
is exempted. However, there is no exemption for waste arising from
clocks or watches on the premises at which they are manufactured or
repaired by a process involving luminous material.

Subsection (2)

The Secretary of State may grant exemptions by way of statutory   **B–99**
instrument. Those statutory instruments currently in force for England
and Wales are:

- The Radioactive Substances (Storage in Transit) Exemption Order
  1962 (S.I. 1962 No. 2646).

- The Radioactive Substances (Phosphatic Substances, Rare Earths,
  etc.) Exemption Order 1962 (S.I. 1962 No. 2648).

- The Radioactive Substances (Lead) Exemption Order 1962 (S.I. 1962
  No. 2649).

- The Radioactive Substances (Uranium and Thorium) Exemption
  Order 1962 (S.I. 1962 No. 2710).

- The Radioactive Substances (Prepared Uranium and Thorium Compounds) Exemption Order 1962 (S.I. 1962 No. 2711).

- The Radioactive Substances (Geological Specimens) Exemption
  Order 1962 (S.I. 1962 No. 2712).

- The Radioactive Substances (Waste Closed Sources) Exemption
  Order 1963 (S.I. 1963 No. 1831).

- The Radioactive Substances (Schools, etc.) Exemption Order 1963
  (S.I. 1963 No. 1832).

- The Radioactive Substances (Electronic Valves) Exemption Order
  1967 (S.I. 1967 No. 1797).

- The Radioactive Substances (Smoke Detectors) Exemption Order
  1980 (S.I. 1980 No. 953), as amended by (S.I. 1991 No. 477).

- The Radioactive Substances (Gaseous Tritium Light Devices) Exemption Order 1985 (S.I. 1985 No. 1047).

- The Radioactive Substances (Luminous Articles) Exemption Order
  1985 (S.I. 1985 No. 1048).

- The Radioactive Substances (Testing Instruments) Exemption Order
  1985 (S.I. 1985 No. 1049).

- The Radioactive Substances (Substances of Low Activity) Exemption
  Order 1986 (S.I. 1986 No. 1002), as amended by (S.I. 1992 No. 647).

383

- The Radioactive Substances (Hospitals) Exemption Order 1990 (S.I. 1990 No. 2512), as amended by (S.I. 1995 No. 2395).

Similar orders have been issued in relation to exemptions for Scotland and Northern Ireland.

Grant of authorisations

B–100   16.—(1) In this section, unless a contrary intention appears, "authorisation" means an authorisation granted under section 13 or 14.

(2) ... the power to grant authorisations shall be exercisable by the [appropriate Agency].

(3) [...]

(4) Any application for an authorisation shall be accompanied by the [charge prescribed for the purpose by a charging scheme under section 41 of the Environment Act 1995].

[(4A) Without prejudice to subsection (5), on any application for an authorisation under section 13(1) in respect of the disposal of radioactive waste on or from any premises situated on a nuclear site in any part of Great Britain, the appropriate Agency—

(a) shall consult the relevant Minister and the Health and Safety Executive before deciding whether to grant an authorisation on that application and, if so, subject to what limitations or conditions, and

(b) shall consult the relevant Minister concerning the terms of the authorisation, for which purpose that Agency shall, before granting any authorisation on that application, send that Minister a copy of any authorisation which it proposes so to grant.]

(5) Before granting an authorisation under section 13(1) in respect of the disposal of radioactive waste on or from premises situated on a nuclear site, the [appropriate Agency] [shall] consult with such local authorities, relevant water bodies or other public or local authorities as appear to [that Agency] to be proper to be consulted by [that Agency].

(6) On any application being made, the [appropriate Agency] shall, subject to directions under section 25, send a copy of the application to each local authority in whose area, in accordance with the authorisation applied for, radioactive waste is to be disposed of or accumulated.

(7) An application for an authorisation [(other than an application for an authorisation under section 13(1) in respect of the disposal of radioactive waste on or from any premises situated on a nuclear site in any part of Great Britain)] which is duly made to the [appropriate Agency] may be treated by the applicant as having been refused if it is not determined within the prescribed period for determinations or such longer period as may be agreed with the applicant.

(8) An authorisation may be granted—

(a) either in respect of radioactive waste generally or in respect of such one or more descriptions of radioactive waste as may be specified in the authorisation, and

(b) subject to such limitations or conditions as the [appropriate Agency] thinks fit.

(9) Where any authorisation is granted, the [appropriate Agency]—

(a) shall furnish the person to whom the authorisation is granted with a certificate containing all material particulars of the authorisation, and

(b) shall, subject to directions under section 25, send a copy of the certificate—

(i) to each local authority in whose area, in accordance with the authorisation, radioactive waste is to be disposed of or accumulated, and

(ii) in the case of an authorisation to which subsection (5) applies, to any other public or local authority consulted in relation to the authorisation in accordance with that subsection.

(10) An authorisation shall have effect as from such date as may be specified in it; and in fixing that date, in the case of an authorisation where copies of the certificate are required to be sent as mentioned in subsection (9)(b), the [appropriate Agency]—

(a) shall have regard to the time at which those copies may be expected to be sent, and

(b) shall fix a date appearing to [it] to be such as will allow an interval of not less than twenty-eight days after that time before the authorisation has effect,

unless in [its] opinion it is necessary that the coming into operation of the authorisation should be immediate or should otherwise be expedited.

[(11) In this section, "the relevant Minister" means—

(a) in relation to premises in England, the Minister of Agriculture, Fisheries and Food, and

(b) in relation to premises in Wales or Scotland, the Secretary of State.]

Definitions

"appropriate Agency": section 47(1).

B–101

"chief inspector": section 47(1).

"disposal": section 47(1).

"local authority": section 47(1).

"nuclear site": section 47(1).

"premises": section 47(1).

"prescribed period for determinations": section 47(1) and (2).

"public or local authorities": section 47(1).

"radioactive waste": section 2.
"relevant water body": section 47(1).

Amendments

B–102 The Environment Act 1995, Sched. 22, para. 205, revoked subsection (3), inserted subsections (4A) and (11), and amended subsections (2), (5), (7), (8) and (10).

General Note

Subsection (2)

B–103 The appropriate Agency has the power to grant authorisations for the disposal and accumulation of radioactive waste. The requirement, formerly in subsection (3), that an authorisation for the disposal of radioactive waste on or from a nuclear site situated in England and Wales (as well as those situated in Northern Ireland, where of course subsection (3) remains in force) must be granted jointly by both the chief inspector and the appropriate Minister, has been revoked.

For an authorisation to be granted the appropriate Agency must consult the statutory consultees which include local authorities, water undertakers, sewerage undertakers or local fisheries committees; in Scotland, a District Salmon Fishery Board and a Water and Sewerage Authority; and in Northern Ireland the Fisheries Conservation Board for Northern Ireland.

Subsection (4)

B–104 See the Environment Act 1995, s.41 for provisions on fees payable.

Subsections (4A) and (11)

B–105 With the revocation in Great Britain of subsection (3) (see note above), the Ministerial role has been expanded to provide a measure of the control in relation to nuclear site radioactive waste disposal which being joint regulatory authority would previously have afforded. The consultation requirement afforded by subsection (4A) is one example (for further discussion of this issue see *(3) Role of Ministers* at B–04 above). It will be noted that the requirement to consult the HSE is confined to the grant of, and limitations and conditions in, authorisations (para. (a)) and is thus more limited in scope than the consultation with the relevant Minister (paras. (a) and (b)).

Subsection (6)

Except to the extent that the Secretary of State has directed that knowledge of the application should be restricted on the grounds of national security pursuant to section 25, the appropriate Agency must furnish relevant local authorities with copies of applications. **B—106**

Subsection (7)

The prescribed period is four months beginning with the day on which the application was received. **B—107**

Subsection (8)

Limitations and conditions may be attached to an authorisation. Additionally, it is open to the Secretary of State to set out general limitations or conditions in regulations made pursuant to section 44 or section 45. Breach of limitations and conditions is a criminal offence pursuant to section 32. **B—108**

Subsection (9)

Except to the extent that the Secretary of State has restricted knowledge of an authorisation on the grounds of national security pursuant to section 25, a copy of the certificate must be sent to the relevant local authorities and where an authorisation is for the disposal of radioactive waste on or from nuclear site premises, to the public and local authorities originally consulted. **B—109**

Person

For a discussion of the meaning of the word "person" see the General Note to section 6. **B—110**

For provisions on the right to appeal where an application is refused, deemed to be refused or has limitations or conditions attached to it see sections 26 and 27.

For the right to a hearing before an authorisation for the disposal of radioactive waste under section 13 is refused or condition or limitation imposed on such an authorisation see section 28.

Revocation and variation of authorisations

B–111   17.—(1) The [appropriate Agency] may at any time revoke an authorisation granted under section 13 or 14.

(2) The [appropriate Agency] may at any time vary an authorisation granted under section 13 or 14—

(a) where the authorisation has effect without limitations or conditions, by attaching limitations or conditions to it, or

(b) where the authorisation has effect subject to limitations or conditions, by revoking or varying any of those limitations or conditions or by attaching further limitations or conditions to the authorisation.

[(2A) On any proposal to vary an authorisation granted under section 13(1) in respect of the disposal of radioactive waste on or from any premises situated on a nuclear site in any part of Great Britain, the appropriate Agency—

(a) shall consult the relevant Minister and the Health and Safety Executive before deciding whether to vary the authorisation and, if so, whether by attaching, revoking or varying any limitations or conditions or by attaching further limitations or conditions, and

(b) shall consult the relevant Minister concerning the terms of any variation, for which purpose that Agency shall, before varying the authorisation, send that Minister a copy of any variations which it proposes to make.]

(3) Where any authorisation granted under section 13 or 14 is revoked or varied, the [appropriate Agency]—

(a) shall give notice of the revocation or variation to the person to whom the authorisation was granted, and

(b) if a copy of the certificate of authorisation was sent to a public or local authority in accordance with section 16(9)(b), shall send a copy of the notice to that authority.

(4) . . .

[(5) In this section, "the relevant Minister" has the same meaning as in section 16 above.]

Definitions

B–112   "appropriate Agency": section 47(1).
"public or local authority": section 47(1).
"relevant Minister": section 16(11).

Amendments

B–113   The Environment Act 1995, Sched. 24, revoked subsection (4); and Sched. 22, para. 206, inserted subsections (2A) and (5).

General Note

Subsection (1)

This provision empowers the appropriate Agency to revoke or vary an authorisation for either the disposal or accumulation of radioactive waste. Where either a revocation or variation has been made, the authorised person may appeal to the Secretary of State pursuant to the provisions of section 26. **B–114**

Subsection (2A)

The requirement for consultation is new. As regards the "relevant Minister" (subs. (5)), it should be seen in the context of amendments to section 16 (see the general introductory note to the Act at B–04 (*Changes effected by the 1995 Act*)). **B–115**

The breach of limitations or conditions attached to an authorisation is a criminal offence pursuant to section 32.

Functions of public and local authorities in relation to authorisations under section 13

18.—(1) If, in considering an application for an authorisation under section 13, it appears to the [appropriate Agency] that the disposal of radioactive waste to which the application relates is likely to involve the need for special precautions to be taken by a local authority, relevant water body or other public or local authority, the [appropriate Agency] shall consult with that public or local authority before granting the authorisation. **B–116**

(2) Where a public or local authority take any special precautions in respect of radioactive waste disposed of in accordance with an authorisation granted under section 13, and those precautions are taken—

    (a) in compliance with the conditions subject to which the authorisation was granted, or

    (b) with the prior approval of the [appropriate Agency] as being precautions which in the circumstances ought to be taken by that public or local authority,

the public or local authority shall have power to make such charges, in respect of the taking of those precautions, as may be agreed between that authority and the person to whom the authorisation was granted, or as, in default of such agreement, may be determined by the [appropriate Agency], and to recover the charges so agreed or determined from that person.

(3) Where an authorisation granted under section 13 requires or

permits radioactive waste to be removed to a place provided by a local authority as a place for the deposit of refuse, it shall be the duty of that local authority to accept any radioactive waste removed to that place in accordance with the authorisation, and, if the authorisation contains any provision as to the manner in which the radioactive waste is to be dealt with after its removal to that place, to deal with it in the manner indicated in the authorisation.

Definitions

B–117 "appropriate Agency": section 47(1).
"disposal": section 47(1).
"disposed of": section 47(1).
"local authority": section 47(1).
"public or local authority": section 47(1).
"radioactive waste": section 2.
"relevant water body": section 47(1).

Amendments

B–118 The Environment Act 1995, Sched. 22, para. 207, amended subsections (1) and (2)(b).

General Note

Subsection (1)

B–119 Under this provision where the disposal of radioactive waste is likely to need "special precautions to be taken" then the appropriate Agency must consult the relevant public or local authorities before granting an authorisation.

Subsection (2)

B–120 Where the public or local authority must take special precautions in disposing of radioactive waste, it has the power to make a charge for taking those precautions.

Subsection (3)

B–121 Where an authorisation to deal with radioactive waste either requires or permits that waste to be removed or disposed of at a local authority refuse site, the local authority is under a statutory duty to accept that waste and to deal with it in a manner required by the authorisation.

Further obligations relating to registration or authorisation

Duty to display documents

19. At all times while—   B–122
    (a) a person is registered in respect of any premises under section 7, or
    (b) an authorisation granted in respect of any premises under section 13(1) or 14 is for the time being in force,
the person to whom the registration relates, or to whom the authorisation was granted, as the case may be, shall cause copies of the certificate of registration or authorisation issued to him under this Act to be kept posted on the premises, in such characters and in such positions as to be conveniently read by persons having duties on those premises which are or may be affected by the matters set out in the certificate.

Definition

"premises": section 47(1).   B–123

General Note

Certificates of registration of premises or authorisation for the disposal or B–124
accumulation of waste must be posted on the premises concerned.

Person

For the meaning of "person," see the General Note to section 6.   B–125

Retention and production of site or disposal records

20.—(1) The [appropriate Agency] may, by notice served on any B–126
person to whom a registration under section 7 or 10 relates or an authorisation under section 13 or 14 has been granted, impose on him such requirements authorised by this section in relation to site or disposal records kept by that person as the [appropriate Agency] may specify in the notice.

(2) The requirements that may be imposed on a person under this section in relation to site or disposal records are—

(a) to retain copies of the records for a specified period after he ceases to carry on the activities regulated by his registration or authorisation, or

(b) to furnish the [appropriate Agency] with copies of the records in the event of his registration being cancelled or his authorisation being revoked or in the event of his ceasing to carry on the activities regulated by his registration or authorisation.

(3) [...]

(4) In this section, in relation to a registration and the person registered or an authorisation and the person authorised—

"the activities regulated" by his registration or authorisation means—

(a) in the case of registration under section 7, the keeping or use of radioactive material,

(b) in the case of registration under section 10, the keeping, using, lending or hiring of the mobile radioactive apparatus,

(c) in the case of an authorisation under section 13, the disposal of radioactive waste, and

(d) in the case of an authorisation under section 14, the accumulation of radioactive waste,

"records" means records required to be kept by virtue of the conditions attached to the registration or authorisation relating to the activities regulated by the registration or authorisation, and "site records" means records relating to the condition of the premises on which those activities are carried on or, in the case of registration in respect of mobile radioactive apparatus, of any place where the apparatus is kept and "disposal records" means records relating to the disposal of radioactive waste on or from the premises on which the activities are carried on, and

"specified" means specified in a notice under this section.

Definition

B–127 "appropriate Agency": section 47(1).

Amendment

B–128 The Environment Act 1995, Sched. 24, revoked subsection (3).

General Note

B–129 This provision sets out the obligation in relation to the retention and production of documentation.

Person

For the meaning of "person", see the General Note to section 6.   B–130

Enforcement notices and prohibition notices

Enforcement notices

21.—(1) Subject to the provisions of this section, if the [appropriate B–131
Agency] is of the opinion that a person to whom a registration under
section 7 or 10 relates or to whom an authorisation was granted under
section 13 or 14—
> (a) is failing to comply with any limitation or condition subject to
>   which the registration or authorisation has effect, or
> (b) is likely to fail to comply with any such limitation or condition,
[it] may serve a notice under this section on that person.
  (2) A notice under this section shall—
> (a) state that the [appropriate Agency] is of that opinion,
> (b) specify the matters constituting the failure to comply with the
>   limitations or conditions in question or the matters making it
>   likely that such a failure will occur, as the case may be, and
> (c) specify the steps that must be taken to remedy those matters
>   and the period within which those steps must be taken.
  (3) [. . .]
  (4) Where a notice is served under this section the [appropriate
Agency] . . . shall—
> (a) in the case of a registration, if a certificate relating to the
>   registration was sent to a local authority under section 7(8) or
>   10(5), or
> (b) in the case of an authorisation, if a copy of the authorisation was
>   sent to a public or local authority under section 16(9)(b),
send a copy of the notice to that authority.

Definition

"appropriate Agency": section 47(1).   B–132

Amendment

The Environment Act 1995, Sched. 24, revoked subsection (3), and part B–133
of subsection (4).

General Note

Subsection (1)

B–134 An enforcement notice may be served in relation to either a registration or an authorisation not only where a condition or limitation is being contravened but also where such contravention appears likely to take place.

Subsection (2)

B–135 The requirement as to the matters to be contained in an enforcement notice are mandatory and it appears that any notice which on its face does not comply with these requirements will be a nullity and could then be ignored or challenged in proceedings.[19] In practice, it is unlikely that many notices will fail to comply with these requirements on their face; rather the question is likely to be whether a notice is bad for failure to specify the required matters with sufficient accuracy and precision. The most widely accepted test as to these matters in the case of planning enforcement notices is whether the notice tells the recipient "fairly what he has done wrong and what he must do to remedy".[20] There seems to be no reason why the same test should not be applied to notices served under section 21.

Provision as to appeals against enforcement notices are contained at sections 26 and 27. Under section 27, the Secretary of State may either quash or affirm the notice and in affirming may do so with modifications. Although it is not explicitly stated, this power may well allow the Secretary of State to correct defects in a notice, but it is questionable whether it would be possible to cure fundamental defects that make the notice a nullity. Reference to the considerable body of case law on defective enforcement notices under the Town and Country Planning Acts may provide some degree of assistance as to the likely attitude of the Courts on this question. In general the Courts have become increasingly reluctant to hold that notices cannot be served.

For provisions on the opportunity to have a hearing, see section 28.

Breach of an enforcement notice is a criminal offence under section 32.

The power to issue enforcement notices was introduced by the EPA 1990 into the regulation of radioactive substances. The HMIP Fifth Annual Report for 1991–1992 (HMSO) reveals that the first of such enforcement notices was issued during that year to Lucas Aerospace for failing to notify the loss of radioactive sources, failure to implement

---

[19] see: *Miller-Mead v. Minister of Housing and Local Government* [1963] Q.B. 196, 226 *per* Upjohn L.J.

[20] see note **19** *ibid.*

security measures that had been previously requested by an inspector, and for having more radioactive sources than was permitted by the company's registration.

Another early enforcement notice was issued to a hospital in Chelmsford which was no longer incinerating its radioactive clinical waste but accumulating it instead. The notice obliged the hospital to recommence incineration.

Prohibition notices

22.—(1) Subject to the provisions of this section, if the [appropriate B–136 Agency] is of the opinion, as respects the keeping or use of radioactive material or of mobile radioactive apparatus, or the disposal or accumulation of radioactive waste, by a person in pursuance of a registration or authorisation under this Act, that the continuing to carry on that activity (or the continuing to do so in a particular manner) involves an imminent risk of pollution of the environment or of harm to human health, [it] may serve a notice under this section on that person.

(2) A notice under this section may be served whether or not the manner of carrying on the activity in question complies with any limitations or conditions to which the registration or authorisation in question is subject.

(3) A notice under this section shall—

    (a) state the [appropriate Agency's] opinion,

    (b) specify the matters giving rise to the risk involved in the activity, the steps that must be taken to remove the risk and the period within which those steps must be taken, and

    (c) direct that the registration or authorisation shall, until the notice is withdrawn, wholly or to the extent specified in the notice cease to have effect.

(4) Where the registration or authorisation is not wholly suspended by the direction given under subsection (3), the direction may specify limitations or conditions to which the registration or authorisation is to be subject until the notice is withdrawn.

(5) [. . .]

(6) Where a notice is served under this section the [appropriate Agency] . . . shall—

    (a) in the case of a registration, if a certificate relating to the registration was sent to a local authority under section 7(8) or 10(5), or

    (b) in the case of an authorisation, if a copy of the authorisation was sent to a public or local authority under section 16(9)(b),

send a copy of the notice to that authority.

(7) The [appropriate Agency] . . . shall, by notice to the recipient, withdraw a notice under this section when [that Agency] is satisfied that the risk specified in it has been removed; and on so doing [that Agency]

shall send a copy of the withdrawal notice to any public or local authority to whom a copy of the notice under this section was sent.

Definitions

B–137 "appropriate Agency": section 47(1).
"chief inspector": section 47(1).
"disposal": section 47(1).
"mobile radioactive apparatus": section 3.
"radioactive material": section 1.
"radioactive waste": section 2.

Amendments

B–138 Subsection (5) and parts of subsections (6) and (7) were revoked by the Environment Act 1995, Sched. 24.

General Note

B–139 The appropriate Agency may serve a prohibition notice even where the activity in question is in full compliance with an authorisation or registration; the relevant issue is whether the activity involves imminent risk of pollution to the environment or harm to human health. The Act does not provide any definitions of pollution of the environment or harm to human health.

Subsection (2)

B–140 There are mandatory requirements as to what matters should be included within the notice and the comments made in relation to enforcement notices in the General Note to section 21 above should apply equally.

The provisions on the rights of appeal against a prohibition notice are set out at sections 26 and 27.

The provisions on the ability to seek a hearing in relation to a prohibition notice are set out at section 28.

Breach of a prohibition notice is a criminal offence under section 32.

The HMIP Annual Report 1991–1992 indicates that no prohibition notices were served in that period. Such notices are considered necessary for use in an emergency situation where there is a risk of serious pollution.

Powers of Secretary of State in relation to applications etc.

Powers of Secretary of State to give directions to [the appropriate Agency]

23.—(1) The Secretary of State may, if he thinks fit in relation to—　　B–141
  (a) an application for registration under section 7 or 10,
  (b) an application for an authorisation under section 13 or 14, or
  (c) any such registration or authorisation,
give directions to the [appropriate Agency] requiring [it] to take any of the steps mentioned in the following subsections in accordance with the directions.

(2) A direction under subsection (1) may require the [appropriate Agency] so to exercise [its] powers under this Act as—
  (a) to refuse an application for registration or authorisation,
  (b) to effect or grant a registration or authorisation, attaching such limitations or conditions (if any) as may be specified in the direction, or
  (c) to vary a registration or authorisation, as may be so specified, or
  (d) to cancel or revoke (or not to cancel or revoke) a registration or authorisation.

(3) The Secretary of State may give directions to the [appropriate Agency], as respects any registration or authorisation, requiring [it] to serve a notice under section 21 or 22 in such terms as may be specified in the directions.

(4) The Secretary of State may give directions requiring the [appropriate Agency] to send such written particulars relating to, or to activities carried on in pursuance of, registrations effected or authorisations granted under any provision of this Act as may be specified in the directions to such local authorities as may be so specified.

[(4A) In the application of this section in relation to authorisations, and applications for authorisations, under section 13 in respect of premises situated on a nuclear site in England, references to the Secretary of State shall have effect as references to the Secretary of State and the Minister of Agriculture, Fisheries and Food.]

(5) In the application of this section to Northern Ireland, references to the Secretary of State shall have effect as references to the Department of the Environment for Northern Ireland.

Definitions

"appropriate Agency": section 47(1).　　B–142
"nuclear site": section 47(1).

Amendment

B–143 The Environment Act 1995, Sched. 22, para. 211, inserted subsection (4A).

General Note

B–144 This section sets out the powers of the Secretary of State to give directions on the conduct of all registrations and authorisations.

Subsection (4A)

B–145 Unlike elsewhere in the United Kingdom, in England directions by the Secretary of State must be given jointly with the Minister of Agriculture, Fisheries and Food if those directions relate to authorisations for the disposal of radioactive waste on or from a nuclear site. This provision was inserted by the 1995 Act to reflect the fact that the Minister is no longer joint authorising authority for the matters in questions (revoked section 16(3)). See further at B–04 (*(3) Role of Ministers*).

Power of Secretary of State to require certain applications to be determined by him

B–146   24.—(1) The Secretary of State may—
   (a) give general directions to the [appropriate Agency] requiring [it] to refer applications under this Act for registrations or authoris-ations of any description specified in the directions to the Secretary of State for his determination, and
   (b) give directions to the [appropriate Agency] in respect of any particular application requiring [it] to refer the application to the Secretary of State for his determination.
   (2) Where an application is referred to the Secretary of State in pursuance of directions given under this section, the Secretary of State may cause a local inquiry to be held in relation to the application.
   (3) The following provisions shall apply to inquiries in pursuance of subsection (2)—
   (a) in England and Wales, subsections (2) to (5) of section 250 of the Local Government Act 1972 (supplementary provisions about local inquiries under that section) but with the omission, in subsection (4) of that section, of the words "such local authority or",
   (b) in Scotland, subsections (2) to (8) of section 210 of the Local Government (Scotland) Act 1973 (power to direct inquiries), and

(c) in Northern Ireland, Schedule 8 to the Health and Personal Services (Northern Ireland) Order 1972 (provisions as to inquiries).

(4) After determining any application so referred, the Secretary of State may give the [appropriate Agency] directions under section 23 as to the steps to be taken by [it] in respect of the application.

[(4A) In the application of this section in relation to authorisations, and applications for authorisations, under section 13 in respect of premises situated on a nuclear site in England, references to the Secretary of State shall have effect as references to the Secretary of State and the Minister of Agriculture, Fisheries and Food.]

(5) In the application of this section to Northern Ireland, references to the Secretary of State shall have effect as references to the Department of the Environment for Northern Ireland.

Definitions

"appropriate Agency": section 47(1).   **B–147**
"nuclear site": section 47(1).

General Note

This section enables the Secretary of State to give either a general **B–148** direction to the chief inspector that all applications of a particular type are referred to him or that an individual application be referred to him.

Subsection (4A)

Unlike elsewhere in the United Kingdom, in England the Secretary of **B–149** State's power to require either particular applications (subs. (1)(b) or kinds of applications (subs. (1)(a)) to be referred to him for his determination is only exercisable jointly with the Minister of Agriculture, Fisheries and Food in relation to the disposal of radioactive waste on or from a nuclear site. This provision was inserted by the 1995 Act to reflect the fact that the Minister is no longer joint authorising authority for the matters in question (revoked s.16(3)). See further at B–04 ((3) Role of Ministers).

Power of Secretary of State to restrict knowledge of applications, etc.

B–150   25.—(1) The Secretary of State may direct the [appropriate Agency] that in his opinion, on grounds of national security, it is necessary that knowledge of [such information as may be specified or described in the directions, being information contained in or relating to—]

    (a) any particular application for registration under section 7 or 10 or applications of any description specified in the directions, or

    (b) any particular registration or registrations of any description so specified,

should be restricted.

(2) The Secretary of State . . . may direct the [appropriate Agency] that in his . . . opinion, on grounds of national security, it is necessary that knowledge of [such information as may be specified or described in the directions, being information contained in or relating to—]

    (a) any particular application for authorisation under section 13 or 14 or applications of any description specified in the directions, or

    (b) any particular authorisation under either of those sections or authorisations of any description so specified,

should be restricted.

(3) Where it appears to the [appropriate Agency] that an application, registration or authorisation is the subject of any directions under this section, the [appropriate Agency] shall not send a copy of [so much of] the application or the certificate of registration or authorisation, as the case may be [as contains the information specified or described in the directions—]

    (a) to any local authority under any provision of section 7 or 10, or

    (b) to any public or local authority under any provision of section 16.

[(3A) No direction under this section shall affect—

    (a) any power or duty of the Agency to which it is given to consult the relevant Minister; or

    (b) the information which is to be sent by that Agency to that Minister.]

(4) In the application of this section to Northern Ireland—

    (a) references to the Secretary of State shall have effect as references to the Department of the Environment for Northern Ireland, and

    (b) in subsection (2), the reference to England shall have effect as a reference to Northern Ireland and the reference to the Minister of Agriculture, Fisheries and Food shall have effect as a reference to the Department of Agriculture for Northern Ireland.

[(4A) In the application of this section in relation to authorisation, and applications for authorisations, under section 13 in respect of premises situated on a nuclear site in England, references to the Secretary of State shall have effect as references to the Secretary of State and the Minister of Agriculture, Fisheries and Food.]

[(5) In this section "the relevant Minister" has the same meaning as in section 16 above.]

Definitions

"appropriate Agency": section 47(1).   B–151
"local authority": section 47(1).
"public or local authority": section 47(1).
"relevant Minister": section 16(11).

Amendments

The Environment Act 1995, Sched. 22, para. 213, amended subsections B–152
(1), (2) and (3); and inserted subsections (3A), (4A) and (5).

General Note

Subsections (1)–(3)

The Secretary of State may prohibit on the grounds of national security B–153
the copying of information contained in or relating to applications or registrations to public or local authorities. This prohibition is subject to subsection (3A).

Appeals

Registrations, authorisations and notices: appeals from decisions of [the appropriate Agency]

26.—(1) Where the [appropriate Agency]—   B–154
    (a) refuses an application for registration under section 7 or 10, or refuses an application for an authorisation under section 13 or 14,
    (b) attaches any limitations or conditions to such a registration or to such an authorisation, or
    (c) varies such a registration or such an authorisation, otherwise than by revoking a limitation or condition subject to which it has effect, or
    (d) cancels such a registration or revokes such an authorisation,
the person directly concerned may, subject to subsection (3), appeal to the Secretary of State.

(2) A person on whom a notice under section 21 or 22 is served may, subject to subsections (3) and (4), appeal against the notice to the Secretary of State.

(3) No appeal shall lie—

(a) [...]

(b) under subsection (1) or (2) in respect of any decision taken by the [appropriate Agency] in pursuance of a direction of the Secretary of State under section 23 or 24.

(4) No appeal shall lie under subsection (2) in respect of any notice served in [...] Northern Ireland by the appropriate Minister in exercise of the power under section 21 or 22.

(5) In this section "the person directly concerned" means—

(a) in relation to a registration under section 7 or 10, the person applying for the registration or to whom the registration relates;

(b) in relation to an authorisation under section 13 or 14, the person applying for the authorisation or to whom it was granted;

and any reference to attaching limitations or conditions to a registration or authorisation is a reference to attaching limitations or conditions to it either in effecting or granting it or in exercising any power to vary it.

[(5A) In the application of this section in relation to authorisations, and applications for authorisations, under section 13 in respect of premises situated on a nuclear site in England, references in subsection (1) to (3) to the Secretary of State shall have effect as references to the Secretary of State and the Minister of Agriculture, Fisheries and Food.]

(6) In the application of this section to Northern Ireland, references to the Secretary of State shall have effect as references to the Department of the Environment for Northern Ireland.

Definitions

B–155 "appropriate Agency": section 47(1).
"appropriate Minister": section 47(1).
"nuclear site": section 47(1).

Amendments

B–156 Subsection (3)(a) was revoked by the Environment Act 1995, Sched. 24. Subsection (4) was amended, and subsection (5A) inserted, by Sched. 22, para. 214, to that Act.

General Note

B–157 This section provides a right of appeal to the Secretary of State from all decisions taken by the appropriate Agency, unless (subs. (3)(b)) that decision was taken in accordance with a direction from him. The same applies in relation to decisions of the chief inspector in Northern Ireland, where, in addition (subs. (4)) no appeal may be made in respect of a notice served by the Department of Agriculture for Northern Ireland in respect of authorisations made or notices given relating to the disposal of radioactive waste on or from nuclear site premises.

Subsection (5A)

Unlike elsewhere in the United Kingdom, in England the Secretary of **B–158** State's appellate function under subsections (1)–(2) is exercisable only jointly with the Minister of Agriculture, Fisheries and Food if exercised in relation to the disposal of radioactive waste on or from a nuclear site. Similarly, in relation to such issues the two Ministers give directions jointly: hence the application of subsection (5A) to subsection (3). This provision was inserted by the 1995 Act to reflect the fact that the Minister is no longer joint authorising authority for the matters in question (revoked section 16(3)). See further at B–04 (*(3) Role of Ministers*).

Procedure on appeals under section 26

27.—(1) The Secretary of State may refer any matter involved in an **B–159** appeal under section 26[, other than an appeal against any decision of, or notice served by, SEPA,] to a person appointed by him for the purpose.

[(1A) As respects an appeal against any decision of, or notice served by, SEPA, this section is subject to section 114 of the Environment Act 1995 (delegation or reference of appeals).]

(2) An appeal under section 26 shall, if and to the extent required by regulations under subsection (7) of this section, be advertised in such manner as may be prescribed.

(3) If either party to the appeal so requests, an appeal shall be in the form of a hearing (which may, if the person hearing the appeal so decides, be held, or held to any extent, in private).

(4) On determining an appeal from a decision of the [appropriate Agency] under section 26 the Secretary of State—

(a) may affirm the decision,

(b) where that decision was the refusal of an application, may direct the [appropriate Agency] to grant the application,

(c) where that decision involved limitations or conditions attached to a registration or authorisation, may quash those limitations or conditions wholly or in part, or

(d) where that decision was a cancellation or revocation of a registration or authorisation, may quash the decision,

and where the Secretary of State does any of the things mentioned in paragraph (b), (c) or (d) he may give directions to the [appropriate

Agency] as to the limitations and conditions to be attached to the registration or authorisation in question.

(5) On the determination of an appeal in respect of a notice under section 26(2), the Secretary of State may either cancel or affirm the notice and, if he affirms it, may do so either in its original form or with such modifications as he may think fit.

(6) The bringing of an appeal against a cancellation or revocation of a registration or authorisation shall, unless the Secretary of State otherwise directs, have the effect of suspending the operation of the cancellation or revocation pending the determination of the appeal; but otherwise the bringing of an appeal shall not, unless the Secretary of State so directs, affect the validity of the decision or notice in question during that period.

(7) The Secretary of State may by regulations make provision with respect to appeals under section 26 (including in particular provision as to the period within which appeals are to be brought).

[(7A) In the application of this section in relation to authorisations, and applications for authorisations, under section 13 in respect of premises situated on a nuclear site in England, references in subsections (1) to (6) to the Secretary of State shall have effect as references to the Secretary of State and the Minister of Agriculture, Fisheries and Food.]

(8) In the application of this section to Northern Ireland, references to the Secretary of State shall have effect as references to the Department of the Environment for Northern Ireland.

Definitions

B–160   "appropriate Agency": section 47(1).
"nuclear site": section 47(1).
"SEPA": section 47(1).

General Note

B–161   This section governs the procedures for appeals made under section 26.

Subsections (1) and (1A)

B–162   The Secretary of State has the power to either determine the appeal himself or to transfer his jurisdiction to a person appointed by him, although it will be noted that in Scotland the only procedure applicable is that under the 1995 Act, s.114, which also provides for specific matters raised in an appeal to be referred to an appointed person.

404

Subsection (3)

The two modes of appeal contemplated are written representations or a **B–163** hearing (which could be either public or private).

Subsection (4)

The Secretary of State is given a wide discretion as to the disposal of **B–164** appeals including substitution of new conditions and modification of notices.

The Radioactive Substances (Appeals) Regulations 1990 (S.I. 1990 No. 2504) have been published under this section (as previously enacted) and set out the relevant procedures in more detail. In particular, notice of appeal should be given to the Secretary of State within two months of the date on which the decision or notice which is the subject of the appeal is sent to the appellant or the date on which application is deemed to have been refused for non-determination. Although a longer period may be allowed by the Secretary of State this would only occur in very compelling circumstances.

Subsection (7A)

By virtue of section 26(5A), and for the reasons outlined in the note to **B–165** that section in England (unlike elsewhere in the United Kingdom) the Secretary of State and Minister of Agriculture, Fisheries and Food are joint appellate authorities in relation to authorisations for the disposal of radioactive waste on or from a nuclear site.

Representations in relation to authorisations and notices when appropriate Minister is concerned

28. ...

General Note

Section 28 was revoked in Great Britain by the Environment Act 1995, **B–166** Sched. 24. It provided for representations to be made in circumstances which now only apply in Northern Ireland, where the role of the appropriate Minister as joint authorising authority meant there could be no appeal against any decision thus made.

Further powers of Secretary of State in relation to radioactive waste

Provision of facilities for disposal or accumulation of radioactive waste

B–167   29.—(1) If it appears to the Secretary of State that adequate facilities are not available for the safe disposal or accumulation of radioactive waste, the Secretary of State may provide such facilities, or may arrange for their provision by such persons as the Secretary of State may think fit.

(2) Where, in the exercise of the power conferred by this section, the Secretary of State proposes to provide, or to arrange for the provision of, a place for the disposal or accumulation of radioactive waste, the Secretary of State, before carrying out that proposal, shall consult with any local authority in whose area that place would be situated, and with such other public or local authorities (if any) as appear to him to be proper to be consulted by him.

(3) The Secretary of State may make reasonable charges for the use of any facilities provided by him, or in accordance with arrangements made by him, under this section, or, in the case of facilities provided otherwise than by the Secretary of State, may direct that reasonable charges for the use of the facilities may be made by the person providing them in accordance with any such arrangements.

(4) In the application of this section to Northern Ireland, references to the Secretary of State shall have effect as references to the Department of the Environment for Northern Ireland.

Definitions

B–168   "disposal": section 47(1).
"local authority": section 47(1).
"public or local authority": section 47(1).
"radioactive waste": section 2.

General Note

B–169   The site at Drigg near Sellafield, owned and occupied by BNFL, was provided pursuant to this power.

Power of [the appropriate Agency] to dispose of radioactive waste

B–170   30.—(1) If there is radioactive waste on any premises, and the [appropriate Agency] is satisfied that—

(a) the waste ought to be disposed of, but

(b) by reason that the premises are unoccupied, or that the occupier is absent, or is insolvent, or for any other reason, it is unlikely that the waste will be lawfully disposed of unless [that Agency] exercises [its] powers under this section,

[the appropriate Agency] shall have power to dispose of that radioactive waste as [that Agency] may think fit, and to recover from the occupier of the premises, or, if the premises are unoccupied, from the owner of the premises, any expenses reasonably incurred by [that Agency] in disposing of it.

(2) In the application of subsection (1) to Northern Ireland, references to the Secretary of State shall have effect as references to the Department of the Environment for Northern Ireland.

(3) For the purposes of this section in its application to England and Wales and Northern Ireland, the definition of "owner" in section 343 of the Public Health Act 1936, and the provisions of section 294 of that Act (which limits the liability of owners who are only agents or trustees), shall apply—

(a) with the substitution in section 294 for references to a council of references to the [Environment Agency] or, in Northern Ireland, the Department of the Environment for Northern Ireland, and

(b) in relation to Northern Ireland, as if that Act extended to Northern Ireland.

(4) For the purposes of this section in its application to Scotland, the definition of "owner" in section 3 of the Public Health (Scotland) Act 1897 and the provisions of section 336 of the Housing (Scotland) Act 1987 shall apply, with the substitution in section 336 of references to [SEPA] for references to a local authority.

Definitions

"appropriate Agency": section 47(1).   B–171
"disposed of ": section 47(1).
"premises": section 47(1).
"radioactive waste": section 2.
"SEPA": section 47(1).

General Note

Subsection (3)

"Owner" is defined by reference to section 343 of the Public Health Act B–172
1936. Under that Act "owner" is defined by reference to the person receiving or entitled to receive the rack rent of the premises. However, it should be noted that in Scotland, although the provisions by which the word "owner" is defined are very similar, they may be interpreted differently by the Scottish Courts.

Rights of entry and inspection

31. ....

General Note

B–173   Section 31 and Schedule 2, to which it gave effect, which provided rights of entry, inspection, etc., for inspectors appointed or authorised for the purposes of this Act, were revoked in Great Britain by the Environment Act 1995, Sched. 24. It remains in force for Northern Ireland. In Great Britain, a broad range of powers and rights comparable to those formerly provided by section 31 are now provided by the 1995 Act, ss.108 and 109.

Offences

Offences relating to registration or authorisation

B–174   32.—(1) Any person who—
   (a) contravenes section 6, 9, 13(1), (2) or (3) or 14(1), or
   (b) being a person registered under section 7 or 10, or being (wholly or partly) exempted from registration under either of those sections, does not comply with a limitation or condition subject to which he is so registered or exempted, or
   (c) being a person to whom an authorisation under section 13 or 14 has been granted, does not comply with a limitation or condition subject to which that authorisation has effect, or
   (d) being a person who is registered under section 7 or 10 or to whom an authorisation under section 13 or 14 has been granted, fails to comply with any requirement of a notice served on him under section 21 or 22,
shall be guilty of an offence.
   (2) A person guilty of an offence under this section shall be liable—
   (a) on summary conviction, to a fine not exceeding £20,000 or to imprisonment for a term not exceeding six months, or both;
   (b) on conviction on indictment, to a fine or to imprisonment for a term not exceeding five years, or both.
   [(3) If the appropriate Agency is of the opinion that proceedings for an offence under subsection (1)(d) would afford an ineffectual remedy against a person who has failed to comply with the requirements of a notice served on him under section 21 or 22, that Agency may take proceedings in the High Court or, in Scotland, in any court of competent jurisdiction, for the purpose of securing compliance with the notice.]

Definition

"appropriate Agency": section 47(1). **B–175**

Amendment

Subsection (3) was inserted by the Environment Act 1995, Sched. 22, **B–176** para. 219.

General Note

This section creates some of the offences under the Act and prescribes the **B–177** various penalties; it also provides the regulator with the alternative of civil enforcement proceedings.

Subsection (1)

For discussion of the word "person" see the General Note to section 6. **B–178**
   For discussion of the elements of the criminal offences see the General Note to section 6.

   Offences are given a maximum fine of £20,000 on summary conviction (in a magistrates' court in England and Wales; a sheriff court in Scotland) which is consistent with the maximum fines available under the Water Resources Act 1991 and the EPA 1990.

   As an illustration of the offences committed and penalties imposed, the HMIP Annual Report for 1991–1992 reveals that three prosecutions were taken and concluded in that year:

(a) AEA Technology were prosecuted for a failure of operating procedures in relation to telephone dials containing tritium. They were fined £3,000 plus costs of £1,395.

(b) BNFL Springfold were found guilty of breaching a condition of authorisation in relation to the disposal of low level radioactive waste. They were fined £7,500 plus costs of £6,140.

(c) Leicester Royal Infirmary were found guilty of unauthorised disposal of radioactive waste and fined £6,000 plus costs of £5,500.

Subsequent prosecutions include:

(a) Plessey GEC Semiconductors Ltd., who pleaded guilty to holding a radioactive source without a certificate of registration and were fined £1,000 plus £2,261 costs in May 1993.

(b) Nicholls Institute Diagnostics, who pleaded guilty to accumulating radioactive waste without authorisation, holding radioactive material without a certificate of registration, and exceeding the limit for the holdings in their registration. They were fined £10,000 and ordered to pay costs of £2,726 in April 1993.

Subsection (3)

B–179 Where the appropriate Agency considers that prosecution for non-compliance with an enforcement or prohibition notice (sections 21 and 22) would be an ineffectual remedy, express provision is made for it to take civil proceedings. The advantage for the Agency in question is that this can be a rapid means of enforcement, particularly in cases of repeated contravention and where criminal courts are not imposing terms of imprisonment: breach of an injunction so obtained constitutes contempt of court, punishable by imprisonment and/or sequestration of assets.

Offences relating to sections 19 and 20

B–180   33.—(1) Any person who contravenes section 19 shall be guilty of an offence and liable—

    (a) on summary conviction, to a fine not exceeding the statutory maximum;

    (b) on conviction on indictment, to a fine.

(2) Any person who without reasonable cause pulls down, injures or defaces any document posted in pursuance of section 19 shall be guilty of an offence and liable on summary conviction to a fine not exceeding level 2 on the standard scale.

(3) Any person who fails to comply with a requirement imposed on him under section 20 shall be guilty of an offence and liable—

    (a) on summary conviction, to a fine not exceeding the statutory maximum or to imprisonment for a term not exceeding three months, or both;

    (b) on conviction on indictment, to a fine or to imprisonment for a term not exceeding two years, or both.

General Note

B–181 This section creates the offence of breaching the provisions in section 19 relating to the posting of certificates of registration or authorisation and also an offence of defacing such displayed documents.

Subsection (3)

B–182 Failure to comply with the requirement to retain or produce documents or records issued pursuant to section 20 is an offence.

The maximum fine for this offence on summary conviction (in a magistrates' court in England and Wales or a sheriff court in Scotland) is the statutory maximum which is currently £5,000 (see section 17(1) of the Criminal Justice Act 1991 which amended the standard scale of fines set out in the Criminal Justice Act 1982).

Disclosure of trade secrets

34.—(1) If any person discloses any information relating to any B–183 relevant process or trade secret used in carrying on any particular undertaking which has been given to or obtained by him under this Act or in connection with the execution of this Act, he shall be guilty of an offence, unless the disclosure is made—
  (a) with the consent of the person carrying on that undertaking, or
  (b) in accordance with any general or special directions given by the Secretary of State, or
  [(bb) under or by virtue of section 113 of the Environment Act 1995, or]
  (c) in connection with the execution of this Act, or
  (d) for the purposes of any legal proceedings arising out of this Act or of any report of any such proceedings.
(2) A person guilty of an offence under this section shall be liable—
  (a) on summary conviction, to a fine not exceeding the statutory maximum or to imprisonment for a term not exceeding three months, or both;
  (b) on conviction on indictment, to a fine or to imprisonment for a term not exceeding two years, or both.
(3) In this section "relevant process" means any process applied for the purposes of, or in connection with, the production or use of radioactive material.
(4) In the application of this section to Northern Ireland, the reference in subsection (1)(b) to the Secretary of State shall have effect as a reference to the Department of the Environment for Northern Ireland.

Definition

"undertaking": section 47(1). B–184

Amendment

Subsection (1)(bb) was inserted by the Environment Act 1995, Sched. 22, B–185 para. 220.

General Note

This section creates an offence of disclosing trade or process secrets B–186 without proper authority. For a note on "person" see the General Note to section 6.

The statutory maximum fine available on summary conviction is £5,000, see the General Note to section 33.

Subsection (1)(bb) adds to the list of acceptable grounds for disclosing trade secrets the disclosure of information pursuant to the 1995 Act, section 113. That section provides for the exchange of information between the Agencies, local enforcing authorities and Ministers, in the furtherance of the Agencies' functions, local enforcing authorities' functions (that is the functions of local authorities under the pollution control, contaminated land and air quality assessment and management systems) and Ministers' environmental functions. It provides an exemption from criminal or civil liability in respect of disclosure under its provisions.

[Offences of making false or misleading statements or false entries

B–187   34A.—(1) Any person who—
  (a) for the purpose of obtaining for himself or another any registration under section 7 or 10, any authorisation under section 13 or 14 or any variation of such an authorisation under section 17, or
  (b) in purported compliance with a requirement to furnish information imposed under section 31(1)(d),
makes a statement which he knows to be false or misleading in a material particular, or recklessly makes a statement which is false or misleading in a material particular, shall be guilty of an offence.
  (2) Any person who intentionally makes a false entry in any record
  (a) which is required to be kept by virtue of a registration under section 7 or 10 or an authorisation under section 13 or 14, or
  (b) which is kept in purported compliance with a condition which must be complied with if a person is to have the benefit of an exemption under section 8, 11 or 15, shall be guilty of an offence.
  (3) A person guilty of an offence under this section shall be liable—
  (a) on summary conviction, to a fine not exceeding the statutory maximum;
  (b) on conviction on indictment, to a fine or to imprisonment for a term not exceeding two years, or to both.]

Commencement

B–188 April 1, 1996 (S.I. 1996 No. 186).

General Note

This section was inserted by the Environment Act 1995, Sched. 19, which   **B–189**
inserted a similar provision into several other Acts. The section makes it a
criminal offence to knowingly or recklessly make statements which are
false or misleading in a material particular, either:

(i) in order to obtain a registration, authorisation or variation
(subs.(1)(a));

(ii) (but see below) in answering questions or supplying information where required to do so by an inspector under section
31(1)(d) (subs.(1)(b)); or

(iii) in an entry in a record required by a registration or authorisation (subs.(2)).

In a serious case, imprisonment could result (subs.(3)).

As regards situation (ii) described here, it will be noted that section 31
now applies only in Northern Ireland. Curiously, as regards Great Britain,
there is no directly comparable provision (although there are a number
of offences provided by section 110 of the 1995 Act which cover some of
the same ground—such as failing to comply with requirements imposed
by Agency officers, which includes requests for information (s.110(2)
(a)), and failing or refusing to provide information which such officers
reasonably require (s.110(2)(b))).

Obstruction

35. . . .

General Note

Section 35 was revoked in Great Britain by the Environment Act 1995,   **B–190**
Sched. 24. It created offences of intentionally obstructing officials in the
exercise of their duties under the Act and refusing or without reasonable
excuse failing to provide assistance to them. These offence provisions still
apply in Northern Ireland. Equivalent and additional offence provisions
are now found in relation to Great Britain in the 1995 Act, s.110. One
notable change is that there is no "reasonable excuse" qualification in
section 110 to the charge of failing to provide assistance, etc., as there was
in section 35(1)(b).

Offences by bodies corporate

36.—(1) Where a body corporate is guilty of an offence under this Act,   **B–191**
and that offence is proved to have been committed with the consent or
connivance of, or to be attributable to any neglect on the part of, any

director, manager, secretary or other similar officer of the body corporate, or any person who was purporting to act in any such capacity, he, as well as the body corporate, shall be guilty of that offence, and shall be liable to be proceeded against and punished accordingly.

(2) In this section "director", in relation to a body corporate established by or under any enactment for the purpose of carrying on under national ownership any industry or part of an industry or undertaking, being a body corporate whose affairs are managed by its members, means a member of that body corporate.

General Note

B–192   This provision is included in numerous statutes with the object of making directors and other senior officers of a company statutory principals in circumstances where they have a certain degree of personal responsibility and culpability.

Consent

B–193   "It would seem that where a director consents to the commission of an offence by his company, he is well aware of what is going on and agrees to it."[21]

Connivance

B–194   This term implies acquiescence in the course of conduct reasonably likely to lead to the commission of the offence. "Where he [the director] connives at the offence committed by the company he is equally well aware of what is going on but his agreement is tacit, not actively encouraging what happens but letting it continue and saying nothing about it."[22]

Neglect

B–195   This term implies "failure to perform a duty which the person knows or ought to know".[23] A director's duty is not absolute and some act or omission constituting neglect must be shown.[24] The director's duties[25] may, in certain circumstances, be properly delegated: "a director may

---

[21] *Huckerby v. Elliot* [1970] 1 All E.R. 189, 194, *per* Ashworth J.
[22] see note **21** *ibid.*
[23] *Re Hughes* [1943] 2 All E.R. 269.
[24] *Huckerby v. Elliot* [1970] 1 All E.R. 189, *per* Ashworth J.
[25] see: *Re City Equitable Fire Insurance Co.* [1925] Ch. 407.

delegate, but each case is one of fact and of the circumstances of the case".[26]

Director, manager, secretary or other similar officer

See *R. v. Boal (Francis)*,[27] in which an employee manager of a bookshop   B–196
was not held to be a "manager" for the purposes of the Fire Precautions
Act 1971. See also *Armour v. Skeen*[28] in which the terms were held to
include a senior officer of a Scottish Regional Council.

Or a person purporting to act in any such capacity

This term will cover directors or officers whose appointment is irregular   B–197
or defective.[29]

Offence due to another's fault

37. Where the commission by any person of an offence under this Act   B–198
is due to the act or default of some other person, that other person may by
virtue of this section be charged with and convicted of the offence
whether or not proceedings for the offence are taken against the
first-mentioned person.

General Note

This section allows a person whose acts or defaults result in the   B–199
commission of an offence by another person, to be charged and
convicted of the offence, whether or not proceedings are taken against
the person who actually committed the offence.[30]

Restriction on prosecutions

38.—(1) Proceedings in respect of any offence under this Act shall not   B–200
be instituted in England or Wales except—
    (a) by the Secretary of State,
    (b) by the [Environment Agency], or

---

[26] *Hirschler v. Birch* (1987) 151 J.P. 396; see also *Re City Equitable Fire Insurance Co.* [1925] Ch.
407.
[27] see: *Tesco Supermarkets v. Natrass* [1972] A.C. 153; and *R. v. Boal (Francis)* [1992] 1 Q.B. 591.
[28] [1977] I.R.L.R. 310.
[29] see: *Dean v. Hiesler* [1942] 2 All E.R. 340.
[30] see: *Olgeirsson v. Kitching* [1986] 1 W.L.R. 304; and *Meah v. Roberts; Lansley v. Roberts* [1978]
1 W.L.R. 1187; for discussion of this type of provision.

(c) by or with the consent of the Director of Public Prosecutions.

(2) Proceedings in respect of any offence under this Act shall not be instituted in Northern Ireland except—

(a) by the head of the Department of the Environment for Northern Ireland, or

(b) by or with the consent of the Attorney General for Northern Ireland.

(3) This section shall be deemed to have been enacted before the coming into operation of the Prosecution of Offences (Northern Ireland) Order 1972.

General Note

B–201   Except where a statute expressly provides to the contrary, any person in England, Wales or Northern Ireland may bring a private prosecution for a criminal offence. The significance of this section is thus that, since it makes such express contrary provision, no private prosecutions may be brought for offences under this Act without the consent of the DPP under subsection (1)(c). Such a bar is of particular relevance in the environmental sphere where environmental groups use private prosecutions as one means of circumventing a perceived passivity on the part of the statutory enforcing authorities. It is perhaps significant that this Act governs a number of high profile environmental subjects—the prolonged series of judicial reviews sought by Greenpeace in relation to the THORP reprocessing facility (see above at B–01) illustrates this.

In Scotland the Lord Advocate's consent is required in any event for a private prosecution, so no specific provision is required.

Public access to documents and records

Public access to documents and records

B–202   39.—(1) The [appropriate Agency] shall keep copies of—

(a) all applications made to [it] under any provision of this Act,

(b) all documents issued by [it] under any provision of this Act,

(c) all other documents sent by [it] to any local authority in pursuance of directions of the Secretary of State, and

(d) such records of convictions under section 32, 33, 34 or 35 as may be prescribed in regulations;

and [the appropriate Agency] shall make copies of those documents available to the public except to the extent that that would involve the disclosure of information relating to any relevant process or trade secret or would involve the disclosure of [information] as respects which the Secretary of State has directed that knowledge should be restricted on grounds of national security.

416

(2) Each local authority shall keep and make available to the public copies of all documents sent to the authority under any provision of this Act unless directed by the [appropriate Agency] ... that all or any part of any such document is not to be available for inspection.

(3) Directions under subsection (2) shall only be given for the purpose of preventing disclosure of relevant processes or trade secrets and may be given generally in respect of all, or any description of, documents or in respect of specific documents.

(4) The copies of documents required to be made available to the public by this section need not be kept in documentary form.

(5) The public shall have the right to inspect the copies of documents required to be made available under this section at all reasonable times and, on payment of a reasonable fee, to be provided with a copy of any such document.

(6) In this section "relevant process" has the same meaning as in section 34.

(7) In the application of this section to Northern Ireland, references to the Secretary of State shall have effect as references to the Department of the Environment for Northern Ireland.

Definitions

"appropriate Agency": section 47(1).   **B–203**
"local authority": section 47(1).
"relevant process": section 34(3).

Amendments

Subsections (1) and (2) were amended by the Environment Act 1995, **B–204**
Sched. 22, para. 223.

General Note

This provision was added to the laws on radioactive substances by the EPA **B–205**
1990 and greatly extends the documents that should be sent to local authorities and in turn made available to the public. The *DoE Circular 21/90* on *Local Authority Responsibilities for Public Access to Information under the Radioactive Substances Act 1960 as Amended by the Environmental Protection Act 1990* as supplemented by *DoE Circular 22/92*, explains the provisions on public access to information and gives guidance to local authorities on their obligations in respect of those provisions.

The local authorities are required to make available to the general public copies of all the documents sent to them by the appropriate Agency unless they have been directed otherwise by the appropriate

Agency that a particular document or part of a document must not be made available for inspection. Such direction would normally involve preventing the disclosure of information relating to any relevant process or trade secret.

The public are entitled to both inspect and also take copies of documents held by the local authority. The local authority is entitled to make a reasonable charge for photocopying.

The Environment Agency's regional offices should hold full sets of documentation relating to the local authorities within its region. The Agency is under the same obligation to allow inspection and photocopying as the local authorities.

Except for special rules relating to the rehabilitation of offenders under the Rehabilitation of Offenders Act 1974, local authorities should keep documents sent to them by the appropriate Agency for a minimum period of four years after the documents cease to have effect.

The Radioactive Substances (Records of Conviction) Regulations 1992 (S.I. 1992 No. 1685) set out the prescribed information to be made available to the public. In relation to each conviction this is: the details of the offence; the name of the offender; the date of conviction; the penalty imposed; and the name of the court. However, spent convictions under the Rehabilitation of Offenders Act 1974 must be removed from the register.

Operation of other statutory provisions

Radioactivity to be disregarded for purposes of certain statutory provisions

B–206   40.—(1) For the purposes of the operation of any statutory provision to which this section applies, and for the purposes of the exercise or performance of any power or duty conferred or imposed by, or for the enforcement of, any such statutory provision, no account shall be taken of any radioactivity possessed by any substance or article or by any part of any premises.

(2) This section applies—
- (a) to any statutory provision contained in, or for the time being having effect by virtue of, any of the enactments specified in Schedule 3, or any enactment for the time being in force whereby an enactment so specified is amended, extended or superseded, and
- (b) to any statutory provision contained in, or for the time being having effect by virtue of, a local enactment whether passed or made before or after the passing of this Act (in whatever terms the provision is expressed) in so far as—
    - (i) the disposal or accumulation of waste or any description of waste, or of any substance which is a nuisance, or so as to

> be a nuisance, or of any substance which is, or so as to be, prejudicial to health, noxious, polluting or of any similar description, is prohibited or restricted by the statutory provision, or
>
> (ii) a power or duty is conferred or imposed by the statutory provision on [the Environment Agency or SEPA or on] any local authority, relevant water body or other public or local authority, or on any officer of a public or local authority, to take any action (whether by way of legal proceedings or otherwise) for preventing, restricting or abating such disposals or accumulations as are mentioned in sub-paragraph (i).

(3) In this section—

"statutory provision"—

> (a) in relation to Great Britain, means a provision, whether of a general or a special nature, contained in, or in any document made or issued under, any Act, whether of a general or a special nature, and
>
> (b) in relation to Northern Ireland, has the meaning given by section 1(f) of the Interpretation Act (Northern Ireland) 1954,

"local enactment" means—

> (a) a local or private Act (including a local or private Act of the Parliament of Northern Ireland or a local or private Measure of the Northern Ireland Assembly), or
>
> (b) an order confirmed by Parliament (or by the Parliament of Northern Ireland or the Northern Ireland Assembly) or brought into operation in accordance with special parliamentary procedure,

and any reference to disposal, in relation to a statutory provision, is a reference to discharging or depositing a substance or allowing a substance to escape or to enter a stream or other place, as may be mentioned in that provision.

(4) The references to provisions of the Water Resources Act 1991 in Part I of Schedule 3 shall have effect subject to the power conferred by section 98 of that Act.

Definitions

"article": section 47(1).

"local authority": section 47(1).

"premises": section 47(1).

"public or local authority": section 47(1).

"relevant water body": section 47(1).

"SEPA": section 47(1).

"substance": section 47(1).

"waste": section 47(1) and (4).

B–207

Amendment

B–208   Subsection (2) (b) (ii) was amended by the Environment Act 1995, Sched. 22, para. 224.

General Note

B–209   This section provides that where any prescribed statutory powers or obligations are being exercised or performed no account is to be taken of any radioactivity of any substances, articles or premises dealt with by that statutory provision. The list of prescribed statutory provisions, other than local enactments, are set out at Schedule 3. (The amendments to paragraphs 2 and 12 of Schedule 3 were made by the Clean Air Act 1993; in the same Schedule, paragraph 9 was amended, 16 substituted and 17A inserted by the Environment Act 1995, Sched. 22, para. 230 and Sched. 17, para. 8.) The previous list at Schedule 1 to the RSA 1960 included at paragraph 8c the Control of Pollution (Special Waste) Regulations 1980. Reference to these regulations, which have now been replaced by the Special Waste Regulations 1996, has been omitted from Schedule 3 to the Act. Special waste under the EPA 1990 is excluded by section 78 of that Act which states that Part II of that Act does not apply to radioactive waste, although the Secretary of State may by regulation bring radioactive waste back into the ambit of that Act. Similar provision is made in relation to Part IIA of that Act (contaminated land) by section 78YC: see further the general note above at B–04 *(6) Contaminated land.*

General

Service of documents

B–210   41.—(1) Any notice required or authorised by or under this Act to be served on or given to any person may be served or given by delivering it to him, or by leaving it at his proper address, or by sending it by post to him at that address

(2) Any such notice may—
   (a) in the case of a body corporate, be served on or given to the secretary or clerk of that body;
   (b) in the case of a partnership, be served on or given to a partner or a person having the control or management of the partnership business.

(3) For the purposes of this section and of section 7 of the Interpretation Act 1978 (service of documents by post) in its application to this section, the proper address of any person on or to whom any such notice is to be served or given shall be his last known address, except that—

   (a) in the case of a body corporate or their secretary or clerk, it shall be the address of the registered or principal office of that body;

   (b) in the case of a partnership or person having the control or the management of the partnership business, it shall be the principal office of the partnership;

and for the purposes of this subsection the principal office of a company registered outside the United Kingdom or of a partnership carrying on business outside the United Kingdom shall be their principal office within the United Kingdom.

(4) If the person to be served with or given any such notice has specified an address in the United Kingdom other than his proper address within the meaning of subsection (3) as the one at which he or someone on his behalf will accept notices of the same description as that notice, that address shall also be treated for the purposes of this section and section 7 of the Interpretation Act 1978 as his proper address.

(5) The preceding provisions of this section shall apply to the sending or giving of a document as they apply to the giving of a notice.

General Note

This provision was not in the RSA 1960 and is set out as in Proposal No. 4 **B–211** of the Memorandum under Consolidation of Enactments (Procedure) Act 1949. It sets out the new standard conditions on service of documents under a statute. The "principal" office of a body corporate or partnership (subs.(3)(a) and (b)) is where its general superintendence and management is carried out.[31]

Application of Act to Crown

42.—(1) Subject to the following provisions of this section, the **B–212** provisions of this Act shall bind the Crown.

(2) Subsection (1) does not apply in relation to premises—

   (a) occupied on behalf of the Crown for naval, military or air force purposes or for the purposes of the department of the Secretary of State having responsibility for defence, or

   (b) occupied by or for the purposes of a visiting force.

(3) No contravention by the Crown of any provision of this Act shall make the Crown criminally liable; but the High Court or, in Scotland, the Court of Session may, on the application of any authority charged with enforcing that provision, declare unlawful any act or omission of the Crown which constitutes such a contravention.

(4) Notwithstanding anything in subsection (3), the provisions of this Act shall apply to persons in the public service of the Crown as they apply to other persons.

[31] *Davies v. British Geon Ltd* [1956] 3 All E.R. 389, C.A.

(5) [...]

(6) Where, in the case of any such premises as are mentioned in subsection (2)—

(a) arrangements are made whereby radioactive waste is not to be disposed of from those premises except with the approval of the [appropriate Agency], and

(b) in pursuance of those arrangements the [appropriate Agency] proposes to approve, or approves, the removal of radioactive waste from those premises to a place provided by a local authority as a place for the deposit of refuse,

the provisions of section 18 shall apply as if the proposal to approve the removal of the waste were an application for an authorisation under section 13 to remove it, or (as the case may be) the approval were such an authorisation.

(7) Nothing in this section shall be taken as in any way affecting Her Majesty in her private capacity; and this subsection shall be construed as if section 38(3) of the Crown Proceedings Act 1947 (interpretation of references in that Act to Her Majesty in her private capacity) were contained in this Act.

(8) In this section "visiting force" means any such body, contingent or detachment of the forces of any country as is a visiting force for the purposes of any of the provisions of the Visiting Forces Act 1952.

(9) In the application of this section to Northern Ireland—

(a) references to the Crown shall include references to the Crown in right of Her Majesty's Government in Northern Ireland, and

(b) the reference in subsection (5) to the Secretary of State shall have effect as a reference to the Department of the Environment for Northern Ireland.

Definition

B–213 "appropriate Agency": section 47(1).

Amendment

B–214 Subsection (5) was revoked by the Environment Act 1995, Sched. 24.

General Note

Subsection (1)

B–215 The general principle underlying this section is that the whole Act binds the Crown with the exception of three matters as set out in subsections (2), (3) and (4).

Subsection (2)

The Act is not binding on the Crown in relation to premises occupied on behalf of the Crown for military or defence purposes, or occupied by or for the purposes of visiting forces.   **B–216**

Subsections (3) and (4)

Contravention of the Act does not render the Crown criminally liable, but the enforcing authority may obtain a declaration of unlawfulness from the High Court or Court of Session (subs.(3)). This protection applies only to the Crown, and not to anyone in the public service of the Crown (subs.(4)).   **B–217**

Subsection (5)

Subsection (5), which has been revoked in Great Britain, provided that the Secretary of State may issue a certificate excluding powers of entry to any Crown premises on the ground of national security. This provision still applies in Northern Ireland, and parallel provision is now made by the 1995 Act, s.115(5).   **B–218**

Subsection (6)

This deals with the disposal of radioactive waste from military premises to which the controls of the Act do not apply. Administrative arrangements may be made that such waste is not to be disposed of except with the approval of the appropriate Agency. Where such arrangements are made, and approval is given or proposed for removal to a local authority refuse site, then the local authority must, by subsections 18(3)–(5), receive the waste. The local authority has statutory rights under these provisions to be consulted in advance as to any special precautions necessary and to recover charges in respect of those precautions.   **B–219**

The *DoE Circular 22/92* on *Local Authority Responsibilities for Public Access to Information under the Radioactive Substances Act 1960 as Amended by the Environmental Protection Act 1990: Monitoring Data and Other Information,* helps to clarify the position on the MOD and visiting forces. Although MOD and visiting forces are exempt from the requirements of the Act, controls similar to those exercised over other radioactive material users are applied on an administrative basis. However, instead of registrations and authorisations, MOD and visiting forces receive "certificates of notification" recording the holding of radioactive material and "certificates of agreement" approving the disposal of radioactive wastes. These documents are to be sent to local authorities, except where there are overriding matters of national security. In this Circular the DoE clarify that the MOD and visiting forces have indicated a wish to comply with the spirit of public access to information provisions and copies of these documents as sent to local authorities are to be made available to the public as described in the General Note to section 39.

Fees and charges

43. . . .

General Note

B–220   Section 43 was revoked as it applies to Great Britain by the Environment Act 1995, Sched. 24. It continues in force in Northern Ireland. It made provision for a charging scheme in relation to fees and charges payable under this Act: this provision is now made by the 1995 Act, sections 41 and 42. Those provisions and others in that Act provide considerable flexibility to the Agencies in relation to charging for their time and services provided, both through charging schemes and otherwise. See the General Note to section 41 of that Act. As at February 1997 the applicable schemes in Great Britain, both made under section 43, were the Radioactive Substances Authorisations and Registrations Fees and Charges Scheme (England and Wales) Revised 1996 and the Radioactive Substances Act 1993 Fees and Charges (Scotland) Scheme, both of which came into force on April 1, 1996.

Regulations and orders: Great Britain

B–221   44.—(1) The Secretary of State may make regulations under this Act for any purpose for which regulations are authorised or required to be made under this Act.

(2) For the purpose of facilitating the exercise of any power under this Act to effect registrations, or grant authorisations, subject to limitations or conditions, the Secretary of State may make regulations setting out general limitations or conditions applicable to such classes of cases as may be specified in the regulations; and any limitations or conditions so specified shall, for the purposes of this Act, be deemed to be attached to any registration or authorisation falling within the class of cases to which those limitations or conditions are expressed to be applicable, subject to such exceptions or modifications (if any) as may be specified in any such registration or authorisation.

(3) Any power conferred by this Act to make regulations or orders shall be exercisable by statutory instrument.

(4) Any statutory instrument containing regulations or an order made

under this Act, other than an order under Schedule 5, shall be subject to annulment in pursuance of a resolution of either House of Parliament.

(5) This section does not extend to Northern Ireland.

General Note

The limitations or conditions referred to in subsection (2) may be attached to registrations or authorisations granted pursuant to sections 7, 10, 13 and 14.

    Breach of any such limitations or conditions is a criminal offence pursuant to section 32.

B–222

Regulations and orders: Northern Ireland

45.—(1) The Department of the Environment for Northern Ireland may make regulations under this Act for any purpose for which regulations are authorised or required to be made under this Act.

(2) For the purpose of facilitating the exercise of any power under this Act to effect registrations, or grant authorisations, subject to limitations or conditions, the Department of the Environment for Northern Ireland may make regulations setting out general limitations or conditions applicable to such classes of cases as may be specified in the regulations; and any limitations or conditions so specified shall, for the purposes of this Act, be deemed to be attached to any registration or authorisation falling within the class of cases to which those limitations or conditions are expressed to be applicable, subject to such exceptions or modifications (if any) as may be specified in any such registration or authorisation.

(3) Any power conferred by this Act to make regulations or orders shall be exercisable by statutory rule for the purposes of the Statutory Rules (Northern Ireland) Order 1979.

(4) Any regulations or orders made under this Act shall be subject to negative resolution within the meaning of section 41(6) of the Interpretation Act (Northern Ireland) 1954.

(5) This section extends to Northern Ireland only.

B–223

Effect of Act on other rights and duties

46. Subject to the provisions of section 40 of this Act, and of section 18 of the Interpretation Act 1978 (which relates to offences under two or more laws), nothing in this Act shall be construed as—

    (a) conferring a right of action in any civil proceedings (other than proceedings for the recovery of a fine) in respect of any contravention of this Act, or

B–224

(b) affecting any restriction imposed by or under any other enactment, whether contained in a public general Act or in a local or private Act, or

(c) derogating from any right of action or other remedy (whether civil or criminal) in proceedings instituted otherwise than under this Act.

General Note

B–225 Common law actions such as personal injury actions, in particular those brought by employees against their employers, often allege "breach of statutory duty" as well as negligence on the part of the defendant. Section 47 of the Health and Safety at Work, etc. Act 1974 expressly provides that breach of health and safety regulations, unless they state otherwise, is actionable where damage results. Many statutes are silent on the point, leading the courts into difficult areas of statutory interpretation when civil actions are founded on their breach.

This section expressly excludes such action based on contravention of this Act (para. (a)), whilst on the other hand making it clear that compliance with its provisions will not of itself serve as a defence to any civil or criminal proceedings other than those for which the Act itself provides (para. (c)).

General interpretation provisions

B–226   47.—(1) In this Act, except in so far as the context otherwise requires—

["the appropriate Agency" means

(a) in relation to England and Wales, the Environment Agency; and

(b) in relation to Scotland, SEPA;]

"the appropriate Minister" means—

[...]

(c) in relation to Northern Ireland, the Department of Agriculture for Northern Ireland,

"article" includes a part of an article,

"the chief inspector" means—

[...]

(c) in relation to Northern Ireland, the chief inspector for Northern Ireland appointed under section 4(7),

"disposal", in relation to waste, includes its removal, deposit, destruction, discharge (whether into water or into the air or into a sewer or drain or otherwise) or burial (whether underground or otherwise) and "dispose of" shall be construed accordingly,

"local authority" (except where the reference is to a public or local authority) means—

(a) in England and Wales, the council of a county, district or London borough or the Common Council of the City of London or an authority established by the Waste Regulation and Disposal (Authorities) Order 1985,

(b) in Scotland, a [council constituted under section 2 of the Local Government etc. (Scotland) Act 1994], and

(c) in Northern Ireland, a district council,

"nuclear site" means—

(a) any site in respect of which a nuclear site licence is for the time being in force, or

(b) any site in respect of which, after the revocation or surrender of a nuclear site licence, the period of responsibility of the licensee has not yet come to an end,

"nuclear site licence", "licensee" and "period of responsibility" have the same meaning as in the Nuclear Installations Act 1965,

"premises" includes any land, whether covered by buildings or not, including any place underground and any land covered by water,

"prescribed" means prescribed by regulations under this Act [...]

"the prescribed period for determinations", in relation to any application under this Act, means, subject to subsection (2), the period of four months beginning with the day on which the application was received,

"public or local authority", in relation to England and Wales, includes a water undertaker or a sewerage undertaker,

"relevant water body" means—

(a) in England and Wales, a water undertaker, a sewerage undertaker or a local fisheries committee,

(b) in Scotland, [...] a district salmon fishery board established under section 14 of the Salmon Act 1986 or [a water and sewerage authority established by section 62 of the Local Government etc. (Scotland) Act 1994], and

(c) in Northern Ireland, the Fisheries Conservation Board for Northern Ireland,

["SEPA" means the Scottish Environment Protection Agency];

"substance" means any natural or artificial substance, whether in solid or liquid form or in the form of a gas or vapour,

"undertaking" includes any trade, business or profession and—

(a) in relation to a public or local authority, includes any of the powers or duties of that authority, and

(b) in relation to any other body of persons, whether corporate or unincorporate, includes any of the activities of that body, and

"waste" includes any substance which constitutes scrap material or an effluent or other unwanted surplus substance arising from the application of any process, and also includes any substance or article which requires to be disposed of as being broken, worn out, contaminated or otherwise spoilt.

(2) The Secretary of State may by order substitute for the period

for the time being specified in subsection (1) as the prescribed period for determinations such other period as he considers appropriate.

(3) In determining, for the purposes of this Act, whether any radioactive material is kept or used on any premises, no account shall be taken of any radioactive material kept or used in or on any railway vehicle, road vehicle, vessel or aircraft if either—

    (a) the vehicle, vessel or aircraft is on those premises in the course of a journey, or

    (b) in the case of a vessel which is on those premises otherwise than in the course of a journey, the material is used in propelling the vessel or is kept in or on the vessel for use in propelling it.

(4) Any substance or article which, in the course of the carrying on of any undertaking, is discharged, discarded or otherwise dealt with as if it were waste shall, for the purposes of this Act, be presumed to be waste unless the contrary is proved.

(5) Any reference in this Act to the contamination of a substance or article is a reference to its being so affected by either or both of the following, that is to say,—

    (a) absorption, admixture or adhesion of radioactive material or radioactive waste, and

    (b) the emission of neutrons or ionising radiations,

as to become radioactive or to possess increased radioactivity.

(6) In the application of this section to Northern Ireland, the reference in subsection (2) to the Secretary of State shall have effect as a reference to the Department of the Environment for Northern Ireland.

Amendments

B–227   The reference in the definition of "relevant water body" to water and sewerage authorities was substituted in by the Local Government, etc. (Scotland) Act 1994; otherwise all amendments were made by the Environment Act 1995, Scheds. 22 and 24, and thus have effect only in relation to Great Britain.

General Note

Disposal

B–228   The DoE Guide to the administration of the Radioactive Substances Act draws the distinction between "storage" and "disposal".

Storage is seen as the placement of radioactive substances in a facility with the intention of taking further action at a later time. That further action may include *in situ* treatment or a declaration that further action is no longer needed and that storage has become disposal.

Disposal is seen as the dispersal of radioactive waste into an environmental medium or its placement in a facility with no intention of further action, apart from precautionary monitoring.

Nuclear Site

Under the Nuclear Installations Act 1965, sites to be used for the **B–229** operation of a nuclear reactor or other prescribed nuclear installations must be licensed by the Health and Safety Executive. Thus, licences under that Act are required for nuclear power stations (such as Sellafield, Dungeness or Bradwell), research reactors, private commercial reactors and processing plants.

Premises

Under United Kingdom statutes generally, "premises" refers only to those **B–230** situated in the United Kingdom. However, under the Continental Shelf Act 1964 and in particular the Continental Shelf (Jurisdiction) Order 1980 (S.I. 1980 No. 184) for the purposes of this Act, any installation within the areas designated under the Continental Shelf Act as English, Scottish or Northern Irish are deemed to be situated within England, Scotland or Northern Ireland respectively.

For revocation and variation of registrations, see section 12.

For exemptions from registration, see section 8.

Waste

See further the General Note to section 2.   **B–231**

Index of defined expressions

48. The following Table shows provisions defining or otherwise **B–232** explaining expressions for the purposes of this Act—

[the appropriate Agency	section 47(1)]
the appropriate Minister	section 47(1)
article	section 47(1)
contamination	section 47(5)
disposal	section 47(1)
licensee (in relation to a nuclear site licence)	section 47(1)
local authority	section 47(1)
mobile radioactive apparatus	section 3
nuclear site	section 47(1)
nuclear site licence	section 47(1)
period of responsibility (in relation to a nuclear site licence)	section 47(1)

premises	section 47(1)
prescribed	section 47(1)
the prescribed period for determinations	section 47(1) and (2)
public or local authority	section 47(1)
radioactive material	section 1
radioactive waste	section 2
relevant water body	section 47(1)
[SEPA	section 47(1)]
substance	section 47(1)
undertaking	section 47(1)
waste	section 47(1) and (4).

Amendments

B–233 References to SEPA and the appropriate Agency were inserted by the Environment Act 1995, Sched. 22, para. 228.

General Note

B–234 The addition of a list of terms defined in the Act is a welcome innovation which will assist in making statutes more user friendly.

Consequential amendments and transitional and transitory provisions

B–235   49.—(1) The enactments specified in Schedule 4 shall have effect subject to the amendments set out in that Schedule, being amendments consequential on the preceding provisions of this Act.

(2) The transitional and transitory provisions contained in Schedule 5 shall have effect.

Repeals

B–236   50. The enactments and instruments specified in Schedule 6 (which include spent enactments) are repealed or, as the case may be, revoked to the extent specified in the third column of that Schedule, but subject to any provision at the end of any Part of that Schedule.

Short title, commencement and extent

B–237   51.—(1) This Act may be cited as the Radioactive Substances Act 1993.

(2) This Act shall come into force at the end of the period of three months beginning with the day on which it is passed.

(3) This Act extends to Northern Ireland.

General Note

The Act came into force on August 27, 1993.   B–238

SCHEDULE 1

Section 1

SPECIFIED ELEMENTS

B–239

Element	Becquerels per gram (Bq g^{-1})		
	Solid	Liquid	Gas or Vapour
1. Actinium	0·37	$7·40 \times 10^{-2}$	$2·59 \times 10^{-6}$
2. Lead	0·74	$3·70 \times 10^{-3}$	$1·11 \times 10^{-4}$
3. Polonium	0·37	$2·59 \times 10^{-2}$	$2·22 \times 10^{-4}$
4. Protoactinium	0·37	$3·33 \times 10^{-2}$	$1·11 \times 10^{-6}$
5. Radium	0·37	$3·70 \times 10^{-4}$	$3·70 \times 10^{-5}$
6. Radon	—	—	$3·70 \times 10^{-2}$
7. Thorium	2·59	$3·70 \times 10^{-2}$	$2·22 \times 10^{-2}$
8. Uranium	11·1	0·74	$7·40 \times 10^{-5}$

SCHEDULE 2

[This schedule was revoked by the Environment Act 1995, Schedule B–240
24. See the note to the revoked section 31.]

SCHEDULE 3

Section 40

ENACTMENTS, OTHER THAN LOCAL ENACTMENTS, TO WHICH SECTION 40 APPLIES

PART I

ENGLAND AND WALES

1. Sections 48, 81, 82, 141, 259 and 261 of the Public Health Act 1936. B–241
2. [Section 16 of the Clean Air Act 1993.]
3. Section 5 of the Sea Fisheries Regulation Act 1966.

4. Section 4 of the Salmon and Freshwater Fisheries Act 1975.

5. Section 59 of the Building Act 1984.

6. The Planning (Hazardous Substances) Act 1990.

7. Part III of the Environmental Protection Act 1990.

8. Sections 72, 111 and 113(6) and Chapter III of Part IV of the Water Industry Act 1991 and paragraphs 2 to 4 of Schedule 8 to that Act so far as they re-enact provisions of sections 43 and 44 of the Control of Pollution Act 1974.

9. Sections 82, 84, 85, 86, 87(1), 88(2), 92, 93, 99, 161, 190, 202, [and 203] of and paragraph 6 of Schedule 25 to the Water Resources Act 1991.

10. Section 18 of the Water Act 1945 so far as it continues to have effect by virtue of Schedule 2 to the Water Consolidation (Consequential Provisions) Act 1991 or by virtue of provisions of the Control of Pollution Act 1974 not having been brought into force.

PART II

SCOTLAND

11. Sections . . . 32, 41, 42 and 116 of the Public Health (Scotland) Act 1897.

12. [*Sections 3, 16 and 17 of the Clean Air Act 1993.*]

13. The Sewerage (Scotland) Act 1968.

14. Sections 56A to 56N and 97B of the Town and Country Planning (Scotland) Act 1972.

15. Section 201 of the Local Government (Scotland) Act 1973.

[16. Sections 30A, 30B, 30D, 30F, 30G, 30H(1), 31(4), (5), (8) and (9), 31A, 34 to 42B, 46 to 46D and 56(1) to (3) of the Control of Pollution Act 1974.]

17. Sections 70, 71 and 75 of the Water (Scotland) Act 1980.

[17A. Pt. III of the Environmental Protection Act 1990.]

PART III

NORTHERN IRELAND

18. Sections 50, 51, 58, 107 and 129 of the Public Health (Ireland) Act 1878.

19. Section 26 of the Public Health Acts Amendment Act 1890.

20. Sections 35, 46, 49 and 51 of the Public Health Acts Amendment Act 1907.

21. Sections 26, 47 and 124 of the Fisheries Act (Northern Ireland) 1966.

22. Sections 5, 7 and 8 of the Water Act (Northern Ireland) 1972.

23. Article 34 of the Water and Sewerage Services (Northern Ireland) Order 1973.

24. The Clean Air (Northern Ireland) Order 1981.

25. The Pollution Control (Special Waste) Regulations (Northern Ireland) 1981.

SCHEDULE 4

Section 49(1)

CONSEQUENTIAL AMENDMENTS

The Continental Shelf Act 1964 (c. 29)

1. In section 7 of the Continental Shelf Act 1964, for "Radioactive B–242 Substances Act 1960" there is substituted "Radioactive Substances Act 1993".

The Nuclear Installations Act 1965 (c. 57)

2. In section 4(1)(d) of the Nuclear Installations Act 1965, for "sections 6 and 8 of the Radioactive Substances Act 1960" there is substituted "sections 13 and 16 of the Radioactive Substances Act 1993".

The Control of Pollution Act 1974 (c. 40)

3. In section 56(6) of the Control of Pollution Act 1974, for "Radioactive Substances Act 1960" (in both places) there is substituted "Radioactive Substances Act 1993".

The Pollution Control and Local Government (Northern Ireland) Order 1978 (S.I. 1978/1049 (N.I. 19))

4. In article 36(4) of the Pollution Control and Local Government (Northern Ireland) Order 1978—
   (a) for "Radioactive Substances Act 1960" there is substituted "Radioactive Substances Act 1993", and
   (b) in paragraph (b) for "1960" there is substituted "1993".

The Atomic Energy (Miscellaneous Provisions) Act 1981 (c. 48)

5. In section 4(1) of the Atomic Energy (Miscellaneous Provisions) Act 1981, in the definition of "radioactive substance", for "has the same meaning as in section 12 of the Radioactive Substances Act 1948" there is substituted "means any substance which consists of or contains any radioactive chemical element, whether natural or artificial".

The Environmental Protection Act 1990 (c. 43)

6. In section 28(2) of the Environmental Protection Act 1990, for "Radioactive Substances Act 1960" there is substituted "Radioactive Substances Act 1993".

7. In section 78 of that Act, for "Radioactive Substances Act 1960" (in both places) there is substituted "Radioactive Substances Act 1993".

8. In section 142(7) of that Act, for "the Radioactive Substances Act 1960" there is substituted "the Radioactive Substances Act 1993".

9. In section 156 of that Act, for "Radioactive Substances Act 1960" there is substituted "Radioactive Substances Act 1993".

The Atomic Weapons Establishment Act 1991 (c. 46)

10. After paragraph 10 of the Schedule to the Atomic Weapons Establishment Act 1991 there is inserted—

"Radioactive Substances Act 1993

10A.—(1) For the purposes of the Radioactive Substances Act 1993, so far as relating to authorisations required under section 13(1) of that Act for the disposal of radioactive waste, a relevant site in designated premises shall be treated as a site in respect of which a nuclear site licence is for the time being in force.

(2) For the purposes of sub-paragraph (1) above, "relevant site" means a site used by a contractor for the purposes of any activity which would, if section 1 of the Nuclear Installations Act 1965 applied to the site, require a nuclear site licence".

The Water Resources Act 1991 (c. 57)

11. In section 98 of the Water Resources Act 1991—
 (a) in subsection (1), for "Radioactive Substances Act 1960" there is substituted "Radioactive Substances Act 1993", and
 (b) in subsection (2)(b) for "1960" there is substituted "1993".

SCHEDULE 5

Section 49(2)

TRANSITIONAL AND TRANSITORY PROVISIONS

PART I

GENERAL TRANSITIONAL PROVISIONS AND SAVINGS

1. The substitution of this Act for the enactments repealed by this Act B–243 does not affect the continuity of the law.

2. Any reference, whether express or implied, in this Act or any other enactment, instrument or document to a provision of this Act shall, so far as the context permits, be construed as including, in relation to the times, circumstances and purposes in relation to which the corresponding provision of the enactments repealed by this Act has effect, a reference to that corresponding provision.

3. Any document made, served or issued after the commencement of this Act which contains a reference to any of the enactments repealed by this Act shall be construed, except so far as a contrary intention appears, as referring or, as the case may require, including a reference to the corresponding provision of this Act.

4. Paragraphs 2 and 3 have effect without prejudice to the operation of sections 16 and 17 of the Interpretation Act 1978 (which relate to the effect of repeals).

5. The power to amend or revoke the subordinate legislation reproduced in the definition of "local authority" in section 47(1) shall be exercisable in relation to the provision reproduced to the same extent as it was exercisable in relation to the subordinate legislation.

6. Subsection (1) of section 80 of the Health and Safety at Work etc. Act 1974 (general power to repeal or modify Acts or instruments) shall apply to provisions of this Act which re-enact provisions previously contained in the Radioactive Substances Act 1960 as it applies to provisions contained in Acts passed before the Health and Safety at Work etc. Act 1974.

7. In the application of paragraph 6 to Northern Ireland—
   (a) the reference to subsection (1) of section 80 of the Health and Safety at Work etc. Act 1974 shall have effect as a reference to paragraph (1) of Article 54 of the Health and Safety at Work (Northern Ireland) Order 1978, and
   (b) the reference to Acts passed before that Act shall have effect as a reference to statutory provisions passed or made before the making of that Order.

PART II

TRANSITORY MODIFICATION OF SCHEDULE 3

8.—(1) If—
  (a) no date has been appointed before the commencement of this Act as the date on which paragraph 8 of Schedule 15 of the Environmental Protection Act 1990 (in this paragraph referred to as "the 1990 provision") is to come into force, or
  (b) a date has been appointed which is later than that commencement,
paragraph 7 of Schedule 3 to this Act shall be omitted until the appointed day.
  (2) In this paragraph "the appointed day" means—
  (a) in the case mentioned in paragraph (a) of sub-paragraph (1) above, such day as may be appointed by the Secretary of State by order, and
  (b) in the case mentioned in paragraph (b) of that sub-paragraph, the date appointed as the day on which the 1990 provision is to come into force.

9.—(1) If—
  (a) no date has been appointed before the commencement of this Act as the date on which the repeal by Schedule 4 to the Control of Pollution Act 1974 of the provisions of the Radioactive Substances Act 1960 specified in sub-paragraph (2) below (in this paragraph referred to as "the 1974 repeal") is to come into force, or
  (b) a date has been appointed which is later than that commencement,
Schedule 3 to this Act shall have effect until the appointed day with the modifications specified in sub-paragraph (3) below.
  (2) The provisions of the Radioactive Substances Act 1960 referred to in sub-paragraph (1)(a) above are—
  (a) in paragraph 3 of Schedule 1, the words "seventy-nine", and
  (b) paragraph 8A of Schedule 1.
  (3) The modifications of Schedule 3 to this Act referred to in sub-paragraph (1) above are as follows—
  (a) in paragraph 1 after "48" there shall be inserted "79", and
  (b) after paragraph 2 there shall be inserted—
    "2A. Sections 2, 5 and 7 of the Rivers (Prevention of Pollution) Act 1961."
  (4) In this paragraph "the appointed day" means—
  (a) in the case mentioned in paragraph (a) of sub-paragraph (1) above, such day as may be appointed by the Secretary of State by order, and

  (b) in the case mentioned in paragraph (b) of that sub-paragraph, the date appointed as the day on which the 1974 repeal is to come into force.

10.—(1) If—

  (a) no date has been appointed before the commencement of this Act for the purposes of paragraph 17 of Schedule 4 to the Planning (Consequential Provisions) Act 1990, or
  (b) a date has been appointed which is later than that commencement,

paragraph 6 of Schedule 3 to this Act shall be omitted until the appointed day.

  (2) In this paragraph "the appointed day" means—

  (a) in the case mentioned in paragraph (a) of sub-paragraph (1) above, such day as may be appointed by the Secretary of State by order, and
  (b) in the case mentioned in paragraph (b) of that sub-paragraph, the date appointed for the purposes of paragraph 17 of Schedule 4 to the Planning (Consequential Provisions) Act 1990.

11. Until the commencement of the repeal by Part II of Schedule 16 to the Environmental Protection Act 1990 of subsection (5) of section 30 of the Control of Pollution Act 1974 (or, if the repeal of that subsection comes into force on different days, until the last of those days) Schedule 3 to this Act shall have effect—

  (a) with the insertion after paragraph 4 of the following paragraph—
  [4B ...], and
  (b) with the insertion after paragraph 17 of the following paragraph—
  [17A ...]

12. Until the commencement of the repeal by Part II of Schedule 16 to the Environmental Protection Act 1990 of section 124 of the Civic Government (Scotland) Act 1982 (or, if the repeal of that section comes into force on different days, until the last of those days) Schedule 3 to this Act shall have effect with the insertion at the end of Part II of the following paragraph—

  "17B. Section 124 of the Civil Government (Scotland) Act 1982."

SCHEDULE 6

Section 50

REPEALS AND REVOCATIONS

PART I

B–244   ACTS OF THE PARLIAMENT OF THE UNITED KINGDOM

Chapter	Short Title	Extent of repeal
11 & 12 Geo. 6 c. 37.	The Radioactive Substances Act 1948.	The whole Act so far as unrepealed.
8 & 9 Eliz. 2 c. 34.	The Radioactive Substances Act 1960.	The whole Act.
1968 c. 47.	The Sewerage (Scotland) Act 1968.	In Schedule 1, paragraph 4.
1973 c. 65.	The Local Government (Scotland) Act 1973.	In Schedule 27, in Part II, paragraph 144.
1979 c. 2.	The Customs and Excise Management Act 1979.	In Schedule 4, in Part I of the Table following paragraph 12, the entry relating to the Radioactive Substances Act 1948.
1980 c. 45.	The Water (Scotland) Act 1980.	In Schedule 10, in Part II, the entry relating to the Radioactive Substances Act 1960.
1984 c. 55.	The Building Act 1984.	In Schedule 6, paragraph 7.
1986 c. 63.	The Housing and Planning Act 1986.	In Part II of Schedule 7, paragraph 1.
1989 c. 15.	The Water Act 1989.	In Schedule 25, paragraph 27.
1990 c. 11.	The Planning (Consequential Provisions) Act 1990.	In Schedule 2, paragraph 7. In Schedule 4, paragraph 17 and the entry relating to it in the Table in paragraph 1(1).
1990 c. 43.	The Environmental Protection Act 1990.	Sections 100 to 105. Schedule 5. In Schedule 15, paragraph 8.

Chapter	Short Title	Extent of repeal
1991 c. 46.	The Atomic Weapons Establishment Act 1991.	In the Schedule, paragraph 5.
1991 c. 60.	The Water Consolidation (Consequential Provisions) Act 1991.	In Schedule 1, paragraph 9.

Note: Except as provided in Part II of this Schedule, the repeal of the Radioactive Substances Act 1948 does not extend to Northern Ireland.

PART II

REPEALS IN RADIOACTIVE SUBSTANCES ACT 1948 EXTENDING TO NORTHERN IRELAND

Chapter	Short Title	Extent of repeal
11 & 12 Geo. 6 c. 37.	The Radioactive Substances Act 1948.	Section 2. Section 5(1)(b). In section 7, in subsection (1), the words "except section two" and in subsection (2)(b), the words "(except section two)". Section 8(7). In section 9(1), the words "or orders". In section 10, the words "or order" in both places. In section 11, the words from the beginning to "under this Act and". In section 12, the definitions of "registered dental practitioner", "registered pharmacist" and "sale by way of wholesale dealing". Section 14(2)(f).

Note: These repeals extend to Northern Ireland only.

PART III

NORTHERN IRELAND LEGISLATION

Chapter or number	Short title	Extent of repeal
1966 c. 17 (N.I.).	The Fisheries Act (Northern Ireland) 1966.	In Schedule 7, the amendments of the Radioactive Substances Act 1960.
1972 c. 5 (N.I.).	The Water Act (Northern Ireland) 1972.	Section 31(1).
S.I. 1973/70 (N.I. 2).	The Water and Sewerage Services (Northern Ireland) Order 1973.	In Schedule 3, paragraph 1.
S.I. 1978/ 1049 (N.I. 19).	The Pollution Control and Local Government (Northern Ireland) Order 1978.	In Schedule 4, paragraph 5.

PART IV

SUBORDINATE LEGISLATION

Number	Title	Extent of revocation
S.I. 1974/ 1821.	The Radioactive Substances Act 1948 (Modification) Regulations 1974.	The whole instrument.
S.R. (N.I.) 1981/252.	The Pollution Control (Special Waste) Regulations (Northern Ireland) 1981.	Regulation 4(2).
S.I. 1985/ 1884.	The Waste Regulation and Disposal (Authorities) Order 1985.	In Schedule 2, paragraph 2.

Number	Title	Extent of revocation
S.I. 1991/ 2539.	The Control of Pollution (Radioactive Waste) (Scotland) Regulations 1991.	Regulation 4.

TABLE OF DERIVATIONS

Notes:

1. The following abbreviations are used in this Table:—

1960	= The Radioactive Substances Act 1960 (c. 34).
1968 c. 47	= The Sewerage (Scotland) Act 1968 (c. 47).
1973 c. 36	= The Northern Ireland Constitution Act 1973 (c. 36).
1973 c. 65	= The Local Government (Scotland) Act 1973 (c. 65).
1974 c. 40	= The Control of Pollution Act 1974 (c. 40).
1978 c. 30	= The Interpretation Act 1978 (c. 30).
1984 c. 55	= The Building Act 1984 (c. 55).
1986 c. 63	= The Housing and Planning Act 1986 c. 63).
1989 c. 15	= The Water Act 1989 (c. 15).
1990	= The Environmental Protection Act 1990 (c. 43).
1990 c. 11	= The Planning (Consequential Provisions) Act 1990 (c. 11).
1991 c. 60	= The Water Consolidation (Consequential Provisions) Act 1991 (c. 60).
S.I. 1976/959	= The Control of Pollution (Radioactive Waste) Regulations 1976 (S.I. 1976/959).
S.I. 1980/1709	= The Control of Pollution (Special Waste) Regulations 1980 (S.I. 1980/1709).
S.I. 1985/1884	= The Waste Regulation and Disposal (Authorities) Order 1985 (S.I. 1985/1884).
S.I. 1989/1158	= The Control of Pollution (Radioactive Waste) Regulations 1989 (S.I. 1989/1158).
M (followed by a number)	= The proposal of that number in the Memorandum under the Consolidation of Enactments (Procedure) Act 1949 (16 Dec 1991, HC 148).

2. The Table does not show the effect of the Transfer of Functions (Wales) (No. 1) Order 1978 (S.I. 1978/272) or the Transfer of Functions (Radioactive Substances) (Wales) Order 1990 (S.I. 1990/2598).

Provision	Derivation
1	1960 ss.18(1)–(3A), 21(2)(a); 1990 s.100(3), Sch. 5 paras.17, 20(a)(i).
2	1960 s.18(4).
3	1960 s.18(5): 1990 Sch. 5 para. 7(2).
4	1960 ss.11A, 12(7B), 20(a), 21(2)(a), (k), (l); 1990 s.100, Sch. 5 paras.13(4), 18(a), 20(a).
5	1960 ss.8(1)(b), (c), 12(7)(a), 20(b), 21(2)(a); 1990 Sch. 5 paras.18(a), 20(a)(i).
6	1960 s.1(1).
7	1960 s.1(2)–(6); 1990 s.100(2), Sch. 5 paras.4(1), 6(1)(a), (b), 11(1).
8	1960 ss.2, 21(2)(a); 1990 s.100(2), (3), Sch. 5 para. 20(a)(i).
9	1960 s.3(1), (2); 1990 Sch. 5 para. 7(1).
10	1960 s.3(3)–(5); 1990 s.100(2), Sch. 5 paras.6(2)(a), (b), 7(1), 11(2).
11	1960 ss.4, 21(2)(a); 1990 s.100(3), Sch. 5 para. 20(a)(i).
12	1960 s.5; 1990 s.100(2), Sch. 5 para. 6(3).
13	1960 s.6(1)–(3), (6); 1990 Sch. 5 para. 7(3).
14	1960 s.7(1)–(3), (5).
15	1960 ss.6(4), (5), 7(4), 21(2)(a); 1990s.100(3), Sch. 5 para. 20(a)(i).
16	1960 ss.8(1)–(5), (6), 20(b), 21(2)(a); 1990 s.100(2). Sch. 5 paras.1, 4(2), 6(4), 11(3), 18(a), 20(a)(i); M2.
17	1960 s.8(7), (8); 1990 Sch. 5 para. 1(5); M2.
18	1960 s.9(3)–(5); 1990 s.100(2), Sch. 5 para. 2(1).
19	1960 s.11(3).
20	1960 s.8A; 1990 Sch. 5 para. 8.
21	1960 s.11B; 1990 s.102.
22	1960 s.11C; 1990 s.102.
23	1960 ss.12A, 21(2)(a); 1990 Sch. 5 paras.12, 20(a)(i).
24	1960 ss.12B, 21(2)(a), (m); 1990 Sch. 5 paras.12, 20(a).
25	1960 ss.1(7), 3(6), 8(5A), 21(2)(a); 1990 Sch. 5 paras.6(1)(c), (2)(c), (4)(c), 20(a)(i).
26	1960 ss.11D(1)–(4), (12), 21(2)(a); 1990 Sch. 5 paras.10, 20(a)(i); M3.
27	1960 ss.11D(5)–(11), 21(2)(a); 1990 Sch. 5 paras.10, 20(a)(i).
28	1960 ss.11(1), (4), 11E, 20(f), 21(2)(a); 1990 Sch. 5 paras.9(2), 10, 18(b), 20(a)(i).
29	1960 ss.10(1)–(3), 21(2)(a); 1990 s.100(3), Sch. 5 para. 20(a)(i).
30	1960 ss.10(4), (5), 20(e), 21(2)(a), (j); 1978 c. 30 s.17(2)(a); 1990 s.100(3), Sch. 5 para. 20(a)(i).

Provision	Derivation
31	1960 s.12(1)–(6A), (7)(b), (7A), (8), (9), 20(b), 21(2) (a); 1990 s.100(3), Sch. 5 paras.2(2), 13, 18(a), 20(a)(i).
32	1960 s.13(1), (2); 1990 Sch. 5 para. 14(2), (3).
33	1960 s.13(4A), (5), (6); 1990 Sch. 5 para. 14(5), (7), (8).
34	1960 ss.13(3), (4), 21(2)(a); 1990 s.100(3), Sch. 5 paras.14(4), 20(a)(i).
35	1960 s.13(5); 1990 Sch. 5 para. 14(6), (7).
36	1960 s.13(8).
37	1960 s.13(9); 1990 Sch. 5 para. 14(10).
38	1960 ss.13(7), 21(3); 1973 c. 36 Sch. 5 para. 7(1); 1990 Sch. 5 para. 14(9).
39	1960 ss.13A, 21(2)(a); 1990 Sch. 5 paras.15,20(a)(i).
40	1960 ss.9(1), (2), (6), 21(2)(i); 1973 c. 36 Sch. 5 para. 1; 1991 c. 60 Sch. 1 para. 9(2).
41	1960 s.160; M4.
42	1960 ss.14, 21(2)(a), (o); 1990 s.104, Sch. 5 para. 20(a).
43	1960 ss.15A, 21(2)(a), (n); 1990 s.101, Sch. 5 para. 20(a).
44	1960 s.15; 1990 s.100(3).
45	1960 ss.15, 21(2)(a), (b), (c); 1990 s.100(3), Sch. 5 para. 20(a)(i).
46	1960 s.19(5).
47	1960 ss.19(1)–(4), 20(a), (c), 21(2)(a), (f), (g); Fisheries Act (Northern Ireland) 1966 (c. 17) Sch. 7; 1973 c. 65 Sch. 27 Part II para. 144; 1978 c. 30 s.17(2)(a); S.I. 1985/1884 Sch. 2 para. 2; Salmon Act 1986 (c. 62) s.14(2); 1989 c. 15 Sch. 25 para. 27(3); 1990 Sch. 5 paras.3, 5, 11(4), 18(a), 20(a).
48–50	—
51(1)–(2)	—
(3)	1960 s.21(2).
Sch. 1	1960 Sch. 3.
Sch. 2	1960 s.20(d), Sch. 2.
Sch. 3 Part I	1960 Sch. I Part I; S.I. 1976/959 reg. 4; 1978 c. 30 s.17(2)(a); 1984 c. 55 Sch. 6 para. 7; 1990 Sch. 15 para. 8; 1990 c. 11 Sch. 2 para. 6; 1991 c. 60 Sch. 1 para. 9(1).
Part II	1960 Sch. 1 Part II; 1968 c. 47 Sch. 1 para. 4; 1980 c. 45 Sch. 10 Part II; 1986 c. 63 Sch. 7 Part II para. 1; 1990 Sch. 5 para. 19(b); Control of Pollution (Radioactive Waste) (Scotland) Regulations 1991 (S.I. 1991/2539) reg. 4.

Provision	Derivation
Part III	1960 Sch. 1 Part III; Fisheries Act (Northern Ireland) 1966 (c. 17) Sch. 7; Water Act (Northern Ireland) 1972 (c. 5) s.31(1); Water and Sewerage Services (Northern Ireland) Order 1973 (S.I. 1973/70 (N.I. 2)) Sch. 3 para. 1; Pollution Control (Special Waste) Regulations (Northern Ireland) 1981 (S.R. (N.I.) 1981/252).
Sch. 4	—
Sch. 5	
paras.1–7	
para. 8	1990 s.164(3), Sch. 15 para. 8.
para. 9	1974 c. 40 s.109(2), Sch. 4.
para. 10	1990 c. 11 Sch. 4 paras.1, 17.
para. 11	S.I. 1980/1709 reg. 3(2); 1990 s.164(3).
para. 12	1990 s.164(3).
Sch. 6	—

APPENDIX C

RADIOACTIVE MATERIAL (ROAD TRANSPORT) ACT 1991 (c. 27)

ARRANGEMENT OF SECTIONS

RADIOACTIVE MATERIAL (ROAD TRANSPORT) ACT 1991 (c. 27)

General Note

This Act came into force on August 27, 1991 except for section 8 which C–01
came into force on June 27, 1991 (the date of Royal Assent).

Extent

The Act applies to England, Scotland and Wales but, with the exception C–02
of section 8, it does not extend to Northern Ireland.

Purpose

The Act makes new provisions relating to the transport of radioactive C–03
material by road. It increases the powers of inspectors or examiners to
enter vehicles used for transporting radioactive packages and for
inspectors, to enter any premises to carry out their functions. Inspectors
and examiners have powers to prohibit the driving of a vehicle used to
transport radioactive packages and inspectors have the power to issue
enforcement notices on any person failing to comply with regulations
made under section 2 of the Act.

Section 1 provides a number of definitions including "radioactive
materials" and gives the Secretary of State power to appoint inspectors.
Section 2 empowers the Secretary of State to make regulations to ensure
the safety of the transport of radioactive material. The section also repeals
section 5(2) of the Radioactive Substances Act 1948 but regulations made
under that subsection which were in force at the commencement of the
1991 Act continue to have effect, as if made under this section.

Section 3 deals with the powers of inspectors or examiners to prohibit
the driving of a vehicle carrying a package that contains radioactive
material while inspectors are empowered to prohibit the transport of any
radioactive package that fails to comply with regulations made under
section 2. There are further provisions under the section relating to
directions and prohibition notices that may be given by inspectors or
examiners.

Section 4 empowers an inspector to issue an enforcement notice on a
person who he believes is failing or likely to fail to comply with regulations
made under section 2. Failure to comply with such a notice is an offence.

Section 5 sets out the powers of inspectors or examiners to enter vehicles used to transport radioactive packages and for inspectors to enter premises. The section also provides the circumstances in which a warrant may be sought by an inspector or examiner for entry, using reasonable force where necessary.

Section 6 sets out the offences and penalties under the Act along with the powers of the court to order the destruction or disposal of radioactive material with any expenses incurred being paid by the convicted person.

Preliminary

C–04 1.—(1) In this Act "radioactive material" means any material having a specific activity in excess of—

(a) 70 kilobecquerels per kilogram; or

(b) such lesser specific activity as may be specified in an order made by the Secretary of State;

and the power to make an order under this subsection shall be exercisable by statutory instrument which shall be subject to annulment in pursuance of a resolution of either House of Parliament.

(2) In this Act—

"examiner" means by examiner appointed under section 68(1) of the Road Traffic Act 1988;

"inspector" means any inspector appointed under subsection (3) below;

"packaging", in relation to radioactive material which has been consigned for transport, means an assembly of packaging components which encloses the material completely;

"packaging components" means components intended for use as part of the packaging of such material, and includes—

(a) receptacles, absorbent materials, spacing structures and radiation shielding; and

(b) devices for cooling, for absorbing mechanical shocks and for thermal insulation;

"radioactive package" means a package comprising radioactive material which has been consigned for transport and its packaging;

"transport" means transport by road.

(3) The Secretary of State may—

(a) appoint as inspectors, to assist him in the execution of this Act and regulations made under it, such number of persons appearing to him to be qualified for the purpose as he may consider necessary; and

(b) make to or in respect of any person so appointed such payments by way of remuneration, allowances or otherwise as he may with the approval of the Treasury determine.

General Note

The definition of "radioactive material" in subsection (1) is expressed in **C–05**
purely quantitative terms or by reference to orders made by the Secretary
of State for Transport. No orders have yet been made under section 1(1).
"Examiner" is defined by reference to section 68(1) of the Road Traffic
Act 1988 and importantly, "transport" only means transport by road. The
Act does not define "road". Under provisions in subsection (3) the
Secretary of State has the power to appoint as many inspectors as he may
require to ensure their effectiveness.

Regulations

2.—(1) The Secretary of State may make such regulations as appear **C–06**
to him to be necessary or expedient—
   (a) to prevent any injury to health, or any damage to property or to
       the environment, being caused by, or by any incident arising out
       of, the transport of radioactive material; and
   (b) to give effect to such international regulations for the safe
       transport of radioactive material as may from time to time be
       published by the International Atomic Energy Agency.
(2) Without prejudice to the generality of subsection (1) above,
regulations under this section may make provision with respect to—
   (a) the design of packaging for radioactive material and the
       manufacture and maintenance of packaging components;
   (b) the preparation, labelling, consignment, handling, transport,
       storage in transit and delivery of radioactive packages;
   (c) the placarding of vehicles used to transport such packages;
       and
   (d) the keeping of records and the furnishing of information.
(3) Regulations under this section may also—
   (a) impose requirements by reference to the approval of the
       Secretary of State or of any person or body specified in the
       regulations;
   (b) make different provision for different cases or different circum-
       stances; and
   (c) provide for such exceptions, limitations and conditions, and
       make such supplementary, incidental, consequential or tran-
       sitional provisions, as the Secretary of State considers necess-
       ary or expedient.
(4) Any person who contravenes or fails to comply with any regulations
under this section shall be guilty of an offence.
(5) The power to make regulations under this section shall be
exercisable by statutory instrument which shall be subject to annulment
in pursuance of a resolution of either House of Parliament.
(6) Subsection (2) of section 5 of the Radioactive Substances Act 1948

shall cease to have effect; and any regulations under that subsection which are in force at the commencement of this Act shall have effect as if made under this section.

Definitions

C–07 "transport": section 1(2).
"radioactive material": section 1(1).
"packaging": section 1(2).
"packaging component": section 1(2).
"radioactive package": section 1(2).

General Note

C–08 Subsection (1) empowers the Secretary of State for Transport to make regulations relating to the safe transport of radioactive material by road.

The International Atomic Energy Agency (IAEA) referred to in subsection (1)(b) is an inter-governmental organisation operating as a specialised agency of the United Nations, based in Vienna, and recommending safety standards for protection of health and minimisation of danger to life and to property. In making regulations under section 2, the Secretary of State gives effect to regulations for safe transport of radioactive material published by IAEA.

Regulations made under subsection (2) may relate to a range of features concerning radioactive packages including the design, manufacture and maintenance of packaging, their preparation, labelling, transport and delivery, the placarding of vehicles used in the transport of radioactive packages and the keeping of records and the furnishing of information.

Subsection (6) repeals section 5(2) of the Radioactive Substances Act 1948, although regulations made under subsection 5(2) continue in force provided they were in force at the commencement of the 1991 Act. The only regulations to which subsection (6) applies are the Radioactive Substances (Carriage by Road) (Great Britain) Regulations 1974 (S.I. 1974 No. 1735) as amended by (S.I. 1985 No. 1729) which have now been revoked by the Radioactive Material (Road Transport) (Great Britain) Regulations 1996 (S.I. 1996 No. 1350), made under this section of the Act.

Offences

C–09 Subsection (4) makes it an offence to contravene or fail to comply with any regulations made under section 2. For penalties, see section 6. In addition to a fine or term of imprisonment or both (s.6(3)), the Court has the power to order the relevant radioactive material to be destroyed or disposed of and the convicted person to pay the reasonable expenses for the destruction or disposal (s.6(4)).

Prohibitions and directions

3.—(1) If it appears to an inspector or examiner, as respects any C–10 vehicle used to transport radioactive packages—

(a) that the vehicle, or any radioactive package which is being transported by it, fails to comply with any regulations under section 2 above;

(b) that the vehicle, or any radioactive package which is or was being transported by it, has been involved in an accident;

(c) that any radioactive package which was being transported by the vehicle, or any radioactive material which was contained in such a package, has been lost or stolen,

he may prohibit the driving of the vehicle.

(2) If it appears to an inspector that any radioactive package or packaging component fails to comply with any regulations under section 2 above, he may prohibit the transport of that package or, as the case may require, the use of that component as part of the packaging of radioactive materials.

(3) A prohibition imposed under this section may apply either absolutely or for a specified purpose and either without any limitation of time or for a specified period.

(4) Where an inspector or examiner imposes a prohibition under subsection (1) above, he may also by a direction in writing require the person in charge of the vehicle to remove it (and, if it is a motor vehicle drawing a trailer, also to remove the trailer) to such place and subject to such conditions as are specified in the direction; and the prohibition shall not apply to the removal of the vehicle or trailer in accordance with the direction.

(5) Where an inspector or examiner imposes a prohibition under this section, he shall forthwith give notice of the prohibition to the person in charge of the vehicle, package or packaging component, specifying the failure to comply or, as the case may be, the accident or other incident in consequence of which the prohibition is imposed and—

(a) stating whether the prohibition applies absolutely or for a specified purpose (and if the latter specifying the purpose); and

(b) stating whether the prohibition applies without limitation of time or for a specified period;

and any direction under subsection (4) above may be given either in such a notice or in a separate notice given to the person in charge of the vehicle.

(6) A prohibition under this section shall come into force as soon as notice of it has been given in accordance with subsection (5) above and shall continue in force—

(a) until it is removed under subsection (7) below; or

(b) in the case of a prohibition imposed only for a specified period, until either it is removed or that period expires, whichever first occurs.

(7) A prohibition under subsection (1) above may be removed by any inspector or examiner, and a prohibition under subsection (2) above may be removed by any inspector, if he is satisfied—

(a) in the case of a prohibition imposed in consequence of a failure to comply with any regulations under section 2 above, that appropriate action has been taken to remedy that failure;

(b) in the case of a prohibition imposed in consequence of an accident or other incident, either that no failure so to comply was occasioned by that accident or incident or that appropriate action has been taken to remedy any such failure which was so occasioned;

and on doing so, the inspector or examiner shall forthwith give notice of the removal of the prohibition to the person in charge of the vehicle, package or packaging component.

(8) Any person who contravenes a prohibition under this section, or fails to comply with a direction under subsection (4) above, shall be guilty of an offence.

Definitions

C–11   "inspector": section 1(2).
"examiner": section 1(2).
"transport": section 1(2).
"radioactive package": section 1(2).
"radioactive material": section 1(1).
"packaging component": section 1(2).
"packaging": section 1(2).

General Note

C–12   This section sets out the powers of inspectors or examiners to impose prohibitions on the driving of vehicles used to transport radioactive packages. Inspectors are further empowered under subsection (2) to prohibit the transport of a radioactive package either because the package or its packaging fails to comply with regulations made under section 2. Where they have imposed a prohibition, inspectors or examiners may also require the removal of the vehicle, by a direction in writing, to a specified place.

Prohibition Notice

In imposing a prohibition, inspectors or examiners must give a prohib- **C–13** ition notice to the relevant person (subs.(5)). The notice must specify the reason for the prohibition, whether it applies absolutely or for a specified purpose and whether it is for a temporary specified period or permanent in its effect. Any direction in writing under subsection (4) may be included in such a notice or it may be given separately. If the latter, it would seem possible that the prohibition notice could be given orally. A prohibition only comes into force once the notice under subs. (5) has been given. The section also sets out the circumstances in which, and the method by which, a prohibition may be removed (subs.(7)).

Offences

The penalty for an offence under subsection (8) is provided in subsection **C–14** 6(3) and may additionally involve the court ordering any radioactive material to be destroyed or disposed of at the expense of the convicted person (s.6(4)).

Enforcement notices

4.—(1) If an inspector is of the opinion that any person is failing or is **C–15** likely to fail to comply with any regulations under section 2 above which make provision for regulating the manufacture, or requiring the mainten- ance, of packaging components, he may serve a notice under this section on that person.

(2) A notice under this section shall—
    (a) state that the inspector is of the said opinion;
    (b) specify the matters constituting the failure to comply with the regulations in question or the matters making it likely that such a failure will occur, as the case may be;
    (c) specify the steps that must be taken in order to remedy those matters and the period within which those steps must be taken.

(3) Any person who fails to comply with a notice under this section shall be guilty of an offence.

Definitions

"inspector": section 1(2).   **C–16**
"packaging component": section 1(2).

General Note

C–17 This section empowers an inspector to serve an enforcement notice where a person has failed or is likely to fail to comply with regulations under section 2 relating to the manufacture, or requiring the maintenance, of packaging components (as defined). The notice must contain a statement of the inspector's opinion concerning the failure to comply, a list of the failures or likely failures and the steps that need to be taken to remedy matters and the time within which the remedy must be applied.

Offences

C–18 The penalties for an offence under subsection (3) are set out in section 6(3).

Powers of entry

C–19   5.—(1) An inspector or examiner shall, on producing, if so required, some duly authenticated document showing his authority, have a right at all reasonable hours—

   (a) to enter any vehicle used to transport radioactive packages for the purpose of ascertaining—

      (i) whether the vehicle, or any radioactive package which is being transported by it, fails to comply with any regulations under section 2 above;

      (ii) whether the vehicle, or any radioactive package which is or was being transported by it, has been involved in an accident; and

      (iii) whether any radioactive package which was being transported by the vehicle, or any radioactive material which was contained in such a package, has been lost or stolen; and

   (b) in the case of an inspector, to enter any premises for the purpose of ascertaining whether there is on the premises any vehicle used for transporting radioactive packages, or any radioactive package or packaging component which fails to comply with regulations under section 2 above.

   (2) If a justice of the peace, on sworn information in writing or, in Scotland, on evidence on oath, is satisfied that there are reasonable grounds for entering any vehicle or premises for any such purpose as is mentioned in subsection (1) above and either—

   (a) that admission to the vehicle or premises has been refused, or a refusal is apprehended, and (in the case of premises) that notice of the intention to apply for the warrant has been given to the occupier; or

(b) that an application for admission, or the giving of such a notice, would defeat the object of the entry, or that the case is one of urgency, or (in the case of premises) that they are unoccupied or the occupier temporarily absent,

he may by warrant signed by him authorise the inspector or examiner to enter and search the vehicle or premises, using reasonable force if need be.

(3) A warrant granted under this section shall continue in force until executed.

(4) An inspector or examiner who enters any vehicle or premises by virtue of this section, or of a warrant issued under it, may seize anything which he has reasonable grounds for believing is evidence in relation to an offence under section 2(4) above.

(5) Any person who intentionally obstructs any person exercising any power conferred by this section, or by a warrant issued under it, shall be guilty of an offence.

(6) If any person who enters any vehicle or premises by virtue of this section, or of a warrant issued under it, discloses any information thereby obtained with respect to any manufacturing process or trade secret, he shall, unless the disclosure was made in the performance of his duty, be guilty of an offence.

(7) In the application of this section to Scotland, any reference to a justice of the peace includes a reference to the sheriff and to a magistrate.

Definitions

"inspector": section 1(2).   C–20
"examiner": section 1(2).
"transport": section 1(2).
"radioactive package": section 1(2).
"radioactive material": section 1(1).
"packaging component": section 1(2).

General Note

Inspectors or examiners are empowered to enter any vehicle used to   C–21
transport radioactive packages. In addition, inspectors are empowered to
enter premises to locate vehicles that fail to comply with regulations made
under section 2. The section also makes provision for obtaining warrants
to effect an entry under specified circumstances from a Justice of the
Peace.

Entry into Vehicles or Premises

C–22 Entry by inspectors or examiners (vehicles) and inspectors only (premises) is permissible provided that it is attempted at a reasonable hour and the inspector or examiner can give written evidence of his authority, if required. "Reasonable hour" would almost certainly include a time during normal business hours[1] but the question of what is a reasonable hour will necessarily depend on the circumstances and is therefore a question of fact.

Whether entry is effected with or without a warrant, an inspector or an examiner is empowered to seize evidence in relation to an offence under section 2(4) *i.e.* a failure to comply with regulations made under section 2.

Premises

C–23 The term is not defined in the Act. In previous cases, it has been construed in such a way as to indicate a whole property which may be subject to a single occupation or a single ownership whichever is applicable in the circumstances.[2]

Intentionally obstructs

C–24 Far from always requiring the threat of physical harm, any behaviour which makes it more difficult for a person to carry out his duty may be construed as obstruction.[3] Subsection (5) requires that the obstruction be intentional *i.e.* deliberate and conscious and not accidental.

Offences

C–25 Two types of offences may be committed under this section. Under subsection (5) it is an offence to obstruct an inspector or an examiner exercising his powers under section 5 or by a warrant issued under subsection (2). Under subsection (6) an inspector or an examiner may commit an offence, if, having obtained entry with or without a warrant, he discloses specified types of information obtained during his entry. No offence is committed under subsection (6) when disclosure occurs in the course of the inspector's or examiner's duty. For penalties, see subsections 6(2) and (3).

---

[1] *Davies v. Winstanley* (1930) 144 L.T. 433.
[2] *Cadbury Bros. Ltd v. Sinclair* [1934] 2 K.B. 389.
[3] *Hinchcliffe v. Sheldon* (1995) 3 All E.R. 406.

Offences and penalties

6.—(1) Where an offence under this Act which has been committed by **C–26** a body corporate is proved to have been committed with the consent or connivance of, or to be attributable to any neglect on the part of—

    (a) any director, manager, secretary or other similar officer of the body corporate; or

    (b) any person who was purporting to act in any such capacity,

he as well as the body corporate shall be deemed to be guilty of that offence and shall be liable to be proceeded against and punished accordingly.

(2) Any person guilty of an offence under section 5(5) above shall be liable on summary conviction to a fine not exceeding level 3 on the standard scale.

(3) Any person guilty of any other offence under this Act shall be liable—

    (a) on conviction on indictment, to a fine or to imprisonment for a term not exceeding two years or to both;

    (b) on summary conviction, to a fine not exceeding the statutory maximum or to imprisonment for a term not exceeding two months or to both.

(4) The court by or before which any person is convicted of an offence under section 2(4) or 3(8) above in respect of any radioactive material may order the material to be destroyed or disposed of and any expenses reasonably incurred in connection with the destruction or disposal to be defrayed by that person.

Definition

"radioactive material": section 1(1).   **C–27**

General Note

This section sets out the offences and penalties that apply to a body **C–28** corporate and to its directors and officers for breaches under the Act. The successful conviction of a director or other officer will first require the successful conviction of the company for that offence and then the demonstration of consent or connivance or neglect on the part of the relevant director or officer. Offences under section 5(5) are triable only in the magistrates' court and currently carry a maximum fine of £1,000 (subs. (2)). All other offences, under the Act, are triable either way. In the Crown court, the maximum penalty is an unlimited fine and imprisonment for up to 2 years or both. In the magistrates court, the maximum penalty is £5,000 and up to 2 months imprisonment or both. In relation to offences under section 2(4) and section 3(8) the court may additionally order radioactive material to be destroyed or disposed of and the expenses incurred be paid by the person convicted of the offence (subs. (4)).

"Consent"

C–29 The courts have taken a common sense view of what consent might mean; effectively, some affirmative action or approval on the part of the individual must have occurred.

"Connivance"

C–30 This behaviour is different from consent in that overt approval of the crime is unnecessary, but there has to be some knowledge amounting to more than inattention, ignorance or negligence. In *Huckerby v. Elliot*,[4] the following reference was made by the judge:

> "Where he [the Director] connives at the offence committed by the company he is equally well aware of what is going on, but his agreement is tacit, not actively encouraging what happens but letting it continue and saying nothing about it."

Such connivance may arise where a director has a suspicion that his employees are taking short-cuts in some of their practices in order to sustain productivity at the expense of health and safety or environmental law. The director does nothing, effectively, "turning a blind eye".

"Neglect"

C–31 This behaviour has been defined as a "failure to perform a duty which the person knows or ought to know".[5] In order to decide whether a director is under a particular duty it is necessary to consider the specific factors of the case including the nature of the company's business and the way that its management is distributed between directors and other officers. Further, while it has never been settled by case law, it would appear that duties resting upon an executive director may be heavier than those upon a non-executive director. It should be noted, however, that duties on a director are not absolute so that he is entitled to delegate part of his workload, provided that this is done in a sensible and responsible manner. In the Canadian case *Her Majesty the Queen v. Bata Industries Ltd and others* (1993), the chairman of the board was held to be entitled to assume that the on-site manager was addressing environmental concerns

---

[4] (1970) All E.R. 189.
[5] *Re Hughes* [1943] 2 All E.R. 269.

at the relevant facility and was also entitled to rely on the systems that had been put in place, provided that he had not become aware that they were in some way defective.

Expenses

7.—Any expenses incurred by the Secretary of State in consequence of the provisions of this Act shall be payable out of money provided by Parliament. C–32

General Note

This section provides for the reimbursement of the Secretary of State for Transport by Parliament for expenses incurred under the Act. C–33

Corresponding provision for Northern Ireland

8.—An Order in Council under paragraph 1(1)(b) of Schedule 1 to the Northern Ireland Act 1974 (legislation for Northern Ireland in the interim period) which contains a statement that it is made only for purposes corresponding to the purposes of this Act— C–34

(a) shall not be subject to paragraph 1(4) and (5) of that Schedule (affirmative resolution of both Houses of Parliament); but

(b) shall be subject to annulment in pursuance of a resolution of either House of Parliament.

General Note

This section is the only one which applies to Northern Ireland and provides by an Order in Council for the introduction of corresponding provisions to the Act to Northern Ireland. C–35

No such order has yet been made.

Short title, repeals, commencement and extent

9.—(1) This Act may be cited as the Radioactive Material (Road Transport) Act 1991. C–36

(2) The enactments mentioned in the Schedule to this Act are hereby repealed to the extent specified in the third column of that Schedule.

(3) Except for section 8 above, this Act shall not come into force until the end of the period of two months beginning with the day on which it is passed.

(4) Except for section 8 above, this Act does not extend to Northern Ireland.

SCHEDULE

C–37

REPEALS

Chapter	Short title	Extent of repeal
1948 c. 37.	The Radioactive Substances Act 1948.	In section 5, subsection (2) and in subsection (6) the words from "and for the purposes of subsection (2)" to the end. In section 7, in subsection (1), the words "or any vehicle, vessel or aircraft", and the words "vehicle, vessel or aircraft", and in subsection (2) the words "vehicle, vessel or aircraft" in both places where they occur.

INDEX

(All references are to paragraph numbers)
(All references with prefix A, B or C are to Appendices)

463

464